Brill's Companion to Greek Land Warfare Beyond the Phalanx

Brill's Companions in Classical Studies

Warfare in the Ancient Mediterranean World

VOLUME 4

The titles published in this series are listed at *brill.com/wamw*

Brill's Companion to Greek Land Warfare Beyond the Phalanx

Edited by

Roel Konijnendijk
Cezary Kucewicz
Matthew Lloyd

BRILL

LEIDEN | BOSTON

On the cover: Departure of warriors on an Attic black-figure terracotta amphora in the Metropolitan Museum of Art, New York (last quarter of the sixth century; MMA 41.85). Attributed to the Group of Toronto 305. Gift of Mrs. Leonard A. Cohn, 1941.

Library of Congress Cataloging-in-Publication Data

Names: Konijnendijk, Roel, editor. | Kucewicz, Cezary, editor. | Lloyd, Matthew T. (Matthew Thomas), 1985– editor.
Title: Brill's companion to Greek land warfare beyond the phalanx / edited by Roel Konijnendijk, Cezary Kucewicz, Matthew Lloyd.
Description: Leiden ; Boston : Brill, [2021] | Series: Brill's companions to classical studies: warfare in the ancient Mediterranean world, 2452-1493 ; 4 | Includes bibliographical references and index.
Identifiers: LCCN 2021048675 (print) | LCCN 2021048676 (ebook) | ISBN 9789004501720 (hardback) | ISBN 9789004501751 (ebook)
Subjects: LCSH: Military art and science—Greece—History—To 500. | Greece—History, Military—To 146 B.C. | Military art and science—History—To 500. | Military history, Ancient.
Classification: LCC U33 .B75 2021 (print) | LCC U33 (ebook) | DDC 355.020938—dc23/eng/20211008
LC record available at https://lccn.loc.gov/2021048675
LC ebook record available at https://lccn.loc.gov/2021048676

Typeface for the Latin, Greek, and Cyrillic scripts: "Brill". See and download: brill.com/brill-typeface.

ISSN 2452-1493
ISBN 978-90-04-50172-0 (hardback)
ISBN 978-90-04-50175-1 (e-book)

Published by Koninklijke Brill NV, Leiden, The Netherlands.
Koninklijke Brill NV incorporates the imprints Brill, Brill Nijhoff, Brill Hotei, Brill Schöningh, Brill Fink, Brill mentis, Vandenhoeck & Ruprecht, Böhlau Verlag and V&R Unipress.

This book is printed on acid-free paper and produced in a sustainable manner.

Contents

Acknowledgments VII
List of Figures and Table IX
Abbreviations and Spellings X

1 Introduction: Beyond the Phalanx 1
Matthew Lloyd, Roel Konijnendijk, and Cezary Kucewicz

2 Men of Iron
Pre-Archaic Greek Warfare in Context 17
Matthew Lloyd

3 The Anatolian Roots of Archaic Greek Warfare 64
Josho Brouwers

4 The War Dead in Archaic Sparta 83
Cezary Kucewicz

5 Women, Diversity, and War off the Battlefield in Classical Greece 122
Jennifer Martinez Morales

6 Worshipping Violence 145
Alexander T. Millington

7 Cavalry and the Character of Classical Warfare 169
Roel Konijnendijk

8 'Not Many Bows'?
Light-Armed Fighters of the Tenth through Fourth Centuries 205
Cezary Kucewicz, Matthew Lloyd, and Roel Konijnendijk

9 Assaults and Sieges
Rewriting the Other Side of Greek Land Warfare 236
Fernando Echeverría

10 The Western Greeks and the 'Greek Warfare' Narrative 266
Joshua R. Hall

11 The First Greek Soldiers in Egypt
Myths and Realities 293
Hans van Wees

Epilogue 345
Matthew A. Sears

Index 353

Acknowledgments

The idea behind this volume originated during a colloquium entitled 'The Phalanx and Beyond: Ways Forward in the Study of Greek Warfare' organised by the three of us, with the assistance of Joshua R. Hall, in April 2014 at the Institute of Classical Studies, University of London. We are grateful to all speakers and attendees for the lively discussions which took place on the day, as well as to the Institute of Classical Studies which generously funded the event. Several of the chapters of this volume are elaborations of the papers delivered that day, and we would like to thank the authors of those papers for their co-operation and patience through the work's long gestation period. We are also grateful to Jennifer Martinez Morales and Matthew Sears, who enriched this collection with separate original contributions.

The volume has been many years in the making and we have benefited greatly from the advice, comments and enthusiasm of many colleagues who helped us along the way. They include Stephanie Craven, Katherine Harrell, Sonya Nevin, Stephen O'Brien, Louis Rawlings, Owen Rees, and Hans van Wees. Although each of them helped to make the volume better, we claim exclusive rights to any mistakes and oversights which remain. The three of us would also like to extend our thanks to the editorial team at Brill: Lee Brice, Giulia Moriconi and the anonymous reviewers of the volume. Lee's eagerness and belief in the project has been particularly helpful, especially as we neared the finish line.

Cezary would like to thank his colleagues at the University of Gdańsk, in particular Bogdan Burliga and Nicholas Sekunda, for their helpful insights and support for the project. Special thanks also to my parents, Ewa and Leszek, and Caroline Musgrove, who read and commented on multiple drafts, both short and long. I am also grateful to the National Science Centre, Poland, which provided funding for my work on this volume (project number 2018/28/C/HS3/00418).

Matthew would like to thank Whitney Dickson, his parents, Mervyn and Sheila, and his in-laws, Jamie and Sharryn, for their support over the years in which this volume was in preparation. Thanks also to Irene S. Lemos and the rest of the Lefkandi excavation team, as well as Josho Brouwers, Joshua R. Hall, Owen Rees, and my fellow editors for keeping me involved in the ancient world through post-DPhil/immigration alienation.

Roel would like to thank the various institutions that have provided him with the time and resources to work on this volume – especially the Institute of Classical Studies and Institute of Historical Research at Senate House, and the European Research Council's EU Horizon2020 programme, whose funding (Grant Proposal 794093) made the historiographical section of the introduction possible. All else is thanks to the kindness and patience of our contributors and my fellow editors, and that of Eri, who stuck with me even when I planned the colloquium to take place on our anniversary.

Figures and Table

Figures

2.1 Attic LG ship battle scene on a krater in the Metropolitan Museum of Art, New York (*ca.* 800–775 BCE; MMA 34.11.2). Attributed to the Workshop of New York MMA 34.11.2 33

2.2 Perfume bottle (aryballos) with lion-head mouth, attributed to the Chigi Painter (British Museum 1889,0418.1) 37

5.1 Slab of the Nereid Monument, showing a woman on the walls of a besieged city *ca.* 390–380 BCE (British Museum 1848,1020.65) 129

8.1 Laconian lead figure of an archer (sixth–fifth century BCE; MMA 24.195.97). Gift of A.J.B. Wace, 1924 217

8.2 Departure of a warrior and archer on an Attic black-figure terracotta neck-amphora in the Metropolitan Museum of Art, New York (last quarter of the sixth century BCE; MMA 41.162.171). Unattributed. Rogers Fund, 1941 218

8.3 Departure of warriors on an Attic black-figure terracotta stamnos in the Metropolitan Museum of Art, New York (late sixth century BCE; MMA 2011.233). Attributed to the Painter of London B 343. Gift of Patricia Stickney, 2011 219

8.4 Departure of warriors on an Attic black-figure terracotta stamnos in the Metropolitan Museum of Art, New York (late sixth century BCE; MMA 2011.233). Attributed to the Painter of London B 343. Gift of Patricia Stickney, 2011 219

10.1 Aristonothos Krater, second quarter of the seventh century BCE. Currently in the Musei Capitolini, Rome 280

Table

9.1 Expeditions and their targets and battles in three case studies from the Classical period 250

Abbreviations and Spellings

Throughout the volume, we favour Hellenised spellings of names of people and places, employing Latinised versions only for the most familiar names, where confusion would be created by using a Hellenised name, and for locations in Italy. All ancient dates cited are BCE unless otherwise indicated.

Abbreviations of journals and works employed in this book are those from *L'Année Philologique Online*, and the *Oxford Classical Dictionary*^{4}, in addition to the following:

ABC	Grayson, A.K. (2000) *Assyrian and Babylonian Chronicles*. Winona Lake, IN.
CAH	*Cambridge Ancient History*, 2nd edn. Vols. 3–4 and 7. (1982–1991) Cambridge.
CHGRW	Sabin, P., M. Whitby, and H. van Wees (eds) (2007) *Cambridge History of Greek and Roman Warfare*, vol. 1. Cambridge.
GSW	Pritchett, W.K. (1971–1991) *The Greek State at War*, vols. 1–5. Berkeley, CA.
GWMR	Van Wees, H. (2004) *Greek Warfare: Myths and Realities*. London.
HCGBE	Hanson, V.D. (ed) (1991d) *Hoplites: The Classical Greek Battle Experience*. London.
MoB	Kagan, D. and G.F. Viggiano (eds) (2013) *Men of Bronze. Hoplite Warfare in Ancient Greece*. Princeton.
NAGRW	Brice, L.L. (ed) (2020) *New Approaches to Greek and Roman Warfare*. Hoboken, NJ.
RINAP 3.1–2	Grayson, A.K. and J. Novotny (2012, 2014) *The Royal Inscriptions of Sennacherib, King of Assyria (704–681 BC)*. Parts 1 and 2. Royal Inscriptions of the Neo-Assyrian Period 3.1 and 3.2. Winona Lake, IN.
RINAP 4	Leichty, E. (2011) *The Royal Inscriptions of Esarhaddon, King of Assyria (680–669 BC)*. Royal Inscriptions of the Neo-Assyrian Period 4. Winona Lake, IN.
RINAP 5.1	Novotny, J. and J. Jeffers (eds) (2018) *The Royal Inscriptions of Ashurbanipal (668–631 BC), Assur-etal-ilani (630–627 BC), and Sin-sarra-iskun (626–612 BC), Kings of Assyria*, Part 1. Royal Inscriptions of the Neo-Assyrian Period 5.1. Winona Lake, IN.
RO	Rhodes, P.J. and R. Osborne (2003). *Greek Historical Inscriptions 404–323 BC*. Oxford.
SAA	*State Archives of Assyria* (1987–). Helsinki. Online at http://oracc.museum.upenn.edu/saao

Sieges	Armstrong, J. and M. Trundle (eds) (2019) *Brill's Companion to Sieges in the Ancient Mediterranean*. Warfare in the Ancient Mediterranean World 3. Leiden.
W&S	Rich, J. and G. Shipley (eds) (1993) *War and Society in the Greek World*. London.
WCW	Campbell, B. and L.A. Tritle, (eds) (2013) *The Oxford Handbook of Warfare in the Classical World*. Oxford.
W&V	Van Wees, H. (ed) (2000) *War and Violence in Ancient Greece*. London.
WWW	Hanson, V.D. (1989) *The Western Way of War: Infantry Battle in Classical Greece*. London.

CHAPTER 1

Introduction: Beyond the Phalanx

Matthew Lloyd, Roel Konijnendijk, and Cezary Kucewicz

When we think about Greek warfare, we tend to think about the hoplite. The word conjures a specific image: a spearman decked out in bronze armour, carrying an iconic round double-grip shield and crested helmet, marching and fighting in a tightly ordered formation – the phalanx. While there may be other types of fighters around this hoplite phalanx, they are the periphery to his centre. We can see this hoplite take pride of place in recent volumes on Greek warfare with titles like *Reinstating the Hoplite*, *The Psychology of the Athenian Hoplite*, *A Storm of Spears*, *Men of Bronze*, and *Hoplites at War*, all of which focus on understanding the hoplite as the key to understanding Greek warfare.[1] Indeed, one of the most influential edited volumes in the subfield is simply titled *Hoplites: The Classical Greek Battle Experience*. We see the same iconic warrior all over representations of the ancient Greek world in popular media, from films like *300*[2] and games like *Assassin's Creed: Odyssey* (2018) to a steady stream of historical novels, popular historical accounts, graphic novels, and YouTube videos on the wars of the Greeks in general and the Spartans in particular.

In both the scholarly and the popular understanding, the hoplite is uniquely Greek. The 'phalanx' way of fighting is treated as a particular feature of warfare in the central mainland area of what is now the Hellenic Republic; there is little interest in Anatolia or northern Greece, the islands, or Italy. Besides the more recent works listed above, classic works in the field have emphasised a universalising 'Greek' way of war, for example, the five volumes of *The Greek State at War* by W.K. Pritchett and *Greek Warfare: Myths and Realities* by Hans van Wees. The assumption is that there is a particular *Greek* way of war, and that this Greek way of war was *hoplite* warfare. This notion is also central to the holistic 'orthodoxy' or 'grand narrative' of Greek warfare, which posits that the rise of the hoplite was fundamental to the rise of Greek society and culture in all its facets – in other words, that the hoplite was uniquely Greek, and by his existence he made Greece uniquely important and interesting to us today.

1 Schwartz (2009); Crowley (2012); Matthew (2012); *MoB*; Bardunias and Ray (2016).
2 dir. Zack Snyder (2006) and its sequel *300: Rise of an Empire*, dir. Noam Murro (2014).

 | DOI:10.1163/9789004501751_002

There is a problem at the centre of this narrative, one that is too often ignored in scholarship: the modern use of 'hoplite' and 'phalanx' has granted them a technical and even ethnic definition that they did not have in the ancient world.[3] When we use the word 'hoplite' to mean something specifically *Greek*, we use it in a way that would be alien to Herodotos, who used it for a Persian at Marathon; it would be alien also to Xenophon, who used it to describe Assyrians and Egyptians. In the surviving literary evidence, the term emerged far later than a putative 'hoplite revolution' *ca.* 700 – its earliest attestation is in the early fifth-century poetry of Pindaros and Aischylos – to describe a generic warrior armed exclusively for close combat. Meanwhile 'phalanx' seems to have had no technical meaning in the Homeric epics, where it is first used as a metaphor for massed infantry. It occurs in the plural to describe separate groups of men; it is not accompanied by any vocabulary that suggests these are ordered formations. The few scattered uses of 'phalanx' in Archaic poetry all follow the Homeric pattern (e.g., Tyrtaios Fr. 12.21). Moreover, the term does not occur in the battle descriptions of authors like Herodotos or Thucydides, whose works are of fundamental importance to any study of Greek warfare. The word 'phalanx' does not function as a specific military term until the fourth century, when Xenophon began to use it to indicate the battle line of massed heavy infantry. The words 'hoplite' and 'phalanx' do not appear together until the fourth century. Scholarship has assigned to these words a specificity that they did not originally have. In modern scholarship, 'hoplite phalanx' is an exclusive term; in the ancient world, *hoplitēs* and *phalanx* were each used more broadly.[4] So where does the idea of the hoplite phalanx as the defining feature of a uniquely Greek warfare come from?

There are a few prominent reasons for the field's intense focus on mainland Greece, on hoplites, and on hoplite battle. Firstly, since the bulk of our literary evidence concerns Athens and Sparta, it is not surprising to find classicists and ancient historians preoccupied with those communities; since most of the surviving texts date to the Classical period, scholarship naturally privileges this period, when hoplites undeniably formed the core of all Greek armies. Secondly,

3 Echeverría (2012).

4 A similar confusion of language can be found in the still-common assertion that hoplites took their name from the *hoplon* shield, e.g., Sears (2019) 32. As explained in Lazenby and Whitehead (1996), this misconception is the result of an attempt by Diodoros to make 'hoplite' a technical term of the same type as 'peltast' – a warrior whose name was derived from his shield, the *pelte*. This assertion is particularly popular among scholars who wish to connect the appearance of the shield with a 'hoplite' style of fighting. However, the word *hoplon* is not commonly used for the shield, which is almost invariably called *aspis* in sources from the Classical period.

the sources themselves sometimes suggest that the hoplite was the central figure, not only of Greek warfare, but also of the socio-political development of Greek society as a whole. The speech that Mardonios used to persuade Xerxes to invade mainland Greece (Hdt. 7.9β.1), in particular, has often been taken at face value as a frank assessment of the Greek way of war: Greeks settled their conflicts with a single decisive engagement between heavy infantry on level ground. Later authors offer much detail on the organisation and importance of hoplite formations, but they rarely discuss other warriors in detail. The hoplite completely dominates art and literature of the Classical period as the generic citizen warrior. Meanwhile, Aristotle already theorised that the mounted aristocracies of early Greece were supplanted by broader-based systems of government when hoplites grew in number, got organised, and began to assert their power (*Pol.* 1297b.16–24). Passages like these not only inspired an image of Greek warfare as hoplite warfare, but also encouraged the formulation of wider theories in which hoplites shaped socio-political systems and defined the values of Greek culture.

The origins of this model in modern scholarship lie in George Grote's *History of Greece* (1846–1856), which noted that the link between hoplite equipment and social class affected the political development of Greek states. Following Aristotle, Grote suggested that political power followed the rise of the hoplite to military supremacy, broadening the base of Greek systems of government and imbuing citizen hoplites with an egalitarian ethos.[5] Within a decade, Grote's work had been translated into German, inspiring many new studies on Greek antiquity including a branch of scholarship dedicated to warfare in particular. The earliest German handbook on Greek warfare saw the phalanx of citizen hoplites as the cornerstone of Greek military methods and presented the fifth century as the era of 'pure hoplite battle'.[6] One German scholar after another, driven by their primary interest in tactical forms and battle narratives, stressed that the disciplined hoplite and his well-ordered phalanx were the essential features of Greek warfare and the highest achievements of Greek military prowess until Epameinondas of Thebes unleashed his tactical innovations in the second quarter of the fourth century.[7] Athenian and Spartan models of military organisation provided templates that were assumed to apply throughout the Greek world; surviving accounts of pitched battles apparently justified the tendency to dismiss non-hoplite warriors as irrelevant and ineffective.

5 Kagan and Viggiano (2013) 2–7.
6 Rüstow and Köchly (1852) 142; Bauer (1893) 325.
7 Jähns (1880); Droysen (1889); Bauer (1893); Delbrück (1900); Lammert and Lammert (1921); Kromayer and Veith (1928).

The first works on naval warfare, cavalry, and light troops did not affect the broad narrative of these general studies.[8] Greek warfare came to be characterised as a contest between citizen-hoplites, who fought rule-bound battles on level ground over limited stakes. The notion of 'hoplite warfare' was born.[9]

Unlike some broader German historical surveys in the mould of Grote, however, early studies of military history rarely credited the rise of the hoplite phalanx with political and cultural significance.[10] Heavily reliant on philological material, some of the early works on Greek warfare declared that the warfare of the Archaic period could not be reconstructed from the fragmentary evidence, while others adhered to Karl Otfried Müller's view that a different style of infantry warfare prevailed in the Dorian parts of the Greek world until after the Persian Wars.[11] Some even suggested a more gradual emergence of phalanx tactics that was not completed until the fifth century.[12] Such interpretations precluded the notion that 'hoplite warfare' had been essential to the development of Greek civilisation from its early origins to its Classical form. Max Weber may have anticipated a century of scholarship with his sweeping socio-economic model in which the arrival of hoplite armour brought about the rise of 'hoplite poleis' dominated by 'citizen-peasant hoplite forces', but historians of warfare do not seem to have noticed.[13]

Around the same time, however, other scholars were independently arriving at similar conclusions. As more iconographic and archaeological material came to be studied, the view became commonplace that hoplite armour (and the phalanx tactics that were assumed to be a prerequisite for its effective use) had spread rapidly throughout the Greek world from the first half of the seventh century onwards, completely transforming the way wars were fought. Links were soon drawn between this technological and tactical transformation,

8 For example, Martin (1887); Assmann (1888); Helbig (1902); Lippelt (1910).

9 For the role of these German scholars in shaping the debate, see Konijnendijk (2018) 7–12.

10 Grote's causal link between hoplite service and the democratisation of city-states is visible in works like Meyer (1893) 561 and Weber (1922) 542–43. However, contemporary works on military history tended to derive the characteristics of military organisation from socio-political conditions, rather than the other way around.

11 Unwilling to discuss the Archaic period: Droysen (1889) 91; Delbrück (1900) 1–2. On 'Dorian' warfare: Müller (1824); Rüstow and Köchly (1852) 17, 50; Jähns (1880) 118–19; Bauer (1893) 304–308, but the theory is disputed in Lammert and Lammert (1921) 445; Nilsson (1929) 242–43.

12 For example, Kromayer and Veith (1928) 22; Lammert (1938) 1629–30.

13 Konijnendijk and Echeverría (2022).

early Archaic demographic and economic growth, and the appearance of the first codified laws and constitutions.[14]

It was the work of George Grundy that did the most to elevate the concept of tightly regulated and exclusive hoplite warfare to a cornerstone of Greek history in the broadest sense. Grundy asserted that hoplite warfare was not just a tactical system, but a way of war mandated by the economic conditions of Classical Greece. Whereas Grote had regarded hoplites as a broad group consisting of farmers, artisans, and shopkeepers, and Eduard Meyer had identified them with a rising merchant class, Grundy (like Weber) tied them intimately to the land from which they drew their sustenance. Greek states, he argued, could not afford to lose a single harvest to devastation, and therefore had to present invaders with a human wall drawn across their fields, ready to physically push the enemy back. He claimed that only hoplites could do this, but also that hoplites – ill-trained amateurs burdened by heavy armour – could *only* do this. The later introduction of mobile troop types and mercenaries notwithstanding, Greek warfare *had* to take its peculiar form.[15] According to this model, it was not strange to focus exclusively on hoplites, but rather to study *anything else*; only the farmer-citizen-hoplite mattered. As Donald Kagan and Gregory Viggiano noted, while Grundy only applied this vision of Greek warfare to the fifth century, later authors like Frank Adcock, drawing on the chronology established by Hilda Lorimer, projected it all the way back to the first introduction of hoplite armour.[16] The model of socio-political development of the Greek states now firmly revolved around the hoplite and his fighting style.

In the 1960s and 1970s, the main topic of controversy in the narrative of Greek warfare was exactly when and how hoplites had come to dominate battle and politics. Ongoing study of archaeological material from the eighth and seventh centuries brought 'gradualist' challenges to the notion of the immediate wholesale adoption of hoplite gear and phalanx tactics, as well as discussion over the theory that hoplites had been instrumental to both the instalment and the removal of tyrants in the Archaic period.[17] In the late 1970s Joachim

14 The key works that established this consensus are Helbig (1911); Nilsson (1929); Lorimer (1947). A detailed early version of the full narrative occurs in Hasebroek (1931) 158–89.

15 Grundy (1911) 242–59.

16 Kagan and Viggiano (2013) 17–18; the work in question is Adcock (1957).

17 Snodgrass (1964, 1965); Greenhalgh (1973). This early criticism and the formation of a 'gradualist' view seems less momentous in hindsight than scholars may have thought at the time; by the early 1980s, some believed it had already been 'convincingly demonstrated' that 'the "hoplite revolution" must be listed among the great non-events of history' (Frost 1984, 292–93).

Latacz went in the opposite direction, proposing to push the advent of the hoplite and his battle tactics back even further, into the 'Homeric Period' of the eighth century and earlier, based on a radical reinterpretation of descriptions of battle in the *Iliad*.[18] But these challenges mainly affected the developmental model insofar as it was used to explain the onset of political change. There were no major critiques of the view that the hoplite defined Greek warfare throughout the Archaic and Classical periods, and no counter-narrative to the notion that his military power allowed him to shape society and its values as he saw fit. Scholars who worked on formerly neglected facets like siege warfare or the role of ritual and religion typically only reinforced the primacy of hoplites and hoplite battle.[19]

While critical reinterpretations were clearly beginning to gain momentum from the 1980s onwards, the established view successfully managed to reassert itself. Some scholars followed Yvon Garlan in challenging Grundy's notion of the economic imperative to hoplite battle, suggesting that the damage done by invaders was moral and political rather than material,[20] but one of their number, Victor Davis Hanson, argued more forcefully than any of his predecessors that Greek warfare remained the exclusive business of socio-economic equals fighting as heavy infantry.[21] Economic factors were not really important, since the foundation of 'hoplite warfare' was ultimately ideological, not material.[22] While there were more frequent objections to Grundy's view that hoplite combat took the form of a concerted mass shove, these only helped to keep hoplites at the centre of scholarly attention.[23] Voices challenging the series of causal connections that led from the introduction of hoplite equipment to the establishment of democracy similarly do not seem to have had an immediate effect.[24] Introductory works from this period show the older view still firmly entrenched in scholarship.[25]

Alternatives to hoplite primacy have only really begun to gain momentum in the last few decades. Ever since Peter Krentz questioned the conventional wisdom that Greeks prior to the Peloponnesian War always marched out to

18 Latacz (1977).
19 See, for example, the chapters in Vernant (1968); Lonis (1979); Garlan (1974, 1989); *GSW*1–5.
20 Garlan (1974) 42–44; Will (1975) 300–304; Hanson (1983); Foxhall (1993).
21 *WWW*; see also Hanson (1983, 1995).
22 Connor (1988); Ober (1991); Dawson (1996) 50–51; Runciman (1998).
23 See especially Cawkwell (1978) 151–53; Krentz (1985, 1994); Goldsworthy (1997).
24 Morris (1987) 196–201; Raaflaub (1993) 80–81, (1997).
25 See, for example, Ducrey (1985); Sage (1995); Mitchell (1996); Hanson (1999).

fight pitched battles on level ground,[26] one scholar after another has worked to expose the limitations and weaknesses of the established model. They have shown that a fresh look at the sources opens all aspects of this model to reinterpretation. Louis Rawlings has challenged the notion of immobile hoplites fully dependent on the phalanx to function; Hans van Wees has cast doubt on the association of hoplites with a particular social class and challenged the view that the phalanx existed throughout the Archaic period; John Dayton has dismissed the idealised vision of rule-bound 'agonal' warfare between citizen-hoplites; Lin Foxhall has shown that the farmer-citizen-hoplite is nowhere to be found on the ground until the end of the sixth century.[27] The primacy of the hoplite has been doubted on ideological, tactical, and philological grounds.[28] The resulting new image of Greek warfare and its relation to the development of Greek societies allows for more nuanced and inclusive readings of the evidence. It is now possible to look beyond the phalanx and see 'Greek warfare' in all its pluriform complexity.

Yet, despite the spreading realisation that we are not beholden to the hoplite and his phalanx in our efforts to characterise Greek warfare, there has been little deliberate effort to explore the consequences of that realisation. In a stream of recent edited volumes, many aspects of ancient warfare have finally received due attention, but none have set out to redefine what the practice of war was about, or where its spatial and temporal borders should be drawn.[29] Not much has been done to integrate the development of warfare before and after the period of the supposed dominance of the hoplite, or to examine how the role of this warrior changed over time.[30] The forces which fought alongside the phalanx could be integrated into the bigger picture instead of being studied separately in recurrent attempts to fill this gap. The study of 'Greek' warfare

26 Krentz (1997, 2000, 2002).

27 Rawlings (2000, 2007); Van Wees (2000, 2001, *GWMR*, 2013); Dayton (2005); Foxhall (2013).

28 Krentz (2002, 2010); Gaebel (2002); Sidnell (2006); Sheldon (2012); Echeverría (2008, 2012); Konijnendijk (2016, 2018, 2021).

29 Some examples from a growing list: Bragg *et al.* (2008); Fagan and Trundle (2010); Brice and Roberts (2011); Meineck and Konstan (2014); Bakogianni and Hope (2015); Contamine *et al.* (2015); Heckel *et al.* (2015); Lee, Whittaker and Wrightson (2015); Armstrong (2016); Caston and Weineck (2016); Clark and Turner (2018); Ñaco del Hoyo and Lopéz Sanchéz (2018); Van Gils *et al.* (2018); Giangiulio *et al.* (2019); *Sieges*; *NAGRW*; Dillon *et al.* (2020); Chlup and Whately (2021). These volumes tend to provide a platform for a wide range of papers across the history of the Ancient Mediterranean and its modern reception, however; since *W&V*, none have set out to challenge our understanding of the basics of Greek warfare.

30 On the gradual end of the hoplite system in the Hellenistic period, see Ma (2000).

might also follow the approach of many other aspects of ancient history and take a Mediterranean perspective. The obsession with the hoplite and the phalanx has led to many misunderstandings of the nature of Greek warfare, particularly in the Greek regions of southern Italy; the narrow spatial approach has prevented a full understanding of the influence of other societies on Greek warfare while overemphasising Greek influence elsewhere.

Our volume makes no pretence of answering all these questions. It is intended as a starting point: a collection of chapters that push 'beyond the phalanx' and show the potential for further research outside the traditional boundaries of 'Greek warfare'. These boundaries have long been stretched in other subfields within the discipline of ancient history. The study of Greek military history is now following suit – a development that is to be celebrated and encouraged. We have collected a cross-section of the latest original research on land warfare in the Greek world, along with survey treatments of key subjects, both of which reflect recent scholarly trends and offer potential new directions.

Our chapters follow a broadly chronological framework, beginning with the Early Iron Age before moving on to the Archaic and Classical periods. By starting with the tail end of the 'prehistoric' era we extend the narrative of 'Greek warfare' beyond the usual eighth-century beginning with the panoply tomb of Argos and the traditional date of the Homeric epics. When we move into the historical era, several chapters look at warfare within Greece, and by Greeks elsewhere, that was not dependent on or defined by the hoplite and phalanx. The final chapters approach 'the Greeks' on the global scale of the Mediterranean world in the first millennium, taking as their starting point the recent view of the first millennium as a more globalising – or 'Mediterraneanising' – world system than has been previously envisioned.

First, Matthew Lloyd considers how warfare in Greece might have been organised and fought after the collapse of the Mycenaean world and how we can see change in the centuries before the return of writing. Building on the work of scholars looking at how Mycenaean palatial-era warfare extends into the post-palatial period,[31] Lloyd looks at how warfare changes throughout the Early Iron Age and how some of these deep roots are still evident in the early Archaic period. While many approaches to early Greek warfare take a 'block-time' approach,[32] dividing the pre-literate Greek world from that which can be reconstructed from written sources with a 'revolution',[33] Lloyd argues

31 Kramer-Hajos (2016); O'Brien (2022).
32 Crellin (2020) 5–7.
33 Viggiano (2013).

that a more fruitful approach to understanding early Greek warfare is to look at change and continuity throughout the Early Iron Age and into the eighth and seventh centuries. In the next chapter, Josho Brouwers casts doubt on the supposed 'Greekness' of Greek warfare as it emerged in the Archaic period by stressing the interconnectedness of the Aegean and Anatolia during that time. He uses archaeological evidence to demonstrate the similarities and frequent exchanges between the ways of war of the peoples of Western Anatolia in particular.

Archaeological evidence also features in the following chapter, which focuses on Spartan military burials in the Archaic era. Recent years have witnessed a growing number of studies exploring how societies – both ancient and modern – remember and honour military conflicts.[34] Commemorating war took many forms in the ancient world, including memorials, buildings, dedications, festivals, and stories of military victories passed down from one generation to another, each illuminating the impact of wars on the cultural life of the Greek poleis, as well as the special place of war commemoration in the process of building political communities. Surveying the ways in which the Spartans interred and commemorated their war dead, Cezary Kucewicz adds to this body of scholarly work, providing new insights into the nature of early Spartan militias and their transformation towards the end of the sixth century.

The Classical period was the age of the hoplite's greatest prominence, but it also provides a wealth of sources to support different approaches to Greek warfare. Broadening the traditional boundaries of the subject also reveals opportunities to draw from models and insights offered by other disciplines, such as cultural studies, economics, psychology, and psychiatry. The latter two have gained particular prominence in discussions which explore the experience of war on the individual, drawing inspiration from the bottom-up 'face-of-battle' approach. These studies focus on aspects like combat trauma and the impact of military conflicts on the home front, redefining our understanding of the full extent of wars' effect on ancient communities.[35] In her chapter, which explores the roles of women in war, Jennifer Martinez Morales points out that our modern picture of women as non-combatants leans too heavily on the assumption that the only combatants are those who engage in battle. She shows the active involvement of women in a range of military activities, and the expectation of Greek authors that they would make themselves useful in the defence of their community. While women often played an active part in

34 For example, Low *et al.* (2012); Arrington (2015); Giangiulio *et al.* (2019); Bérard (2020); Schröder (2020); Kucewicz (2021); Proietti (2021).

35 See Meineck and Konstan (2014); Cecchet *et al.* (2019); Rees (2020); Tritle (2020).

the latter, success in military undertakings was also heavily reliant on securing divine favour, reflecting the integral role of religion in Greek warfare.[36] Taking a broader perspective on the Greek perception of war as a religious phenomenon, Alexander Millington examines the presence of Ares in a pair of Athenian tragedies to show that there was a great deal of ambiguity in the way Greeks thought about this god and his realm. The notion of a deity representing the violence and chaos of war – one whose support must be secured, but who could not be celebrated – reflects a society where war was often not imposed from above, but agreed upon from below.

The next three chapters focus on blind spots in the 'hoplite warfare' paradigm. The last few years have seen renewed and laudable efforts to integrate non-hoplite warriors into our picture of war in the Greek world[37] and to underline the centrality of sieges in ancient warfare from an early period.[38] The contributions to this volume bolster and encourage these efforts. Roel Konijnendijk argues that the scholarly view of the hoplite as the quintessential Greek warrior is a misjudgement of the tactical system of the Classical period: horsemen were a greater threat than hoplites in most tactical situations, and Greek states went to great lengths to secure cavalry superiority. The editors use a combination of archaeological and literary approaches to construct an unprecedented general survey of light-armed infantry from the tenth to the fourth century, showing how the importance of these troops is consistently underplayed by the elites who produced the surviving sources. Fernando Echeverría shows that attacks on towns and fortified places were not an incidental and poorly developed aspect of Greek warfare, as scholars have traditionally argued, but one of the most common features of their campaigns in enemy territory. Contrary to received wisdom, these invasions were open-ended; invaders were likely to seek targets of opportunity, and did not necessarily expect or intend to provoke a formal pitched battle.

A final pair of chapters considers the Greeks in their Mediterranean context. On this point, too, the last decade has seen a good deal of ground-breaking work.[39] The key element of such work is the recognition that the practices of Greek warriors varied depending on local conditions and on the range of non-Greek peoples they fought or served. Joshua Hall shows how little the warfare of the Western Greeks conformed to our traditional picture of 'hoplite

36 Recent works on religion and warfare include Ulanowski (2016); Nevin (2017); Dillon *et al.* (2020).

37 Sekunda and Burliga (2014); Wrightson (2019); Blome (2020).

38 Echeverría (2017); *Sieges*; Hülden (2020); Seaman (2020).

39 For example, Hale (2013); Raaflaub (2013); Verčik (2014); Stoyanov (2015); Rop (2019); Manning (2021); Lewis (forthcoming).

warfare' and its predilections; he argues that we should not base our understanding of the warfare of communities outside of mainland Greece on what we assume to be the common 'Greek' way of war. Hans van Wees looks in detail at the evidence for Greeks serving the Egyptian pharaoh Psammetichos I – Herodotos' account of which is the origin of the famous moniker for hoplites, 'men of bronze'. He concludes that Greek mercenaries cannot have been a substantial force in Egyptian armies until a generation later than scholars have traditionally assumed. His reconstruction of events implies that these Greeks were not hired as specialist heavy infantry, but simply as readily available manpower for ambitious kings, primarily in Lydia and Egypt.

These chapters present points of departure for a more comprehensive study of warfare in ancient Greece and the Mediterranean. They build on the latest scholarship, embracing the opportunities offered by our rapidly changing understanding of a contentious subject. They illuminate underappreciated aspects of the period, present familiar evidence from new points of view, and suggest new ways to integrate the study of warfare into our understanding of ancient Greek and Mediterranean history. In short, they offer glimpses of a history of Greek warfare beyond Greece, beyond the hoplite, and beyond the phalanx.

Bibliography

Adcock, F.E. (1957) *The Greek and Macedonian Art of War*. Berkeley, CA.

Armstrong, J. (ed) (2016) *Circum Mare: Themes in Ancient Warfare*. Leiden.

Arrington, N.T. (2015) *Ashes, Images, and Memories: The Presence of the War Dead in Fifth-Century Athens*. Oxford.

Assmann, E. (1888) 'Seewesen', in *Denkmäler des klassischen Altertums* III, ed. A. Baumeister. Munich: 1593–639.

Bakogianni. A. and Hope, V.M. (eds) (2015) *War as Spectacle: Ancient and Modern Perspectives on the Display of Armed Conflict*. London.

Bardunias, P.M. and Ray, F.E., Jr. (2016) *Hoplites at War: A Comprehensive Analysis of Heavy Infantry Combat in the Greek World, 750–100 BCE*. Jefferson, NC.

Bauer, A. (1893) *Die griechischen Kriegsaltertümer*. Munich.

Bekker-Nielsen, T. and Hannestad, L. (eds) (2002) *War as a Cultural and Social Force: Essays on Warfare in Antiquity*. Copenhagen.

Bérard, R.-M. (2020) 'La politique du cadaver: Traitements funéraires et usages civiques des morts à la guerre en Grèce archaïque et Classique'. *Annales HSS* 75.1: 3–38.

Blome, D.A. (2020) *Greek Warfare Beyond the Polis: Defense, Strategy, and the Making of Ancient Federal States*. Ithaca.

Bragg, E., Hau, L.I., and Macaulay-Lewis, E. (eds) (2008) *Beyond the Battlefields: New Perspectives on Warfare in the Graeco-Roman World.* Newcastle.

Brice, L.L. and Roberts, J.T. (eds) (2011) *Recent Directions in the Military History of the Ancient World.* Claremont, CA.

Caston, V. and Weineck S.-M. (eds) (2016) *Our Ancient Wars: Rethinking War Through the Classics.* Ann Arbor, MI.

Cawkwell, G. (1978) *Philip of Macedon.* London.

Cecchet, L., Degelmann, C., and Patzelt, M. (eds) (2019) *The Ancient War's Impact on the Home Front.* Newcastle.

Chlup, J.T. and Whately, C. (eds) (2021) *Greek and Roman Military Manuals: Genre and History.* Abingdon.

Clark, J.H. and Turner, B. (eds) (2018) *Brill's Companion to Military Defeat in Ancient Mediterranean Society.* Warfare in the Ancient Mediterranean World 2. Leiden.

Connor, W.R. (1988) 'Early Greek Land Warfare as Symbolic Expression'. *P&P* 119: 3–29.

Contamine, P., Jouanna, J., and Zink, M. (eds) (2015) *La Grèce et la Guerre.* Paris.

Crellin, R. (2020) *Change and Archaeology.* Abingdon.

Crowley, J. (2012) *The Psychology of the Athenian Hoplite: The Culture of Combat in Classical Athens.* Cambridge.

Dawson, D. (1996) *The Origins of Western Warfare: Militarism and Morality in the Ancient World.* Boulder, CO.

Dayton, J.C. (2005) *The Athletes of War: An Evaluation of the Agonistic Elements of Greek Warfare.* Toronto.

Delbrück, H. (1900) *Geschichte der Kriegskunst im Rahmen der Politischen Geschichte* I: *Das Altertum.* Berlin.

Dillon, M., Matthew, C., and Schmitz, M. (eds) (2020) *Religion & Classical Warfare: Archaic and Classical Greece.* Barnsley, UK.

Droysen, H. (1889) *Heerwesen und Kriegführung der Griechen.* Freiburg.

Ducrey, P. (1985) *Guerre et Guerrier dans la Grèce Antique.* Paris.

Echeverría, F. (2008) *Cuidadanos, Campesinos y Soldados: El Nacimiento de la 'Pólis' griega y la Teoría de la 'Revolución Hoplita'.* Madrid.

Echeverría, F. (2012) 'Hoplite and Phalanx in Archaic and Classical Greece: A Reassessment'. *CPh* 107.4: 291–318.

Echeverría, F. (2017) 'Greek Armies against Towns: Siege Warfare and the *Seven against Thebes*', in Torrance (2017) 73–90.

Fagan, G.G. and Trundle, M. (eds) (2010) *New Perspectives on Ancient Warfare.* Leiden.

Foxhall, L. (1993) 'Farming and Fighting in Early Greece', in *W&S* 134–45.

Foxhall, L. (2013) 'Can We See the "Hoplite Revolution" on the Ground? Archaeological Landscapes, Material Culture, and Social Status in Early Greece', in *MoB* 194–221.

Frost, F.J. (1984) 'The Athenian Military before Cleisthenes'. *Historia* 33: 283–94.

Gaebel, R.E. (2002) *Cavalry Operations in the Ancient Greek World.* Norman, OK.

Garlan, Y. (1974) *Recherches de Poliorcétique Grecque*. Paris.

Garlan, Y. (1989) *Guerre et Économie en Grèce Ancienne*. Paris.

Giangiulio, M., Franchi, E., and Proietti, G. (eds) (2019) *Commemorating War and War Dead: Ancient and Modern*. Stuttgart.

Goldsworthy, A.K. (1997) 'The *Othismos*, Myths and Heresies: The Nature of Hoplite Battle'. *War in History* 4.1: 1–26.

Greenhalgh, P.A.L. (1973) *Early Greek Warfare: Horsemen and Chariots in the Homeric and Archaic Ages*. Cambridge.

Grundy, G.B. (1911) *Thucydides and the History of his Age*. London.

Hale, J.R. (2013) 'Not Patriots, Not Farmers, Not Amateurs: Greek Soldiers of Fortune and the Origins of Hoplite Warfare', in *MoB* 176–93.

Hamilton, C.D. and Krentz, P. (eds) (1997) *Polis and Polemos: Essays on Politics, War, and History in Ancient Greece in Honor of Donald Kagan*. Claremont.

Hansen, M.H. (ed) (1993) *The Ancient Greek City-State. Symposium on the Occasion of the 250th Anniversary of The Royal Danish Academy of Sciences and Letters, July, 1–4 1992*. Copenhagen.

Hanson, V.D. (1983) *Warfare and Agriculture in Classical Greece*. Pisa.

Hanson, V.D. (1995) *The Other Greeks: The Family Farm and the Agrarian Roots of Western Civilization*. New York.

Hanson, V.D. (1999) *The Wars of the Ancient Greeks*. New York.

Hasebroek, J. (1931) *Griechische Wirtschafts- und Gesellschaftsgeschichte bis zur Perserzeit*. Tübingen.

Heckel, W., Müller, S., and Wrightson, G. (eds) (2015) *The Many Faces of War in the Ancient World*. Newcastle.

Helbig, W. (1902) *Les ἱππεῖς Athéniens*. Paris.

Helbig, W. (1911) 'Über die Einführungszeit der geschlossenen Phalanx'. *Sitzungsberichte der Königlich Bayerischen Akademie der Wissenschaften* 12: 3–41.

Hülden, O. (2020) *Das griechische Befestigungswesen der archaischen Zeit: Entwicklungen – Formen – Funktionen*. Vienna.

Jähns, M. (1880) *Handbuch einer Geschichte des Kriegswesens von der Urzeit bis zur Renaissance*. Leipzig.

Kagan, D. and Viggiano, G.F. (2013) 'The Hoplite Debate', in *MoB* 1–56.

Konijnendijk, R. (2016) 'Mardonius' Senseless Greeks'. *CQ* 66.1: 1–12.

Konijnendijk, R. (2018) *Classical Greek Tactics: A Cultural History*. Leiden.

Konijnendijk, R. (2021) 'Playing Dice for the Polis: Pitched Battle in Greek Military Thought'. *TAPhA* 151.1: 1–33.

Konijnendijk, R. and Echeverría, F. (2022) 'Max Weber, the Rise of the Polis, and the "Hoplite Revolution" Theory'. *Journal of the History of Ideas* (forthcoming).

Kramer-Hajos, M. (2016) *Mycenaean Greece and the Aegean World: Palace and Province in the Late Bronze Age*. Cambridge.

Krentz, P. (1985) 'The Nature of Hoplite Battle'. *ClAnt* 4: 50–61.
Krentz, P. (1994) 'Continuing the *Othismos* on *Othismos*'. *AHB* 8: 44–49.
Krentz, P. (1997) 'The Strategic Culture of Periclean Athens', in Hamilton and Krentz (1997) 55–72.
Krentz, P. (2000) 'Deception in Archaic and Classical Greek Warfare', in *W&V* 167–200.
Krentz, P. (2002) 'Fighting by the Rules: The Invention of the Hoplite Agôn'. *Hesperia* 71: 23–39.
Krentz, P. (2010) *The Battle of Marathon*. London.
Kromayer, J. and Veith, G. (1928) *Heerwesen und Kriegführung der Griechen und Römer*. Munich.
Kucewicz, C. (2021) *The Treatment of the War Dead in Archaic Athens: An Ancestral Custom*. London.
Lammert, E. and Lammert, F. (1921) 'Schlachtordnung', in *Pauly-Wissowa* II A, col. 436–94.
Lammert, F. (1938) 'Phalanx', in *Paulys Real-Encyclopädie der classischen Altertumswissenschaft* 19.2, col. 1625–46.
Latacz, J. (1977) *Kampfparänese, Kampfdarstellung und Kampfwirklichkeit in der Ilias, bei Kallinos und Tyrtaios*. Munich.
Lazenby, J.F. and Whitehead, D. (1996) 'The Myth of the Hoplite's *Hoplon*'. *CQ* 46.1: 27–33.
Lee, G., Whittaker, H., and Wrightson, G. (eds) (2015) *Ancient Warfare: Introducing Current Research, Volume I*. Newcastle.
Lewis, D. (forthcoming) 'The Cretan Way of War: Status, Violence, and Values from the Classical Period to the Roman Conquest'. *JHS* 143.
Lloyd, A.B. (ed) (1996) *Battle in Antiquity*. London.
Lonis, R. (1979) *Guerre et Religion en Grèce a l'Époque Classique*. Paris.
Lorimer, H.L. (1947) 'The Hoplite Phalanx with Special Reference to the Poems of Archilochus and Tyrtaeus'. *ABSA* 42: 76–138.
Low, P., Oliver, G., and Rhodes, P.J. (eds) (2012) *Cultures of Commemoration: War Memorials, Ancient and Modern*. Oxford.
Ma, J. (2000) 'Fighting *Poleis* of the Hellenistic World', in *W&V* 337–76.
Manning, S. (2021) *Armed Force in the Teispid-Achaemenid Empire: Past Approaches, Future Prospects*. Stuttgart.
Martin, A. (1887) *Les Cavaliers Athéniens*. Paris.
Matthew, C. (2012) *A Storm of Spears: Understanding the Greek Hoplite at War*. Barnsley, UK.
Meineck, P. and Konstan, D. (eds) (2014) *Combat Trauma and the Ancient Greeks*. New York.
Meyer, E. (1893) *Geschichte des Alterthums* II. Stuttgart.

Mitchell, L.G. and Rhodes, P.J. (eds) (1997) *The Development of the Polis in Archaic Greece*. London.
Mitchell, S. (1996) 'Hoplite Warfare in Ancient Greece', in Lloyd (1996) 87–107.
Morris, I. (1987) *Burial and Ancient Society: The Rise of the Greek City-State*. Cambridge.
Müller, K.O. (1824) *Die Dorier*. Breslau.
Ñaco del Hoyo, T. and López Sánchez, F. (eds) (2018) *War, Warlords and Interstate Relations in the Ancient Mediterranean*. Leiden.
Nevin, S. (2017) *Military Leaders and Sacred Space in Classical Greek Warfare: Temples, Sanctuaries and Conflict in Antiquity*. London.
Nilsson, M.P. (1929) 'Die Hoplitentaktik und das Staatswesen'. *Klio* 22: 240–49.
Ober, J. (1991) 'Hoplites and Obstacles', in *HCGBE* 173–96.
O' Brien, S. (2022) 'Warfare and "Social Complexity" from LH III to the Early Iron Age', in Shelton and Kvapil (2022) forthcoming.
Proietti, G. (2021) *Prima di Erodoto: Aspetti della memoria delle Guerre persiane*, Stuttgart.
Raaflaub, K.A. (1993) 'Homer to Solon: The Rise of the Polis: The Written Sources', in Hansen (1993) 41–105.
Raaflaub, K.A. (1997) 'Soldiers, Citizens and the Early Greek Polis', in Mitchell and Rhodes (1997) 49–59.
Raaflaub, K.A. (2013) 'Early Greek Infantry Fighting in Mediterranean Context', in *MoB* 95–111.
Rawlings, L. (2000) 'Alternative Agonies: Hoplite Martial and Combat Experiences beyond the Phalanx', in *W&V* 233–59.
Rawlings, L. (2007) *The Ancient Greeks at War*. Manchester.
Rees, O. (2020) 'We Need to Talk about Epizelus: 'PTSD' and the Ancient World'. *Medical Humanities* 46: 46–54.
Rop, J. (2019) *Greek Military Service in the Ancient Near East, 401–330 BC*. Cambridge.
Runciman, W.G. (1998) 'Greek Hoplites, Warrior Culture, and Indirect Bias'. *Journal of the Royal Anthropological Institute* 4.4: 731–51.
Rüstow, W. and Köchly, H. (1852) *Geschichte des Griechischen Kriegswesens von der Ältesten Zeit bis auf Pyrrhos*. Aarau.
Sage, M.M. (1996) *Warfare in Ancient Greece: A Sourcebook*. London.
Schröder, J. (2020) *Die Polis als Sieger: Kriegsdenkmäler im archaisch-klassischen Griechenland*, Berlin.
Schwartz, A. (2009) *Reinstating the Hoplite: Arms, Armour and Phalanx Fighting in Archaic and Classical Greece*. Stuttgart.
Seaman, M.G. (2020) 'Early Greek Siege Warfare', in *NAGRW* 29–38.
Sears, M.A. (2019) *Understanding Greek Warfare*. New York.
Sekunda, N.V. and Burliga, B. (eds) (2014) *Iphicrates, Peltasts and Lechaeum*. Gdánsk.

Sheldon, R.M. (2012) *Ambush: Surprise Attack in Ancient Greek Warfare*. Barnsley, UK.

Shelton, K. and Kvapil, L. (eds) (2022) *Brill's Companion to Warfare in the Bronze Age Aegean*. Warfare in the Ancient Mediterranean World 5. Leiden.

Sidnell, P. (2006) *Warhorse: Cavalry in Ancient Warfare*. London.

Snodgrass, A.M. (1964) *Early Greek Armour and Weapons: From the End of the Bronze Age to 600 BC*. Edinburgh.

Snodgrass, A.M. (1965) 'The Hoplite Reform and History'. *JHS* 85:110–22.

Stoyanov, T. (2015) 'Warfare', in Valeva, Nankov and Graninger (2015) 426–42.

Torrance, I. (ed) (2017) *Aeschylus and War: Comparative Perspectives on* Seven against Thebes. New York.

Tritle, L.A. (2020) 'Soldiers' Home: Life After Battle', in *NAGRW* 53–64.

Ulanowski, K. (ed) (2016) *The Religious Aspects of War in the Ancient Near East, Greece, and Rome*, Leiden.

Valeva, J., Nankov, E., and Graninger, D. (eds) (2015) *A Companion to Ancient Thrace*. Malden, MA.

Van Gils, L.W., de Jong, I.J.F., and Kroon, C.H.M. (eds) (2018) *Textual Strategies in Ancient War Narrative: Thermopylae, Cannae and Beyond*. Leiden.

Van Wees, H. (2000) 'The Development of the Hoplite Phalanx: Iconography and Reality in the Seventh Century', in *W&V* 125–66.

Van Wees, H. (2001) 'The Myth of the Middle-Class Army: Military and Social Status in Ancient Athens', in Bekker-Nielsen and Hannestad (2001) 45–71.

Van Wees, H. (2013) 'Farmers and Hoplites: Models of Historical Development', in *MoB* 222–55.

Verčik, M. (2014) *Die barbarischen Einflusse in der griechischen Bewaffnung*. Leidorf.

Vernant, J.-P. (ed) (1968) *Problèmes de la Guerre en Grèce Ancienne*. Paris.

Viggiano, F.G. (2013) 'The Hoplite Revolution and the Rise of the Polis', in *MoB* 112–33.

Weber, M. (1922) *Wirtschaft und Gesellschaft*. Tübingen.

Will, E. (1975) 'Le territoire, la ville et la poliorcetique grecque'. *RH* 235: 297–318.

Wrightson, G. (2019) *Combined Arms Warfare in Ancient Greece: From Homer to Alexander the Great and his Successors*. Abingdon.

CHAPTER 2

Men of Iron

Pre-Archaic Greek Warfare in Context

Matthew Lloyd

In the eighth century, the Greek world changed. Settlements reveal new modes of burial suggesting a reorganisation of their political systems. Interregional sanctuaries familiar from later history see a massive increase in dedications. Connections overseas become more evident and Greek settlements begin to appear in the bay of Naples and Sicily. The alphabet, and eventually literature, provide an entirely new perspective on ancient Greece. Among these changes is evidence of new armour and weapons, and an apparently new way of fighting.

Theories of change play a major part in historical and archaeological thinking. How and why things change is a fundamental part of the discipline. They are complicated, however, by the way we divide history into periods. As such, it is tempting to look at the changes of the eighth century and view them as a radical breaking point in the history of Greece.[1] Archaeologist Rachel Crellin has called this kind of approach to change 'block-time': the idea that history consists of long periods of stasis punctuated by revolutionary changes, when in fact things are always changing, as we can see when we look at the world in the present.[2] Historically, it makes sense that this approach has been applied to the eighth century – much of the Early Iron Age (EIA) was unknown a century ago and the study of early Greek history extends long before that; the invention of writing drastically changes our evidence; and the eighth century saw many changes, as listed above. John Papadopoulos has described the position of the EIA as an 'epistemological divide' between Aegean prehistory and Classical archaeology as it serves as something of a coda to the former and prologue to the latter.[3] But as our understanding of the EIA grows ever more substantial, it is necessary to integrate it into our understanding of the changes at the start of history.

1 Morris (2009); Viggiano (2013).
2 Crellin (2020) 5–7.
3 J. Papadopoulos (2015) 181.

 | DOI:10.1163/9789004501751_003

In this chapter, I will take a long-term view of Greek warfare from the collapse of the Mycenaean palaces *ca.* 1200, through the post-palatial Bronze Age and EIA, into the early Archaic period to *ca.* 600. In order to understand change and continuity in early Greek warfare, we must look to the centuries preceding the eighth to see how things change and develop, and how they do not. By broadening the perspective from which we approach early Greek history, we can work towards a more nuanced understanding of the development of pre-Archaic warfare and society.

Crellin lists seven 'hurdles' for understanding change, many of which are relevant to the period in question.[4] These are (1) block-time approaches, (2) progressive and (3) teleological narratives that focus on things becoming more like the present, (4) a focus on origins and revolutions, (5) determinism in changes beyond human control, (6) singular causation, and (7) anthropocentrism in assuming humans always do have control. In what follows I will focus on the period in question in two 'blocks', 1200 to 800 and 800 to 600, but I will emphasise changes throughout these periods as well as within them. I will show how multiple factors both within and beyond human control affect change, based on pre-existing conditions.

In the eighth century, the Greek world changed; but the Greek world was always changing. In order to understand the changes that took place in the eighth century and afterward, we must look back to the preceding period to understand where we are coming from.

1 Greek Warfare from 1200 to 800

Our understanding of warfare in the post-palatial Bronze and Early Iron Ages relies on archaeological evidence. Most of that evidence consists of weapons found in burials. The second largest source is pictorial pottery, but this is limited after the twelfth century – the period from around 1100 to 800 has been described as a 'long, pictureless hiatus' in which figurative images are almost entirely absent from Greek pottery.[5] A few examples of fortifications and settlement destruction date to this period, although the latter are also predominantly twelfth century. Finally, we may turn to the Homeric epics, but the questions regarding the date of their composition as well as whether or not they are relevant to any historical period make this source less reliable; our dating of any particular element of the poems to a historical time period relies

4 Crellin (2020) 5–17.

5 Coldstream (2006); an explanation for the drastic change in the decoration of pottery is offered by S. Murray *et al.* (2020).

first on observing evidence suggesting 'Homeric' practices in the archaeological record.[6] Thus we must tease our understanding out of diverse and complex sources of evidence.

1.1 *Burials and Weapons*

Burials with weapons were once unambiguously called 'warrior burials', but more recent research, particularly osteological analysis, has called this definition into question.[7] Rather than being the graves of 'warriors' – that is, those who fought and died in battle – these burials represent the construction of a social role through the creation of the burial.[8] Where burials with weapons are known, we need to think less in terms of *warriors* and more of a *warrior culture* in which 'being a warrior' held social value and thus individuals laid claim to that status. They do not directly inform us about warfare, but rather about the emergence of a symbolic system in which objects connected to warfare are valued and significant. Furthermore, both 'warrior burials' and 'burials with weapons' lay emphasis on one component of this symbolic system – the weapons, armour, and iconography of warfare. Other aspects of this symbolic system are also significant to our understanding of the social structures of EIA Greece, in particular the treatment of objects from the eastern Mediterranean and Near East.[9]

The pattern of burials with weapons in Greece changes over time. In the post-palatial twelfth century, burials with bronze weapons are known in significant numbers from Achaia in the northwest Peloponnese and from East Crete, as well as sporadically elsewhere.[10] Many of these sites are coastal: Achaia is on the Corinthian gulf; a few weapons are known from the Perati cemetery in Attika, which is on the Euboian gulf; among the Aegean islands, one or two burials with weapons are known from each of Kos, Naxos, and Rhodes.[11] In the eleventh and tenth centuries, burials with iron weapons are known from new cemeteries at Athens, Tiryns in the Argolid, and across Crete; from the tenth

6 Van Wees (1992, 262–63) acknowledged some time ago the similarities between Homeric and Archaic society, suggesting that Homer reflects the early Archaic period rather than the Early Iron Age.

7 For example, Härke (1990); Smith (2009); Harrell (2014); Georganas (2018); Lloyd (2020a). Nor is the term 'burial with weapons' unproblematic, continuing to focus interpretation of such burials on the weapons rather than the place of the weapons within the burial context as a whole. In EIA Greek burials with weapons this emphasis is largely justified by the treatment of the weapons (Lloyd 2015); but their significance is also connected to changes in the burial ritual – notably the use of cremation (Whitley 2002).

8 Whitley (2002) 219–20.

9 Morris (2000) 195–256; Antonaccio (2002).

10 Deger-Jalkotzy (2006); Giannopoulos (2008).

11 Deger-Jalkotzy (2006) 154–67.

century they are known from sites such as Lefkandi on Euboia and throughout northern Greece from Epirus to Macedonia. With the exception of East Crete, burials with weapons are not known in the same areas in the twelfth century as they are from the eleventh century onwards. While several of the Iron Age burials are from coastal sites – Lefkandi, Tiryns, and Knossos – others are inland. Significantly, in Attika the twelfth-century Perati cemetery is abandoned as the cemeteries in the area of Athens begin to grow, indicating a shift in settlement inland in this region. Nevertheless, from the twelfth century the settlement pattern across Greece shifts toward coastal sites, particularly as an indication of prosperity – and, significantly, archaeological visibility.[12] Burials with weapons are one such signifier of prosperity.

As for the weapons in these burials, there is comparatively little change between the Late Bronze and Early Iron Ages except that in the former most are bronze and in the latter most are iron.[13] Bronze Naue type II swords appear in Greece in the late thirteenth century,[14] becoming widespread in the twelfth, and, in iron, this is the only known sword-type in the Aegean from the eleventh to the seventh century. Spearheads are sometimes discovered singly, and sometimes in pairs or with other weapons, which may indicate that they were javelins – one to throw, and one to hold on to. Groups of arrowheads are found in a few burials, and an arrowhead may have been the cause of death in at least one Athenian EIA burial.[15] With regards to armour, LBA shields are known only from iconography and thus it is difficult to trace their continuity through the EIA; bronze *phalara*, known from a handful of graves from the twelfth to the seventh centuries and from sanctuaries from the eighth, have been identified as shield-bosses, although their identification as a single class of object is contentious.[16] In all probability, there were a variety of different shields used in EIA combat, as there are in the LBA, the *Iliad*, and eighth- and seventh-century iconography. While it is contentious to suggest that the weapons deposited in burials represent those used in battle, the consistent appearance of these types in burials across Greece, and later in sanctuaries and in iconography, suggests

12 Snodgrass (2000) 374–76; more detailed explanations of this phenomenon of coastal prosperity include Broodbank (2013) 500–502, Crielaard (2006), and Dickinson (2006) 212–15.

13 For a more detailed account of this evidence, see O'Brien (2022) and Lloyd (2020b).

14 Probably from Italy: Jung and Mehofer (2013) 178–83.

15 Lagia (2007) 277, 279. The burial (Grave 147) belonged to a child aged 7–12 years; the arrowhead was found in the shoulder of the skeleton.

16 Snodgrass (1973); Lloyd (2020b) 514–16; as O'Brien (2022) notes, there is convincing evidence that they served as metal armour attachments for perishable materials, cf. Bouzek (1988); Douzougli and J. Papadopoulos (2010) 23–35.

that they were the most prominent. It is certainly the case, however, that these patterns cannot be taken at face value. In order to place these weapons into their context, a wider set of evidence must be taken into account.

1.2 *Pictorial Pottery*

Pictorial pottery is prominent in the post-palatial Bronze Age pottery phase Late Helladic IIIC (LH IIIC, *ca.* 1200–1050), when it is usually found in settlements. It is almost entirely absent in the 'long, pictureless hiatus' that lasts to around 800. The majority of LH IIIC pictorial pottery is known from the Argolid and the Euboian gulf, particularly Lefkandi, although this is in part because of the extensive publication of the material from these sites. Thus, while it is notable that the Argolid and Euboian gulf are regions in which twelfth-century burials with weapons are not known,[17] unpublished pictorial pottery from Achaian sites makes it clear that these two kinds of evidence are not mutually exclusive.

Warfare is the subject of a sizable portion of twelfth-century pictorial pottery. Warriors – that is, men in helmets and armour, carrying spears, shields, and swords – are depicted in procession, in chariots, and perhaps also in battle.[18] These scenes appear mainly on kraters, vessels used to mix large quantities of wine and water, indicating that these images were on display at communal gatherings.[19] That twelfth-century pictorial pottery is almost entirely known from domestic settings indicates that it was genuinely used for this function, in contrast to the more ambiguous nature of weapons from burials. As Margaretha Kramer-Hajos argues, this points to the centrality of warrior identity in the social life of the twelfth century, at least where these vases are known.[20] As with burials with weapons, these are symbols of a culture in which the image of a warrior and warfare were significant.

Some of the most significant twelfth-century depictions are the sea battles on kraters from Kynos on the Euboian gulf coast of East Lokris, among the first such depictions in the Mediterranean.[21] The scenes, of which at least five are known, usually show warriors on oared galleys, carrying shields, raising their

17 It is also noteworthy that twelfth-century burials are not widely known from either of these sites, and thus burials with weapons may come to light as a result of future excavations.
18 A. Papadopoulos (2009) 73–76.
19 Kramer-Hajos (2016) 154–55.
20 Kramer-Hajos (2016) 152–56.
21 Emanuel (2015); A. Papadopoulos (2009) 75–76, fig. 9.10; Kramer-Hajos (2008) 91–96; Kramer-Hajos (2016) fig. 7.2.

javelins to throw at an opposing group of warriors on another ship;[22] one, however, shows warriors in combat on a single ship as if on land. Kramer-Hajos has argued that these depictions reflect an emerging 'sailor-warrior' society in the post-palatial Euboian gulf that arose in response to the collapse of the palaces.[23]

1.3 *Feasting Sailor-Warriors in the Post-Palatial Period and Beyond*

The significance of communal gatherings to the elite male self-presentation as 'sailor-warriors' offers a way into the question of social power in the post-palatial period. Social events, particularly those involving fine pottery and the consumption of wine, were occasions at which alliances could be brokered, maintained, and developed. Feasts are designed to put people in a positive mood through appeals to the senses and to the emotions;[24] in the twelfth century the evidence for feasts is discovered primarily in settlements, while in the tenth and ninth it is more apparent in cemeteries, where reaffirmation of the social structure is a necessity. Feasts are also situations in which those with limited social power might seek to establish themselves in the community – aggrandisers.[25] It is possible that military connections could be used to strengthen the position of these aggrandisers.

Kramer-Hajos proposes that the 'sailor-warriors' of the Euboian gulf were intentionally reacting against palatial domination in their iconography and settlement layout;[26] but this argument assumes that the palaces had a broad level of control over the Aegean. If, as Stephen O'Brien has argued,[27] palatial elites gained access to forces under the control of local elites, then a site such as Lefkandi may have been providing the nearest palace, Thebes, with warriors who were now left without that centralising force.[28] Local elites may have maintained or strengthened their control into the post-palatial period, explaining why there is such strong continuity in the weapons and armour of the period. The interrelation of settlements may have broken down or

22 A. Papadopoulos (2009, 75) comments, 'the full armour of the warriors certainly suggests full scale combat'.

23 Kramer-Hajos (2012), (2016) 149–65.

24 Hayden (2009).

25 On the concept of social aggrandisers, see Hayden (2009) 35–37.

26 Kramer-Hajos (2016) 150–52.

27 O'Brien (2022).

28 On the possible relationship between the palace at Thebes and Euboia, see Lemos (2020) 789–90.

changed, but the internal organisation of those settlements may have been maintained – at least for a while.[29]

Kramer-Hajos sums up the effect of the Kynos kraters on the men viewing them:

> For LH IIIC inhabitants of Kynos, ship-board battles were an important iconographical theme, which in turn suggests that these elite LH IIIC Mycenaeans valued their own (perceived) role as warriors: they wished to promote a self-image of fearless sailor-warriors who took their ships to fight on foreign coasts or, if necessary, on board of their ships.[30]

But Kramer-Hajos does not fully address what happened to these Mycenaean elites at the end of the post-palatial period, once pictorial pottery disappears in the eleventh century. In the Euboian gulf, pictorial pottery disappears after the middle phase of LH IIIC pottery around 1100, more than a century before the rise of burials with weapons in the Toumba cemetery at Lefkandi in the middle of the tenth century.[31] Connectivity by sea continues to play an important role in the Euboian gulf throughout the EIA, but evidence for a 'warrior' identity is limited for over a century. The inhabitants of the Euboian gulf appear to reject the sailor-warrior culture of the twelfth century, before reviving it in a different form in the middle of the tenth.

The archaeological evidence from the Euboian gulf in the twelfth to ninth centuries is abundant compared to many other regions. Indeed, in looking at warfare in the Early Iron Age it is integral that we recognise the significance of regional variation in not only the quantity of the evidence, but also its expression. Regional variation in social organisation and material culture is a widely accepted feature of the EIA that continues into the eighth century and beyond.[32] Although we must be careful not to assume that different forms of social organisation necessitate corresponding forms of warfare,[33] we can

29 As Foxhall (1995) argues for agriculture; see further below.

30 Kramer-Hajos (2016) 155.

31 Popham *et al.* (1980) 101. A single burial with a dagger is known from the Lefkandi-Skoubris cemetery in the eleventh century (Early Protogeometric pottery phase).

32 Snodgrass (2000, 24) states this most simply, but it permeates his analysis, beginning with pottery and following through into the grave and metals. It is also implicit in his emphasis on 'advanced' and 'backward' regions of Greece (2000, xxiv–xxv, 374–75), terminology that we should avoid in the twenty-first century CE. J. Papadopoulos (2015, 186) criticises 'regionality' on the basis that it is usually dependent on pottery, but nevertheless characterises post-palatial and EIA settlement as 'a more fragmented but independent network of settlements' than in the palatial period.

33 O'Brien (2022).

accept that in a period in which multiple forms of social organisation existed, there too existed variety in warfare, or at least in its social significance and how that significance was expressed.

The archaeological evidence gives us some idea about how local elites maintained or strengthened their position. As mentioned above, in the twelfth century pictorial decoration is usually found on kraters, vessels that indicate a shared dining ritual within an exclusive group. The significance of the krater continues into the EIA. Around the middle of the tenth century, toward the end of the Middle Protogeometric pottery phase (MPG, *ca.* 1000–950), a man, a woman, and four horses were buried in a monumental building in the Toumba cemetery at Lefkandi. These burials were marked by a huge krater, while the cremated remains of the man were buried in a bronze krater-like vessel, probably a Cypriot import.[34] Another huge krater is known from the mid-ninth century burial of the 'warrior-trader', Toumba Tomb 79, which also contained several drinking vessels,[35] and similar drinking and eating equipment is found in some contemporary cremation pyres.[36] The implication is that funerals in the EIA could have been elaborate feasting parties.[37]

The significance of funerals as feasts is a widespread human phenomenon, grounded in the emotional resonance of the funeral combined with the physical consequences of eating and drinking heavily – these could be opportunities for strengthening allegiances between elites in order to maintain their positions, for creating in- and out-groups, or for aggrandisers to challenge those alliances and groups.[38] The increasing importance of cremation in some regions of EIA Greece is also significant in this context: while at Lefkandi funeral pottery particularly indicates funeral feasting, cremation is a ritual that requires resources and time, as well as being physically imposing and visually impressive.[39] At sites where burials with weapons are connected to

34 R. Catling and Lemos (1990) 25–26; Popham *et al.* (1993) 81–96, 98.

35 Popham and Lemos (1995).

36 Toumba Pyres 13 (Popham *et al.* 1982, 227–29) and 14 (Popham and Lemos (1996) plates 20, 86–9, 100, 103, 109–10, 122, 146), both of which date to the later Subprotogeometric period (SPG III *ca.* 850–825), contained fragments of spits, kraters, and/or pendent semi-circle skyphoi. These graves and Tomb 79 all also contained a bronze grater, which has been linked to drinking rituals described in the *Iliad*: Ridgway (1997).

37 The connection between weapons, feasting equipment, and high status in burial is widely attested in the Iron Age Aegean. See also Xydopoulos (2017) 78 (Macedonia); Courbin (1957) 322–86, (1974) 40–1 (the Argos 'Panoply Tomb').

38 Hayden (2009).

39 Whitley (2002); Maria Liston estimated that the cremation of the woman in Tomb 15 at the Athenian Agora would have taken six or seven hours, and the burial contained bones from at least seventy kilograms of meat (Liston and J. Papadopoulos 2004, 15–16).

cremation – including Athens, Lefkandi, Knossos, and Thessaly – it is likely that funerary rites were a way of strengthening military alliances between local elite groups. The high-running emotions of funeral rituals would not just solidify a system already in place, but also offer opportunities to change it; they did not just strengthen ties, but also challenged them. Aggrandisers might use the upheaval of the funeral to enforce their own interests, especially through the creation of military alliances, resulting in ongoing changes in the balance of power within existing communities.[40]

Other aspects of the social order must not be ignored. The burials at Lefkandi, as well as those at Knossos and elsewhere, contain many objects from distant locations across the sea.[41] Access to these objects, whether through personal exploration or long-distance trade, appears to have been restricted; or, rather, to have been a symbol of social power in the communities in which they are found. At Lefkandi, Antonaccio argued, access to Near Eastern objects was controlled by well-connected individuals;[42] that is, those who had formed alliances at social occasions throughout the eastern Mediterranean. It is also possible that some of these objects were obtained through violence, such as raids.[43] In a period in which overseas travel had declined, evidence of connections overseas was a symbol of power.

Just as we cannot ignore regional differences in EIA Greece, we must also acknowledge change over time. While Athens maintains a fairly consistent pattern of burials with weapons from 1100 to 800, at Lefkandi the evidence is particularly strong between 950 and 850, after which the known cemeteries are abandoned. In certain regions, the EIA was a dynamic period in which social organisation could often change; military structures were probably similarly fluid. However, it is likely that the military organisation and combat of the post-palatial period set the tone for combat in Greece for the next several centuries: local elites strengthened their control over warriors, forming war-bands that fought on land and sea; the sea itself could be a source of danger, or a source of profit. These war-bands forged and broke alliances, perhaps especially at funerals. As such, the most prosperous regions of Greece attached significant social value to the status of 'warriors' between 1200 and 800, reflected in pictorial pottery and burials.

40 Hayden (2009) 37–38.

41 Morris (2000) 195–256; Antonaccio (2002). While the focus has typically been on the Near Eastern objects, Lemos (2001, 17) notes that 'in almost every tomb with imports from the East, there are also imports from the North'.

42 Antonaccio (2002) 16.

43 For the eighth-century examples at Eretria, discussed further below, see Luraghi (2006) 38–40.

1.4 *The Question of Battle*

It is more difficult to reconstruct EIA battle. In a period without literature, pictorial pottery, or other visual media, we have no contemporary guide to structure our interpretation of the use of weapons. One exception to the absence of pictures during the period is a depiction of confronted archers on an MPG hydria from Lefkandi.[44] Alongside the arrowheads found in burials,[45] this hydria hints at a significant role for the bow-and-arrow in EIA battle, at least along the Euboian gulf. Javelins also indicate the importance of long-range weapons; but the prominence of swords suggests that close combat held social significance in the EIA.[46] Furthermore, different patterns of deposition across Greece suggest regional variations: at Lefkandi, swords and spearheads are found together, but there is no grave with more than one of either; meanwhile, in Athens, swords and daggers are found together and spears are found in pairs. However, tactics cannot be deduced from these groupings.

Sparse evidence across Greece between 1200 and 800 suggests that the chariot remained in use throughout the period, at least in Euboia and Crete. Chariots with warriors appear on Argive LH IIIC pictorial pottery and burials of horse-groups are known at Lefkandi, Knossos, and Prinias during the 'pictureless hiatus'. These burials need not indicate the use of the chariot in battle. The four horses buried in the Toumba Building at Lefkandi are associated as much with the female inhumation as with the male cremation-with-weapons,[47] while another horse burial in the same cemetery shows no connection to burials with weapons. At both Knossos and Prinias, horse burials are associated with those of dogs, which may indicate that in central Crete the chariot was associated more with hunting than with warfare.[48] Fighting from horseback is similarly controversial: some, but not all, of the thirteenth- and twelfth-century figurines of horse-riders known from the Argolid are armed;[49] a twelfth-century krater from a tholos tomb at Mouliana on Crete depicts a warrior on horseback, but not in combat.[50] An exceptional Early Geometric (EG, *ca.* 900) burial with horse-bits and weapons from the Areopagus has been

44 Popham *et al.* (1980) 127, fig. 4.

45 At Lefkandi, Toumba Tombs 26 (*ca.* 950–900) and 79 (*ca.* 850) contained arrowheads, while Toumba Pyre 1 may have contained a bow; at Athens, Kerameikos Graves 28 (*ca.* 950–900) and 147 (*ca.* 1000; Lagia 2007, 277, 279) each included a single arrowhead.

46 For swords as particularly significant weapons in the EIA, see Lloyd (2015).

47 Harrell (2014) 101, who also notes that the female burial contains a dagger-like object of her own and that she could also be considered a 'burial with weapons'.

48 Rizza (1979).

49 Kelder (2012).

50 A. Papadopoulos (2009) 74, fig. 9.8.

dubiously interpreted as evidence that cavalry existed in EIA Athens.[51] It certainly cannot be argued that cavalry was prevalent in EIA Greece; most likely a true cavalry was not widespread until the later Archaic period.[52] As O'Brien has argued, if chariots were in use in EIA warfare, it looks much more complex than is usually assumed.[53]

In the century following the destruction of the Mycenaean palaces, several sites across the Aegean suffer repeated destruction and rebuilding, including Tiryns, Lefkandi, and Koukounaries on the island of Paros.[54] In the eleventh century, such evidence is absent; then in the middle of the tenth century, a building at Mitrou, near Kynos, collapsed and was subsequently set on fire.[55] To the south, at Lefkandi, a long wall had been constructed across the isthmus connecting the promontory Xeropolis, on which the settlement was located, to Euboia. While this wall would not have served as much of a fortification, it marks the only point from which Xeropolis could have been approached by land and thus may have served a defensive function.[56] As noted above, the significance of the Euboian gulf as a sea-facing region should not be ignored. It is possible that, with the destruction of Mitrou and the revival of burials with weapons at Lefkandi, we can see a similar pattern of sailor-warrior violence along this gulf as in the twelfth century, but the evidence is tenuous.

While such holistic narratives of this transformative period may be superficially seductive, the evidence simply does not support a complete narrative of combat in the EIA with the detail to which many historians are accustomed. To try to reconstruct it in detail is a clear example of what Anthony Snodgrass has called the 'positivist fallacy' – the belief that what is archaeologically visible is historically significant.[57] Rather, the evidence better supports analysis of social systems that value warriorhood as an expression of elite, male identity.

2 Greek Warfare from 800 to 600

The eighth and seventh centuries see an abundance of new evidence that we can use to build a narrative of warfare in the period. An increase in dedications at sanctuaries, particularly interregional sanctuaries, includes greater

51 Sidnell (2006) 25; D'Onofrio (2011) 653.
52 Brouwers (2007) 315–16.
53 O'Brien (2022).
54 See Deger-Jalkotzy (2008) 394–95 for a discussion with references.
55 Van de Moortel and Zachou (2004).
56 Frederiksen (2011) 107.
57 Snodgrass (1987) 36–66.

numbers of weapons and armour. Figurative decoration reappears in the Middle Geometric pottery phase (MG, *ca.* 850–750), becoming more prominent in the Late Geometric (LG, *ca.* 750–700) and seventh-century styles, although it remains diverse regionally. Eighth-century Attic burials are marked with monumental vases, on which figurative scenes including battles on land and sea are depicted.[58] Burials with weapons disappear from Attika but become much more widely known across the Aegean including burials with armour in Argos. As well as the burials of individuals or small groups, there are also the Parian *polyandria*, a unique example of a pair of mass graves belonging to 117 men aged sixteen to sixty, including some weapons. Finally, the creation of the Greek alphabet in the eighth century leads to the recording of the *Iliad* and the *Odyssey*, as well as early lyric poetry on the theme of war by Archilochos and Tyrtaios that dates to the seventh century.[59]

We must be cautious about treating these changes as evidence of a major break from the preceding period. As we have seen, there were changes throughout the EIA in the use of visual culture and the frequency of burials with weapons. In the eighth century, certain changes appear more dramatic, but this is not a move from stasis to activity. Although it may move at different speeds, change is constant. The prevailing character of EIA elites is that of masculine warriorhood. The question to which we now turn is how this affected Greek culture from the eighth century.

2.1 *Warfare in Epic and Lyric Poetry*

Written texts add a new wealth of data to our archaeological evidence, opening up new avenues of historical investigation. But it is also significant because of the distorting effect that data has on our understanding of the period. As it becomes possible to explore certain ideas in written texts from the eighth century onward, it *looks* like there is a sharp contrast between the eighth and seventh centuries and what went before. This is not necessarily the case. There is also a tendency to view these literary accounts of warfare as representing a relative chronology of battle – 'Homeric' warfare, dated to the eighth century and perhaps earlier, is replaced by new modes of warfare represented by the armour and weapons connected to 'hoplite' warfare as described by Archilochos and Tyrtaios in the seventh.[60] The degree to which 'Homeric' war-

58 Ahlberg (1971), although no longer complete, is still the fundamental study of these scenes.

59 For a recent account of the creation of the Greek alphabet, see Steele (2020).

60 For example, Lorimer (1947) 111–15; *WWW* 40–44. As Hall (2014, 167–69) notes, this idea is as indebted to Aristotle (*Pol.* 1297b.16–24) as it is to Homer.

fare aligns with that of the eighth century and the distinctness of these fighting styles are questionable.

Many scholars date the Homeric epics to the eighth century, the moment that we know a Greek alphabet existed – some even posit that the alphabet was invented to record these epics.[61] But we must recognise that even a partially literate ancient society is likely to have been largely an oral culture, particularly in this early stage,[62] and that the Homeric epics probably continued to develop as oral poems until they became widespread in the sixth century, a factor that is significant to our understanding of the poems' historical relevancy.[63] As they continue to develop, they retain certain elements of their past forms, stretching back into the Bronze Age, but as they become more well-known they are less susceptible to major changes. Snodgrass noted that this approach 'poses a direct and obvious threat to any theory of consistent and coherent Homeric battle tactics, belonging to a single period'.[64]

Even if we could securely date the composition of the Homeric epics to *ca.* 700, then the date of the warfare they depict is another matter. There are three broad viewpoints on this topic. The first is that Homeric warfare is 'pre-hoplite' as it focuses on individual champions dominating melee fighting, and thus it must date to the eighth century and perhaps earlier.[65] The second is the view that Homeric warfare is contemporary with the life of the poet in the seventh century, as it most closely resembles scenes of combat depicted in vase painting at this time.[66] Finally, some simply dismiss Homeric warfare as ahistorical.[67]

If one does not dismiss Homeric warfare as ahistorical, the Homeric evidence on its own is insufficient and we must turn once again to the other contemporary evidence: archaeology. As mentioned above, Homeric warfare corresponds much more closely to the depictions of warfare that we see on pottery of the seventh century than that of the eighth.[68] But it is worth establishing first what 'Homeric warfare' actually looks like, and it is neither simple

61 Morris (2001) 81–82. For the *Iliad* as a seventh-century poem, e.g., West (2011).
62 O. Murray (1993) 182–83.
63 Nagy (1997) 206–207; Dué and Ebbott (2010) 14–30.
64 Snodgrass (2013) 89–90 (quotation 89).
65 Lorimer (1947); Kagan and Viggiano (2013) 2–4; Viggiano (2013) 113–14; Morris (2009, 75) admits a lack of clarity on when such a vision might have been applicable to historical warfare.
66 Van Wees (1992, 1994a, 1994b).
67 Snodgrass (1993, 2013). Here, Snodgrass is largely responding to Latacz (1977), who argued that Homeric warfare is hoplite warfare, and that it dates to the eighth century; cf. Snodgrass (2013) 86–87.
68 Van Wees (1994a, 1994b).

nor homogenous. While the narrative focuses on 'super-warriors', in the background there is much more going on.[69] The *Iliad* includes scenes of ambush and counter-ambush (e.g., *Il.* 10.335–377; 18.520–534), sieges against the walls of both the Achaian camp and Troy (e.g., *Il.* 12.445–466; 16.702–704), the formation of shield-walls (e.g., *Il.* 13.125–135), as well as one-on-one duels (e.g., *Il.* 3.343–380), and battles over the dead (e.g., *Il.* 16.755–764); not to mention the skirmish that ends the *Odyssey* (e.g., *Od.* 22.1–389). There are also scenes of mass combat, in which large groups of warriors work together to defeat the enemy (e.g., *Il.* 13.339–43).[70] As we will see in the next section, eighth- and seventh-century vase painting is similarly varied with regards to scenes of warfare.

Alongside the longstanding tradition of oral epic poetry, in the seventh century other genres of poetry and named poets become known – notably the Parian Archilochos in the middle of the century and Tyrtaios of Sparta in the second half; although Alkaios of Mytilene at the end of the seventh century is also relevant. What survives of these lyric poets is largely, although not exclusively, that which is cited by other ancient sources for its continuing relevance.[71] This selective nature of survival means that, as with epic, we must be cautious using lyric poetry to understand seventh-century developments independently of securely dated archaeological evidence. Our understanding of Archaic Greek warfare is based on partial, selective evidence which has been subject to a number of factors affecting its preservation.

Approaches to the written evidence for warfare in this period typically takes a typological approach; that is, seeking to show how the kind of warfare apparent in Homer is replaced by the type of warfare apparent in Archilochos and Tyrtaios. Similarities are usually attributed to the lyric poets mimicking epic style or even to interpolations into the epic text, and differences to new styles of warfare.[72] One of our fragments of Archilochos, however, points to a different explanation. In this fragment (fr. 3 West), Archilochos announces the preference of the 'spear-famed' Euboians for swords over bows and slings, suggesting that the choice of fighting style is regional, not chronological.[73]

69 *GWMR* 153–65.

70 On the distinction between *mass* combat – warriors working together – and *massed* combat – warriors fighting in organised, tight formation, typically requiring a command structure – see Brouwers (2013) 92–94.

71 Van Wees (2000a) 149.

72 Lorimer (1947); Snodgrass (1964) 181. On the use of epic style in lyric poetry, see Silk (2010) 426–31; West (2011) 15 notes that ancient dates sometimes made Homer and Archilochos contemporaries.

73 Lloyd (2017).

In terms of chronology, preference for swords better matches LG pottery than the *Iliad*, as does a preponderance of bows elsewhere. In the *Iliad*, swords represent fewer than ten percent of the weapons used, while spears, both cast and thrust, represent eighty percent;[74] Archilochos and Tyrtaios also emphasise spears over swords. The *Iliad* and Tyrtaios describe the practice of light-armed warriors hiding behind the shields of their comrades for cover (*Il.* 8.266–272; Tyrtaios fr. 11 West). In Tyrtaios, these men are called *gymnetes* (sing. *gymnes*), meaning 'naked-men', in contrast to the 'all-armed' *panoploi* who fight with spear and shield. While some take Tyrtaios' calls for unity among the fighting warriors as an indication that he represents organised massed fighting, this appeal to the light-armed suggests that there must have been considerable room for manoeuvre between them.[75] Archilochos sings of throwing away his shield when he flees battle, suggesting a heavy object that could not be flung across the back on a *telamon* (fr. 5 West); Alkaios also loses his shield, which is dedicated by the Athenians to Athena (Alkaios fr. 401b Voigt).

By the turn of the sixth century, when Alkaios sings of arming men in his hall, he lists crested bronze helmets, greaves, and linen corselets, as well as swords (fr. 140 Voigt). There are no spears, which perhaps men provided for themselves, nor anything for the light-armed, although the greaves 'keep strong arrows off' (trans. West). This distinction may be because of the social circle of those whom Alkaios addresses, as archers were still significant enough to be defended against.

2.2 *Pictorial Pottery Resumes*

The resumption of pictorial decoration on pottery is much more precisely dated evidence for warfare from *ca.* 800, and it covers a much wider chronological range than the poems. As visual evidence, however, it lacks the detail and specificity that is so useful in the Homeric and lyric descriptions of battle. Furthermore, pictorial decoration is not a photograph – it, like poetry, is affected by the cultural needs at the time of its creation.

Battles appear among the earliest narrative scenes in Attic MG pottery and become prominent in the LG style. Human figures on Geometric pottery start out as silhouettes with added details – helmet plumes, swords, shields – although as the style develops more details are left in reserve. Pictorial decoration continues into seventh-century styles, such as Protocorinthian and Protoattic. There is furious debate over what is depicted in seventh-century

74 Van Wees (1994b) 144.

75 *GWMR* 173.

iconography.[76] There is more detail in outline and black-figure techniques than there is in Geometric pictorial, including armour, weapons and perhaps tactics, but it remains an open question how much of our interpretation of seventh-century iconography is derived from it and how much is imposed upon it from pre-existing scholarly ideas.

In MG and LG Attic pictorial pottery the most common weapon shown in use is the sword, followed by the bow. Spears represent about a quarter of weapons used,[77] and they are typically held aloft as if they are about to be thrown.[78] From the earliest LG representations of battle the Dipylon shield is prominent. This round, convex shield with crescents on either side appears in LG iconography across the Aegean and was carried by a *telamon* which allowed it to be slung across the warrior's back. In the later eighth century the Dipylon shield is less prominent, as rectangular and, most significantly, round shields become more common.[79] In LG depictions, the victors in these conflicts become much more clearly marked – Ahlberg notes that the victorious warriors usually approach from the left.[80] Often, the naked dead are stacked alongside the battlefield, stripped of their arms; a scene on an Attic oinochoe in the Louvre has been interpreted as captured men being stripped of their weapons and armour.[81] Several of these scenes show battle taking place at the shore beside a ship, the amphibious context contrasting with their twelfth-century predecessors, where combat was at sea (Figure 2.1).

While the evidence for cavalry is meagre before the eighth century, pictorial pottery gives us a better idea of the role of horses in warfare. The clearest depictions of horse-riders in combat are two LG amphorai from the eighth-century *polyandria* from Paroikia on Paros. The scene on one of these amphorai shows a raid on a group of herdsmen in which figures on horseback hold their spears aloft as if to throw them at the unarmed figures below; on the other, a pitched battle between confronted warriors over a dead warrior, with horse-riders stood at the rear of the lines.[82] On one LG vase from Athens and

76 An overview of the debate is given in Viggiano and Van Wees (2013); referenced in particular are Lorimer (1947) and Van Wees (2000a). For more on this debate, see Lloyd (2017) 239–43.

77 Van Wees (1994b) 144.

78 On the 'throwing' stance for spears: Matthew (2012) 15–38, especially 19 and 29–30.

79 Rombos (1988) 132–46.

80 Ahlberg (1971) 53, cat. A5.

81 The Lambros oinochoe, Louvre CA 2509; for this interpretation of the decoration, see Ahlberg (1971) 21–25.

82 Zaphiropoulou (2006), figs 1–4, 5–8; Lloyd (2017) 232–33. Williams (2015) 262 insists that the latter amphora must be an Attic import.

FIGURE 2.1 Attic LG ship battle scene on a krater in the Metropolitan Museum of Art, New York (ca. 800–775 BCE; MMA 34.11.2). Attributed to the Workshop of New York MMA 34.11.2
PUBLIC DOMAIN

a seventh-century relief pithos from Paros, horse-riders are depicted raising javelins, but the context of these figures is unclear.

Horse-drawn chariots also appear in a handful of LG battle scenes, although they are more common in scenes of procession, transport, or races. Chariots were probably used throughout the EIA; depictions of them in the eighth century closely resemble those of the post-palatial period, except that LBA chariots are always drawn by two horses while those in the eighth century vary between one and four.[83] Fragments of an MG vessel from the sanctuary of Aphaia on Aigina show two chariots, one with an archer and the other with a spear-thrower, apparently engaged in combat.[84] An LG oinochoe from a burial in the Athenian Agora is usually, though debatably, interpreted as a combat scene in which the mythical conjoined twins, the Aktorione-Molione, escape in a chariot.[85] A chariot is running down a man with a round shield in the Parian raid amphora scene. Finally, a Sub-Geometric (early seventh century) Boiotian

83 Crouwel (1981) 97; (1992) 29–30.
84 Williams (2015) 261–62, fig. 6.
85 Ahlberg (1971) 12–13, cat. A1. Dahm (2007) criticises this interpretation as the twin-figures are hidden beneath the handle of the oinochoe, hardly the place for the figures who define the scene. Dahm suggest that these two figures bookend a scene, distancing the chariot from combat. On figures of conjoined twins in early Greek art in general, see Snodgrass (1998) 26–32.

amphora depicts a warrior being carried from the battlefield in a chariot. The depiction of chariot fighting in the *Iliad* perhaps provides an explanation for these scenes. In the *Iliad* the chariot functions as a status symbol, which the heroes ride around the battlefield but rarely use in combat.[86] It is likely that the same function applied to the horses seen in later Corinthian vase painting described below. It is possible, then, that horse-owning individuals rode their horses to and around the battlefield in the eighth and seventh century, and perhaps even later.[87]

By the early seventh century, duels appear to be the favoured scene, without indication of the victor. On the Early Protoattic (EPA, *ca.* 700–675) Hymettos amphora from Aigina,[88] most figures fight with spears, but some have Naue type II swords. It is also one of the earliest depictions of the inside of an *aspis*, the convex, circular, double-grip shield with the *porpax* and the *antilabe* that becomes the standard Greek shield for centuries to come. Many of the figures on EPA vases wear Corinthian helmets and greaves, although body-armour is not always evident. It is unclear how far the introduction of this new armour was contemporary with the stylistic change and how far its depiction was facilitated by it. Certainly, seventh-century changes in tactics are not clear on Attic pottery, although there is a notable absence of horses, chariots, and ships.

In the Early Protocorinthian (EPC) style, depictions of warriors and combat are somewhat different to their Attic contemporaries. Two vessels have been interpreted as 'unrealistic "heroic" scene[s]' because they feature nudity and supposedly anachronistic armour – the Dipylon shield, or perhaps its successor, the Boiotian shield.[89] On an aryballos from Lechaion,[90] a group of warriors with Dipylon/Boiotian shields raise their spears as if to throw them at a figure with a double-grip *aspis*. Behind this figure, who also wears a bell-cuirass, crouches an unarmoured archer – not unlike the tactic described in the *Iliad* and by Tyrtaios. A slightly later aryballos from Perachora shows a similar scene, although in this case the warrior with the double-grip shield is supported by a warrior with a Dipylon/Boiotian shield. It is noteworthy that on both of these

86 *GWMR* 158–60, with references to the *Iliad*. Chariots in combat: *Il.* 4.293–309, 11.289, 150–54, 15.352–54.

87 Worley (1994) 21–58; Van Wees (2013) 241–42.

88 Lorimer (1947) fig. 4.

89 Viggiano and Van Wees (2013) 63–65 (quotation 63), fig. 2.4, 2.5. Lorimer (1947) 94–95, whom Viggiano and Van Wees follow, only discusses the Perachora aryballos. While Viggiano and Van Wees state that Lorimer's interpretation of the scene was based, in part, on 'heroic nudity' (2013 64–65), Lorimer does not mention that nudity is heroic while discussing this vase.

90 Corinth Museum CP-2096; Viggiano and Van Wees (2013) fig. 2.4.

vessels the warrior with the double-grip shield approaches from the left, as did the victorious warriors in Attic Geometric vase painting.

Our understanding of these scenes depends, in part, on how we understand the Dipylon/Boiotian shield. While it is usually accepted that the Dipylon shield was a real object,[91] the Boiotian shield, which is sometimes depicted with the interior double-grip known principally on the *aspis*, is more controversial. It has been suggested that the Boiotian shield is a symbol of archaism indicating that the scene in question is of the 'heroic' past.[92] The basis of this argument seems to be that with the advent of the *aspis*, the Dipylon shield was redundant and thus can only have survived as a reference to 'pre-hoplite' tactics – even though it is often depicted alongside the *aspis*. In a classic article on early Greek warfare, Lorimer argued that it was impossible for the Dipylon/Boiotian shield to be used alongside the double-grip *aspis*: 'That no such combination of incompatible equipment and tactics could ever have been practised in real life hardly needs stating; the hoplite phalanx was of its essence one and indivisible'.[93] This viewpoint explains away elements of the ancient vase paintings in order to fit a modern assumption about 'hoplite warfare', rather than adjusting modern interpretations to match the ancient evidence. As Krentz states, 'we have to guard against letting assumptions about how hoplites fought prejudge what equipment they used'.[94] In order to avoid prejudging we must base our understanding on the evidence from the period. That evidence is scenes of combat in which the Dipylon/Boiotian shield was in use. As I do not believe there is a convincing reason to doubt the existence of the Boiotian shield, I assume that it was a genuine piece of equipment in the seventh and sixth centuries. On balance, it seems strange that a shield should become useless in combat, and yet remain a symbol of heroism – often apparently the only such symbol in a scene – for centuries to come.[95]

Nevertheless, the nudity in these scenes poses a problem; as Jeffrey M. Hurwit notes, 'in combat nakedness was suicidal'.[96] Hurwit argues that the concept of 'heroic nudity' seems to be new to Greece in the seventh century as it is not present in the Homeric epics. Yet it is surely significant that in many of these early seventh-century depictions it is light-armed warriors who are

91 Although not always – see Webster (1955); Snodgrass (1980) 74–75.

92 Lorimer (1947) 94–95; Ahlberg (1971) 63; Greenhalgh (1973) 64; Hurwit (1985).

93 Lorimer (1947) 95.

94 Krentz (2013) 136.

95 Krentz (2013) 136–37, 148 offers both an argument for how the shield might have been used, and connects it to 'aristocratic *promachoi*' to explain its connection to heroes.

96 Hurwit (2007), quotation 45–46. See also Bonfante (1989) 547–50.

'naked' – warriors we would usually assume are not heroic.[97] Here, we must turn to other contemporary evidence for clues. In the later seventh-century lyric poems of Tyrtaios, the word for light-armed warriors is *gymnetes*, meaning 'naked-men', in contrast to the 'all-armed' *panoploi* (fr. 11.35–38 West). Rather than heroism, in these early depictions of light-armed troops their nakedness is symbolic of their vulnerability in contrast to the highly protected fully-armoured warrior behind whose shields they crouch. This nudity is new – archers in LG vase painting are sometimes shown with shields and armour. If this nakedness cannot be called heroic, then the only marker of heroism in these scenes is the 'obsolete' Dipylon/Boiotian shield; but if we cannot be certain that this shield was obsolete, we cannot assume that it only appears in heroic scenes. Rather, we should take these scenes at face-value as attempts to signal contemporary warfare, which also contain various symbolic elements.

The Middle to Late Protocorinthian (MPC, *ca.* 690–650; LPC, *ca.* 650–625) vessels with decoration attributed to the Chigi Painter are considered especially significant to our understanding of developments in warfare. In particular, the upper register of the Chigi Olpe from Veii in Etruria is said to be the earliest certain depiction of hoplite phalanges;[98] however, the two opposing lines of warriors in heavy armour, about to clash, are in uneven ranks and armed with a pair of spears, one held aloft as if to throw,[99] which much more closely resembles warfare in Homer. On other vessels attributed to this artist, we see four disordered warriors on a fragmentary oinochoe from Erythrai in Asia Minor, rout and pursuit on the MacMillan aryballos in the British Museum (Figure 2.2), and a similar stage of battle alongside two other confronted groups of warriors on the Berlin aryballos.[100] We see here some developments toward more orderly combat, but there is continued use of javelins and uneven ranks.

Contemporary depictions by other Corinthian painters also tend to focus on either single combat or multiple groups of warriors. All carry the double-grip shield and wear Corinthian helmets and bell-cuirasses; the most common weapon is the spear, usually held aloft, and many carry a second spear. Certainly, if one focuses solely on the Corinthian repertoire, it may appear that

97 Brouwers (2013) 60 notes that there is some evidence to show that archers were not perceived as un-heroic at this time, although this is in circumstances where they wear armour.

98 The Chigi Olpe has been discussed extensively in literature on early Greek warfare; see especially: Lorimer (1947) 80–83, fig. 2; Van Wees (2000a) 134–39; Lloyd (2017) 240–42.

99 Matthew (2012, 15–38) argues that this stance is ineffective for thrusting a spear; however, in some depictions it seems as if the intention of the artist can only have been to depict thrusting.

100 Berlin 3773; Viggiano and Van Wees (2013) fig. 2.6.

FIGURE 2.2 Perfume bottle (aryballos) with lion-head mouth, attributed to the Chigi Painter (1889,0418.1)

these seventh-century painters are struggling to depict something resembling the Classical Greek phalanx. But we must remember that we are coming at these images with the phalanx in mind; the artists themselves may have been more-or-less accurately depicting contemporary warfare that would eventually change into the phalanx. If we put the idea of the phalanx aside, these scenes can be interpreted as moments of *mass* combat – warriors working together – rather than *massed* combat – men fighting together in close formation.[101]

101 Van Wees (1994a) 15, n. 7; Brouwers (2013) 92–94.

In contrast to contemporary Athenian vase painting horses are common in the Protocorinthian 'knight-and-squire' motif. These scenes show a youth on horseback, labelled *hippostrophos* ('horse-turner'), accompanying a fully-armed warrior, always on foot, labelled *hippobatas* ('horse-fighter'). Late seventh-century depictions show the *hippostrophos* accompanied by a second horse, suggesting that the *hippobatai* rode to battle before dismounting to fight leaving the *hippostrophos* to look after his master's steed – no known seventh-century scene of combat includes horses.[102] Additionally, eighth- and seventh-century figurines of horse-riders are known from the Argolid, the Corinthia and Boiotia, some of whom wear helmets or carry shields, and both the bell-cuirass and double-grip shield have been interpreted as developments benefitting horse-riders.[103] Horses, even if they were not in combat, were central to developments in eighth- and seventh-century warfare.

2.3 *Arms and Armour*

Metal armour reappears in the Greek archaeological record early in the eighth century. Conical helmets, known from Eleutherna on Crete and on figurines from the Argolid and Boiotia, appear to have been influenced by Assyrian types; so too does the *kegelhelm*, known from Argive tombs and dedications at Olympia, which was constructed from a conical cap, neck- and cheek-pieces and a forehead guard.[104] The Corinthian helmet becomes dominant from the early seventh century, perhaps because its construction, from a single sheet of hammered bronze, removed the structural weaknesses of the *kegelhelm*. But the Corinthian helmet was uncomfortable, limited the senses, and offered little protection to the wearer.[105] These limitations raise the question why the helmet was used at all. It is likely that the image of the 'man of bronze' encased in armour was intended to be intimidating, particularly in combination with a

102 Greenhalgh (1973) 84–95; Brouwers (2013) 76–78, with illustrations.

103 Bell-cuirass: Snodgrass (1971) 45–46; shield: Brouwers (2007). See further below.

104 Dezsö (1998). The *kegelhelm* in Argos Tomb 45 is illustrated, along with the bell cuirass, in Snodgrass (2000) 272, fig. 95.

105 *WWW* 71–75; O. Murray (1993) 128. Hanson argues that these weaknesses of the Corinthian helmet necessitated the development of close-order tactics, but fails to make any positive argument for the adoption of this kind of helmet in the first place. Matthew (2012, 96–101, with plate 22.1) argues that the limitations to vision and hearing have been overstated, although he uses an unusually open-faced helmet design to make his argument; his conclusion that the helmet did not reduce the immediate field of vision can be reconciled with Brouwers' observation (2013, 109) that the helmet would be more difficult to use in close-order, but fine if one needs only focus on a few enemies at a time, with one's compatriots never too distant. Scholars' approaches to the Corinthian helmet often depend on the point that they are trying to prove.

helmet crest and if the armour was well-polished, reflecting the bright Greek sun.[106] The effect, however, would be somewhat muted if the Corinthian helmets were worn by all Greeks engaged in combat in the late eighth to sixth century, when the helmet is dominant among dedications and in vase painting.[107]

The bronze bell-cuirass is known from around 700 among dedications at Olympia and from Argos Tomb 45, where it is accompanied by a *kegelhelm*. The name 'bell-cuirass' describes the flaring lower-rim of the breast-plate, which may have allowed the wearer to sit more comfortably on horseback.[108] In contrast to the eastern Mediterranean inspiration behind eighth-century helmets, the closest parallels for body-armour come from the north and west, including a group of ninth- and eighth-century cuirasses from Marmesse (Haute Marne) in France.[109] The tradition of European body-armour represented by this group stretches back to the twelfth century and was itself apparently inspired by Mycenaean armour. It is likely that this reappearance of body-armour in the archaeological record represents its reintroduction into Greece, although there are those who doubt that it truly disappeared.[110] The essential question is whether it is more feasible that the lack of evidence for armour in twelfth-to-eighth-century Greece is because there was no metal armour, or because little survives. The limited evidence implies that body armour could have consisted of perishable materials to which metal pieces were attached;[111] in this case, the bell-cuirass would have been a significant new adoption. On the other hand, the changing nature of the disposition of weapons and armour could indicate that there were no contexts in which a bell-cuirass would have survived from the EIA.[112] On the balance of the evidence, the former argument seems more likely, but the latter cannot be completely ruled out.

Aside from the continuing tradition of bronze *phalara* in eighth- and seventh-century graves and sanctuaries, the development of the shield is better known from iconography. Several decades after it is depicted on the EPA Hymettos amphora in the latter half of the seventh century, shield-facing and blazons are known from Olympia. The inspiration for the double-grip shield is

---

106 *GWMR* 53; but see also Van Wees in this volume on who found the 'men of bronze' impressive.

107 Van Wees (2013) has argued that only the propertied class would have been able to afford body armour: see also Van Wees and Fisher (2015) 12. Limitations on the availability of the Corinthian helmet may explain its value as an intimidating visage.

108 Snodgrass (1971) 45–46.

109 For the Marmesse armour, see Mohen (1987).

110 Reappearance: Snodgrass (1964) 83, (1971). Doubts: H. Catling (1965).

111 See O'Brien (2022); Douzougli and J. Papadopoulos (2010) 30–34.

112 Van Wees (1998) 340.

contentious. It has been variously argued that the shield functions optimally in the phalanx and so must have been invented for it, that it was a necessary precursor of the phalanx, that it is simply a matter of personal protection, and that the shield was designed, like the bell-cuirass, for ease when riding on horseback.[113] It is, of course, this shield that forms the basis of the 'hoplite revolution' in tactics and sometimes even in its social or political sense.

Changes in weaponry are far less profound than those in armour. The Naue type II remains largely unchallenged as the Greek sword of choice, both in depictions on pottery and in excavated contexts. Spearheads show more variety, with those excavated from eighth-century graves maintaining the javelin-like shape of EIA spearheads, while some from seventh-century sanctuary contexts are so large that they can only have been thrusting-lances or for show. Evidence for other weapons remains scarce in the eighth century. Arrowheads occasionally appear as dedications at sanctuaries, but are no longer found in graves. Nevertheless, archery is common in eighth-century pictorial pottery and is mentioned in both epic and lyric poetry.

2.4 *Other Archaeological Evidence*

Other archaeological evidence points towards a change in the scale and organisation of combat towards the end of the eighth century. Both in and around the many vases in the two Parian *polyandria* were buried the cremated remains of 117 men and one woman aged between sixteen and sixty, many of whom suffered skeletal trauma – some remains contained fragments of iron weapons.[114] Besides the Parian *polyandria*, the only pre-Classical *polyandrion* known is a late seventh-century burial from Akragas (modern Agrigento) in Sicily containing one-hundred and fifty vases and many inhumed corpses.[115] While the evidence is limited, these *polyandria* appear to indicate not only greater casualties in combat, but broader care for those killed.[116]

Another indication of an increase in scale are the measures taken to protect settlements, and the evidence for those that failed to protect themselves. Eighth-century fortifications, consisting of a stone socle with a mud-brick superstructure, might encircle an entire settlement, like those of Old Smyrna,

113 Optimal use in a phalanx: Lorimer (1947) 76–77; Hanson (2013) 267; Schwartz (2013), especially 161–64; Viggiano (2013) 118–19; a necessary precursor of 'hoplite tactics': O. Murray (1993) 129; personal protection: Van Wees (2000a) 126–31; Krentz (1985, 1994); horseback: Brouwers (2007). A century's worth of argumentation is summarised in Krentz (2013) 137–40.

114 Zaphiropoulou (2002); Agelarakis (2017).

115 *GSW* 4 126; Griffo (1946).

116 Agelarakis (2017).

or only part of the city, as at Emporion on Chios. At Zagora on Andros, as at EIA Lefkandi, a single wall was constructed to protect the only approach to the settlement by land. The significance of fortification is debateable. Sites like Asine in the Argolid, where fortifications were constructed just prior to the settlement's destruction toward the end of the eighth century, may signify an atmosphere of violence. However, when Zagora was abandoned at about the same time as Asine, the town appears to have been resettled at Palaiopolis, an undefended coastal site with better harbours.[117] Palaiopolis was not fortified, suggesting that the protective but inaccessible site of Zagora was no longer considered worthwhile. On the other hand, the seventh century sees a much wider adoption of fortification than the eighth. Siege of a walled city is a common theme in Archaic poetry: note the siege of Troy and the city on Achilles' shield in the *Iliad* (*Il.* 18.509–40), the casual mention of the destruction of a city by Hesiod (*Works and Days* 246–7), and the tower defended by Archilochos (fr. 98 West). But city walls serve ideological as well as practical functions: they mark the territory of the settlement, delineating the separation between the natural and social orders.[118]

The archaeological identification of settlement destruction is fraught with difficulties, especially the 'positivist fallacy' – equating archaeological visibility with historical significance.[119] Abandoning a city is a complex, multicausal process that may involve warfare, but just as often involves environmental and social factors. Evidence for the destruction of a building at Lefkandi, *ca.* 700, has been linked to the Lelantine War between Chalcis and Eretria, to which Thucydides alludes (1.15), but about which virtually nothing certain is known.[120] The 'destruction' of the site is limited to one known building, and while the hypothesis of settlement destruction is supported by the subsequent abandonment of the site, the example of Zagora reminds us that abandonment may have been the result of relocation. The historical and archaeological evidence seem more aligned at Asine, which was destroyed and abandoned *ca.* 700.[121] The proposed aggressor, following Strabo and Pausanias, is Argos (Strabo 8.6.11; Pausanias 2.36.4–5). While these sources are late, there are reasons for supporting the attribution. Argos is local, one of the larger known settlements near to Asine; an arrowhead was found in the destruction layer; after the destruction, the sanctuary of Apollo Pythaeus on the Barbouna hill was

117 Hall (2007) 42, (2014) 78–79; Morgan (2012).
118 Brouwers (2013) 55; Seaman (2020) 29–33. See also Echeverría, this volume.
119 Snodgrass (1987) 36–66.
120 Hall (2014) 1–8.
121 Frödin and Persson (1938) 437.

rebuilt and the same cult appears to have been adopted in Argos at this time.[122] While this evidence is hardly conclusive, it does support an interpretation of the destruction as violent and the aggressor as Argos.

The sack of Troy is depicted on a relief pithos from Mykonos, dated *ca.* 675, showing Greeks exiting the wooden horse at Troy and proceeding to kill children and, presumably, abduct and rape women.[123] While the scene is firmly mythological, the slaughter of the children and the abduction of women may reflect the reality of seventh-century warfare in the Aegean.[124] Wars and raids by seventh-century Greeks may have ended with the massacre of men and the enslavement of their dependents, and occasionally even the destruction of entire communities, a rarity for nearly four centuries beforehand.[125]

2.5 *Changes in Greek Warfare from 800 to 600*

Warfare and warriorhood remained a fascination of eighth- and seventh-century Greeks in their poetry, pottery, and even their religious activity. Mass fighting seems to become more common, and thus, perhaps, the sack and destruction of cities; but raids and the desire for booty remained significant. Warriorhood was identified through the depiction and dedication of new kinds of armour, contrasted against the 'nakedness' of the light-armed who stood lower in social standing. The abundance of evidence for warfare suggests that it continued to play a significant role in the identification of individuals as warriors. Furthermore, the Parian *polyandria* with their one-hundred-and-seventeen cremated men indicates that the status of 'warrior' may have begun to encompass a much wider group than in the preceding centuries. The significance of this social change will be examined in the next section.

It is also important to recognise that there were significant regional variations in these developments.[126] The *polyandria* may represent a specifically Parian development, while burials of individuals or small groups with weapons remain common elsewhere into the early seventh century. Horses, chariots, and mass combat are depicted on vases from a diverse selection of settlements and are rarely found together. Fortifications and destroyed settlements are more prominent in the coastal areas of the Aegean and the first depiction of the sack of Troy comes from an island, Mykonos. Archilochos seems to suggest knowledge of regional differences when he describes a way of fighting

122 Hall (1995) 581–82.
123 Ervin (1963).
124 Ebbinghaus (2005) 68. See also Martinez Morales, this volume.
125 *GWMR* 121–26; see Echeverría in this volume for attacks on settlements in the Classical period.
126 Lloyd (2017).

with swords and limited long-range weaponry particularly associated with the Euboians (fr. 3 West); meanwhile Tyrtaios' Spartans incorporate both light- and heavy-armoured warriors into their combat (fr. 11 West). Different regions of Greece depict a variety of combat in the eighth and seventh centuries.

3 Early Greek Warfare in Context

In turning to the explanations for the changes outlined in the previous section, it is worth returning for a moment to Crellin's hurdles, discussed in the introduction. Of central importance to understanding early Greek warfare is her sixth hurdle: singular causation. The major changes in the eighth and seventh century were the result of a number of intersecting factors within which cause and effect cannot always be disentangled – did new methods of warfare influence new political organisations, or vice versa? – but which resulted in a new dynamic in the following centuries. These are population growth, climate change, agricultural and political organisation, and the increasing connectivity of the Mediterranean and the spread of ideas across it.

3.1 *Population Growth, Climate Change, and Agriculture*

Population has been a key factor in our understanding of both the beginning and the end of the post-palatial Bronze and Iron Ages. At the beginning, following the collapse of the Mycenaean palaces, the diminishing evidence in the subsequent centuries, especially graves, has led to the conclusion that the population dropped sharply, reaching a low point *ca.* 1000.[127] With the increase in evidence from the eighth century, population growth has been seen as a key factor in the concurrent revival in Greek fortunes.[128]

The extent of eighth-century population growth has been controversial. Initially, Snodgrass interpreted an increase in the number of graves in Athens and Argos as evidence for an eighth-century population explosion.[129]

127 Morris (2009) 66–67. Dickinson (2006, 93–98, quotation 97) discusses the difficulty of estimating the size of the population in EIA Greece, nevertheless concluding that it is 'safest simply to accept that the population was low, without commitment to precise figures'.

128 It is worth noting that the radiocarbon dates from Sindos in Macedonia, which suggest a much shorter Protogeometric period and much longer Geometric period, would lead to a more even distribution of the material evidence recovered from these periods, implying both a less sharp drop in population and a much slower increase in the eighth century: Gimatzidis and Weniger (2020).

129 Snodgrass (1980) 18–24.

This hypothesis was rejected by Ian Morris on the grounds that a surge in sub-adult burials exaggerates the increase, which should be interpreted not as an increase in mortality but as a change in the visibility of graves resulting from changing social customs.[130] Morris later argued that changing patterns of housing at Zagora on Andros may indicate population growth over the eighth century, with family needs changing as they grew larger; but the limited evidence for houses in the eighth and seventh centuries makes it difficult to project a conclusion from Zagora, a site abandoned *ca.* 700, to the rest of Greece.[131] In both these cases we see evidence for changing social customs – changes which themselves need explanation. Population growth will not be the only cause, but it is likely one of them.

We must also ask what caused the growth in population. On the one hand, if we are correct in believing the population slump to have reached its nadir around 1000, the sustained population growth of the subsequent two centuries will go some way to explaining this change. Walter Scheidel's models of the growth cycle show an interesting point of conversion: the eighth century is approximately the point at which the population of EIA Greece reaches that proposed for the thirteenth century, before the Mycenaean collapse.[132] It is possible that the eighth century marked a tipping-point in population growth – one that should have an impact on agricultural production, as we will see further below.

One of the problems with population growth as a factor is that it begs the question: what caused the population growth? While cultural factors may have had an impact on a slower increase, there is some evidence that climate change led to a cooler, wetter climate across the Mediterranean and Europe. As variability in rainfall was a big agricultural problem and summer intestinal problems were major killers in the Mediterranean, this change in climate would have increased lifespans and, it has been argued, allowed women to produce many more children in their longer lives.[133]

---

130 Morris (1987) 171–210. Dickinson (2006, 177) does not agree with Morris on the notion of exclusion from formal burial, which would require a complex social structure, but emphasises instead the limited scope of the archaeological record. The publication of the Early Iron Age burials from the Athenian Agora questions Morris' definition of 'sub-adult' before the eighth century, see Liston (2017) 515–16. Nevertheless, the general conclusion that we cannot be certain that an increase in evidence means an increase in population remains sound.

131 Morris (1999) 308–309; Foxhall (2009) 498–500.

132 Scheidel (2003) 125.

133 Morris (2009) 66–67. Langdon (2008, 130–43) observed a shift in the lavishness of graves of Athenian women *ca.* 750 away from those of older women toward those of young, perhaps unmarried, girls. It is possible that this reflects a move toward a more patriarchal

In the context of climate change, two more of Crellin's hurdles are worth considering. Her seventh hurdle is anthropocentrism – the belief that human action and agency play a central role in change.[134] Crellin argues that we must be careful not to underplay the role of non-human factors in shaping change, whether the natural environment or non-human factors such as animals. On the flip side of this is hurdle five: environmental and technological determinisms.[135] While we may be able to see that the environment is playing a role in the changes of the eighth century, we must not assume that the actions humans take as a result of these changes are predetermined. Rather, they are rooted in the circumstances that predate the changes and intersect with one another to result in new patterns.

Changes in both climate and population should result in changes in agricultural production: more people need more food, and the means of producing it change with the climate.[136] Our understanding of Early Iron Age agriculture is hampered by the limited textual evidence, but it is likely that there was a degree of continuity in practice from the Late Bronze Age. While it has been suggested that the collapse of the Mycenaean palaces would have left farmers 'directionless',[137] it is probable that the impact was less on practice and more to do with the everyday economic decisions farmers had to make having lost a major consumer.[138] With the drop in population, there was likely a surplus of land and a lack of labour; a common response to which is to diversify agricultural practice and avoid labour-intensive investments.[139] When the population rose again, practices would change once more. Morris has suggested that visible evidence for this phenomenon would be the intensification of agriculture and the exploitation of marginal lands.[140] Unfortunately, the focus of archaeological research on built environments – cities, sanctuaries, and cemeteries – rather than agricultural sites means that we do not have evidence of increased irrigation, manuring, terracing, or even archaeobotanical evidence to back this hypothesis up.[141]

society evident in Classical Athens, as the greater loss was felt with the death of girls who did not reach child-bearing age, rather than women who had lived successful lives on their own terms.

134 Crellin (2020) 16–17.

135 Crellin (2020) 13–14.

136 An agricultural revolution has been seen as a key factor in the military and political changes of the eighth century; see especially Hanson (1995).

137 Hanson (1995) 30–32.

138 Foxhall (1995).

139 Nakassis (2020) 277.

140 Morris (2009) 67–68.

141 Nakassis (2020) 277.

The evidence from archaeological survey does not back up the claim that marginal land was further exploited in the eighth century. Rather, the pattern of settlement from the eighth century to the sixth supports the idea that farmers in several regions of Greece lived in nucleated settlements close to the best agricultural land, on which cultivation focused. Exploitation of marginal land is suggested by the survey evidence only from the mid-sixth century, and its appearance across the fragmented landscape of Greece is variable into the third century.[142] Archaeological survey does not offer the precision of excavation,[143] and it is likely that further research will reshape our knowledge of agriculture in early Greece. From our present position, however, we cannot build an argument based on the hypothesis that sudden population growth or climate change sparked radical new agricultural practices. Rather, we should focus on the consequences of several centuries of *sustained* population growth, and the impact of an apparent increase in agricultural labour on the structure of contemporary political organisations.

3.2 *Political Changes and the Interconnected Mediterranean*

With a growing potential workforce, landowners will have increasingly found themselves in the position where they could derive their wealth from the labour of others. This non-working elite has been described as a leisure class, after the leisure time afforded to them by other's labour.[144] Hans van Wees has described these landowners as 'gentlemen' farmers, who emerged in the eighth century and consolidated their power during the seventh.[145] Not all of these gentlemen farmers hold political power: the poet Hesiod writes in the persona of an elite landowner who employs at least two labourers and enslaves four more, and yet he is excluded from the politics of the *basileis*; he is also ambitious, wanting to accumulate wealth and achieve glory in poetry competitions.[146] Nevertheless, there appears to have been a conscious self-definition on the part of these leisured elites through vigorous exercise, competition in sport and poetry, and military endeavours – perhaps including raids

142 Foxhall (2013) 215–17. This evidence corresponds far better with Van Wees' proposal (2013, 235–36) that the 'agricultural revolution' dates to the end of the sixth century, on which see the following discussion.

143 Foxhall (2013) 196–99.

144 On the 'leisure class': Van Wees and Fisher (2015) 38, at 50, n. 104, they define 'leisure class': 'a class of people who derive most or all of their income from the labour of others (as opposed to self-employment or employment by others) and are thus in principle able to live in leisure, whether or not they do exempt or exclude themselves from work'.

145 Van Wees (2013) 236–45.

146 Van Wees and Fisher (2015) 9.

on other settlements;[147] leisure time would have also given these individuals additional time to spend raising their all-important horses. Politics was also a leisure activity, as Hesiod himself notes (*Op.* 30–32, trans. Daryl Hine):

> You have no business getting in fights and disputing in public,
> Not if you haven't sufficient for life laid up in your storeroom,
> Seasonal fruits of the earth, ripe grain of abundant Demeter.

Therefore, when we begin to see evidence for a growing leisure class, we should expect to see political changes too.

Archaeological evidence hinting at 'leisure' activities appears as early as the tenth century.[148] Material associated with the elite graves of Early Iron Age cemeteries in places like Lefkandi, Athens, and Knossos include weapons and perhaps armour, but also indications of overseas connections and the ability to travel, or at least communicate, with distant elites. In the eighth century, as the quantity of material retrieved from graves decreases, similar material evidence begins to appear as dedications in sanctuaries, especially interregional sanctuaries such as Olympia in Elis and Kalapodi in Phokis.[149] The prominence of interregional sanctuaries over urban or extra-urban sanctuaries again points to the significance of communicating with a wider community, one that could not easily be reached in burials. It can also be seen as an attempt to restrict attendance at significant social engagements. Hesiod describes the difficulty that he faced travelling across the straits of Euboia from Boiotia to attend a funeral (*Op.* 655–659); we should not imagine that the journey to interregional sanctuaries in the eighth and seventh centuries was any less daunting. In order to make such a journey, an individual would require transport, whether on land or sea, and the leisure time to make the journey.

Contemporary burials also hint at the emphasis on travel: one example is the amphibious battles on gigantic Athenian grave markers (Figure 2.1); another is Tomb 45 at Argos, where the deceased is buried with his bell-cuirass – a shape adopted from Europe that is suitable for horse-riding – and firedogs in the shape of ships. These firedogs not only hint at the significance of ships through their form, but their resemblance to ship-shaped firedogs from East

147 Van Wees and Fisher (2015) 38.

148 Van Wees and Fisher (2015) 21–25 and 27 argue that leisure is part of elite behaviour in the *Iliad* and *Odyssey*, and thus is not in opposition to the concept of a 'warrior elite' in Homer that earlier scholars dated prior to the Archaic period, e.g., Donlan (1980) 49–64; O. Murray (1993) 81.

149 Snodgrass (1980), especially 54–55; Morris (2009) 77–78; on weapons in particular: Baitinger (2011).

Crete and Cyprus reinforces the role of such overseas connections in the burials of wealthy individuals.[150]

It is likely that such movements were limited to those with access to ships or horses, as indicated by dedications of bronze models of horse-drawn chariots at eighth-century Olympia.[151] Although Hesiod makes little mention of horses,[152] the leisured, gentlemen farmers that owned them are likely to be connected to the *hippobatai* of the seventh century discussed above.[153] Those who rode their horse or chariot to the battlefield may also have ridden to inter-regional sanctuaries.

Sanctuaries were connected to other leisure activities, too. Notable among early dedications are bronze tripod cauldrons, objects connected to victory in leisure activities such as poetry competitions and athletic contests.[154] Sanctuaries provided a largely exclusive location where elites could compete among themselves for social status. While we should not assume that there was no social mobility, the locations chosen for elite behaviour suggest an attempt to restrict that behaviour to those with the freedom to travel. It is also clear that travel did not include an individual in the sphere of power if they were traders, as indicated by Euryalos' challenge to Odysseus on Scheria in which he accuses the stranger of being a merchant captain, ignorant of athletics and concerned only for wealth and cargo (*Od.* 8.159–164).[155] These individuals travelled for work, and were perhaps threatening to the status of the leisured elite who viewed them as potential aggrandisers.

Warfare will have also played a role in creating and maintaining overseas connections. The prominence of amphibious warfare in Athenian Late Geometric vase painting, the coastal location of many sites that are destroyed

150 Coldstream (2003) 146.

151 Morgan (1990, 91–92) suggested that chariot models at Olympia supported the idea of the sanctuary as a meeting place of local chiefs; see also Eder (2006) 565–66 on the significance of chariots as a conspicuous display of wealth and prestige.

152 On the other hand, he appears to know much about owning a ship: *Op.* 618–46. Van Wees and Fisher (2015) 9 suggest Hesiod's persona owns a ship, although if so, he does not use it himself *Op.* 650.

153 Van Wees (2013, 241–42) and earlier Worley (1994, 51–56) note that Aristotle's succession of constitutions from kings to horsemen to hoplites would fit a model in which the seventh and sixth centuries were dominated by these 'horsemen' (*Pol.* 1297b.16–24). As Van Wees says, however, the historical applicability of Aristotle's model is dubious.

154 Hesiod wins his tripod at a funeral and a tripod is the prize for the wrestling at Patroklos' funeral, suggesting that we should not draw too sharp a distinction with the change in depositional practices: Hes. *Op.* 655–59; *Il.* 23.700–702, see also the simile at *Il.* 22.162–166. In the Homeric poems they are also given as gifts and rewards between elites: *Il.* 8.285–291; *Od.* 13.13–15.

155 O. Murray (1993) 202–203.

or fortified, as well as the first appearance of *polyandria* on an island, Paros, indicate that this was the case. Several of the early surviving mentions of conflict are overseas, whether the mythical siege of Troy or the seventh-century Parian conflict with the Thracian Saians on Thasos (Archil. Fr. 5). Theoretically, these opportunities were open to anyone willing to abandon their home life and to join a ship's crew; in practice, it is unclear how widely available they were. John R. Hale has suggested that the possibilities for mercenary service in the Near East were so prominent that it formed the catalyst for the changes in Greek warfare in the late eighth and seventh century; with so few known battles in the Aegean in the seventh century, he goes as far as to suggest that there were more Greek 'hoplites' fighting in the Near East than in the emerging poleis.[156] Furthermore, the round, hollow Argive shield is buoyant and lacks sharp corners to damage the ship.[157] Hale suggests that 'aristocrats' are behind this move: 'No one else would have had the resources needed to acquire the raw materials, compensate the shipbuilders, protect and maintain the finished vessel, and assemble the crew of rowers that was required to propel the ship on its voyages'.[158] We might add that leisure time to engage in less obviously productive activities was also necessary. Whether or not we agree with Hale on the significance of mercenaries,[159] it is certainly the case that ships are expensive to build and maintain and it is unlikely that there were many places available for people from the Aegean to board these vessels – presumably as rowers – to earn a fortune overseas, whether as a mercenary or as a pirate. But Hale, like many others looking at seventh-century warfare, sees a continuity of elites gathering their fellows around them to form war-bands. The inspiration of overseas contexts to eighth- and seventh-century Greek warfare remains largely exclusive.

The nature of the weapons and armour deposited at sanctuaries also indicates that certain elements of warfare were restricted to those enjoying the leisured elite lifestyle. In graves, armour had been scarce, while weapons, especially swords, daggers and spearheads, were common; in sanctuaries armour is much more prominent.[160] Armour also appears in a handful of later burials with weapons at Argos. These changes may reflect the new availability of armour discussed above, but they may also reflect the changing relationship between the dedicant and dedicated armour. These weapons are either

156 Hale (2013) 185–86. See also Brouwers and Van Wees in this volume.

157 Hale (2013) 189; Schwartz (2013) 159, citing Arr. *An.* 1.19.4.

158 Hale (2013) 187.

159 See Van Wees, this volume, for an alternate assessment of the role of early Greek mercenaries.

160 Jarva (1995) 111–12, 124–28.

buried with or dedicated by the individual who fought with them, or they are the equipment of another warrior, whose weapons and armour were stripped from their body after they had been killed in combat. Pictorial evidence indicates that the stripping of war dead was practised in the eighth century, and inscriptions identify seventh-century armour from Afrati on Crete and later dedications of armour at Olympia as battlefield booty.[161] It is unclear whether weapons or armour in graves were ever spoils of battle, but multiple uses of captured armour are known: the armour from Afrati may have been on display in a communal dining hall;[162] when Alkaios loses his armour to the Athenians, they dedicate it to Athena (Alkaios fr. 401b Voigt); in the *Iliad* captured armour is often re-used (e.g., *Il.* 17.188–197). These different contexts indicate that captured armour could be removed from circulation or remain in it, but in both instances, it could be put on display. If the stripping of the war dead and dedication of the spoils is a new development in the eighth century, this still suggests that warfare was a means of social aggrandisement.

The question remains who was capturing armour and choosing where to deposit it. At Eretria on Euboia there are two examples of possible booty from overseas that indicate a context to which access was restricted. A bronze spearhead from Tomb 6 at the West Gate Cemetery may have been a European Urnfield object, perhaps taken in a conflict in Italy,[163] while a ninth-century North Syrian horse-blinker dedicated at the sanctuary of Apollo Daphnephoros is believed to have been booty from an eighth-century campaign in which Greek 'mercenaries' fought on the side of the Assyrian empire.[164] If the interpretation of these objects as 'booty' is correct, then at the end of the eighth century in Eretria booty was used both as a grave gift and as a dedication in a sanctuary. The West Gate Cemetery at Eretria is a small, exclusive burial plot in use *ca.* 710–690 that includes some of the latest examples of EIA burials

161 Pictorial evidence: Ahlberg (1971) 21–25. Inscriptions: Hoffmann (1972) 15–16; *GSW*3 227. On the stripping of the war dead, see also Kucewicz this volume.

162 Viviers (1994) 244–49. Erickson (2011, 386–87) points out that the unpublished finds from the building point toward a cultic context and thus that Viviers' reinterpretation of this building as an *andreion* rather than a temple is unfounded. The rather less substantial evidence of an iron spearhead and bronze helmet crest from an early sixth-century building at Azoria may be a better example of weapons on display in an *andreion*, but there is no evidence to suggest that these weapons were captured (Haggis *et al.* 2004, 389; Erickson 2011, 387–88).

163 Bettelli (2001); Lloyd (2015) 25–28; although this was originally interpreted as a Mycenaean heirloom or antique, Bérard (1970) 16–17.

164 Luraghi (2006) 38–40. A similar object is known from the Heraion on Samos that Luraghi suggests was booty from the same campaign. On early Greek mercenaries, see Brouwers and Van Wees, this volume.

with weapons in central Greece.[165] Tomb 6, dated to the late eighth century, is an elaborate cremation in two bronze cauldrons buried with four killed iron swords and five iron spearheads, in addition to other metal objects;[166] it is part of an on-going tradition of metal urn cremations connecting the Mediterranean elite including the burials in the Toumba Building and of the 'warrior trader' at Lefkandi more than a century earlier.[167] The North Syrian horse-piece also has a role in defining the Mediterranean elite: such dedications are only available to those with the resources to travel, and perhaps fight, overseas.[168] It suggests that captured armour and weapons were displayed as a means to reinforce the power of those in a position to capture such objects and then dispose of them – the already powerful leisured elite.

Burial and dedications are both forms of competitive consumption – the removal of valuable objects from circulation – a practice that increases across Greece from *ca.* 800.[169] By removing objects such as weapons and armour from circulation, members of the elite were ensuring that they remained scarce, and thus valuable. But warfare is also a way in which valuable objects can be obtained, not least through the capture of weapons and armour from the battlefield. If warfare was largely restricted to or dominated by an elite group then access to these spoils would also have been restricted to those already in a position to arm themselves with the most impressive armour and weapons. However, there is another aspect to this process. Evidence such as the Parian *polyandria* and the increasing scale of warfare indicate that greater numbers of individuals may have become involved in warfare, perhaps as the newly 'naked' light-armoured fighters. This would have given certain men the opportunity to advance their social standing through warfare.

Above, I discussed how in the EIA 'aggrandisers' used social events such as funerals and feasts to bolster their social position through the formation of 'alliances', both political and military. By moving these practices to locations that necessitated travel time, the leisured elite was also restricting access for aggrandisers from other social classes. With the decline in significance of the local funeral as a social gathering and the rise of the interregional sanctuary, even the ability to attend these events and integrate oneself into the

165 Bérard (1970).

166 Bérard (1970) 13–17; Blandin (2007) 43–45. Elsewhere (2015, 26–28), I have argued that the swords can also be interpreted as battlefield spoils as they, like the bronze spearhead, are intentionally destroyed. Crielaard (2016, 62) independently makes the same suggestion based on the number of the weapons.

167 Crielaard (2016), especially 62–64.

168 See also the discussion on Hale (2013) above.

169 Morris (2000) 278–79; Lloyd (2015) 26–28.

social elite may have become more restricted. Van Wees and Fisher note that the competition that we can see in texts for political privileges is typically between the wealthy, and thus that wealth did not guarantee access to political privileges;[170] by focusing on competition at some distance from the growing civic centres, those competing for political privileges further restricted access to the very act of competing for power to those with the ability to travel.

Local politics were a way for individuals to become part of the local elite, and thus possibly the interregional elite. At the turn of the sixth century, the poet Alkaios plans to equip a small army against the tyrant Pittakos at a gathering in his hall, perhaps a feast (Alkaios fr. 140 Voigt); a conflict between two groups of elite aggrandisers aiming to control the social world of Mytilene.[171] Alkaios' example shows how these gatherings were central to the formation of political and military alliances: as they had done for centuries, elites gathered their followers in order to fight amongst themselves.[172] The individuals competing for power in these instances are different members of a wealthy elite for whom military success is important, some of whom defend their powerful positions, others who wish to claim that position for themselves. The absence of bows and slings from this repertoire indicates the exclusion of the *gymnetes* from this gathering.

Competition between members of local elites such as Alkaios and Pittakos, and the exclusion from power of men like Hesiod, raises the question of how elite power was legitimised.[173] The significance of military excellence as part of this legitimisation seems to be present in the Aegean from at least the tenth century, and perhaps as early as the twelfth – at least in those regions where prominent men are buried with weapons. Military excellence seems also to have played a legitimising role in the Homeric epics, although not exclusively.[174] We can safely presume that legitimisation in the EIA was more

170 Van Wees and Fisher (2015) 9.

171 Van Wees and Fisher (2015, 21–22) emphasise the similarity of Homeric feasts to Archaic symposia; later (27), they criticise O. Murray's assertion (1993, 82) that 'The transformation of the Homeric warrior feast of merit into the aristocratic symposium [was] one of the most significant changes in Greek aristocratic life' on account of this continuity. If Alkaios fr. 140 Voight does take place at a feast, this would further suggest that leisured gatherings could have a significant 'warrior' component.

172 Hall (2014) 144–52.

173 On 'legitimising' elite values (as opposed to 'differentiating' values, into which the leisured lifestyle falls) see Van Wees and Fisher (2015) 15–16.

174 Van Wees and Fisher (2015) 16–21. Against 'an historical Homeric society': Snodgrass (1974, 2013); Van Wees and Fisher (2015, 4) emphasise that the Homeric account is 'at best an idealised version of reality'; nevertheless, their discussion of Archaic and Classical 'aristocratic' values seems to presume an Homeric society preceding these periods.

complex too, although our evidence is not always substantial enough to show how. Yet male identity as a 'warrior' appears to have remained a significant part of elite male identity into the Archaic period, as indicated by the presence of weapons and armour, possibly battlefield spoils, among dedications in sanctuaries, the prominence of fully-armoured men on pottery and their distinction from the light-armed 'naked men', and the significant role warfare plays in surviving Archaic lyric poetry.

Surviving written accounts from the Archaic period show us the perspectives of those claiming political legitimacy through their superior *moral* values. The corpus of poems attributed to Theognis take this approach regarding Megara, in which the poet claims that his group are more courageous, trustworthy, just, *et cetera*, than their rivals for power.[175] While it is unclear whether military excellence was part of this claim to superiority, Van Wees argues that the conflict for political power remained violent in Theognis' Megara.[176] The same is true of the conflict on Mytilene recorded in Alkaios' poetry, which rests on an ability to arm one's allies. As noted above, however, Alkaios' gathering includes no weapons for the light-armed – certain groups remain even further distanced from power. In some ways these processes are *more* socially restrictive than those that preceded them.[177]

The availability of written evidence complicates the question of change here. Simply put, we have a new class of evidence in which we can understand the social dynamics of the period that we did not have for the preceding period. What we can see is that many of the values we know to have existed in the EIA, whether through archaeological evidence or derived from Homeric sources, persist into the changed contexts of the historical period. It is possible that many of the values that do not have a material component, and thus are unknown before the eighth century, nevertheless existed then too.

In the EIA, Greece supported a diverse array of forms of social organisation. Even with the increasing connectivity of the eighth century it is unlikely that one sweeping change could have affected all of them equally. Indeed, the varying circumstances of different regions may have left their populations with varying degrees of willingness to adopt new ideas. In the eighth century, the

175 Van Wees and Fisher (2015) 4; Van Wees (2000b), especially 59, where he also notes that, according to Theognis (383–92), the poor can never become good: 'To him, only men born into prosperity *can* ever be righteous and achieve excellence'.

176 Van Wees (2000b), especially 60–63.

177 As Van Wees and Fisher (2015) 35–36 note: 'The growing number of burials (…) in the late eighth century, for example, may in this light be interpreted as reflecting, not the broadening of an existing elite (let alone mere population growth), but the first emergence of stratified communities in the Early Iron Age'.

replacement of the funeral as the locus of elite interaction with the sanctuary gives some idea of how negotiations of power were kept exclusive. Only a select few leisured elites could afford both the travel time and the resources to reach interregional sanctuaries and make dedications. Individuals with leisure and resources became the mediators between the local and the outside world, including the world of the gods. The political changes of the eighth century resulted in the consolidation of power by a larger, but more rigid elite group best defined as a leisure class. These were the men who could ride to battle on horseback, equip themselves and their allies with the latest bronze armour, and dedicate spoils at interregional sanctuaries.

4 Conclusion

Following the collapse of the Mycenaean palaces, the concept of the warrior resurfaced in the Aegean. In the twelfth century, groups along the Corinthian and Euboian Gulfs, across the Aegean islands to Crete, and in the Argolid represented themselves at feasts and in burial as warriors, perhaps reflecting the conditions of their daily lives in the post-palatial world. These men, who at first were perhaps local leaders free of their obligations to the palaces, seem to have formed alliances at social gatherings that strengthened their social position among local and more distant groups. Some of these groups may have been responsible for repeated destructions at sites including Tiryns, Lefkandi, and Koukounaries on Paros.

After a century or so of 'sailor-warriors', elements of this culture are abandoned and changed; the scale on which these elite individuals operated seems to have shrunk. In the Early Iron Age, pictorial pottery almost entirely disappears, and gatherings with wine and feasting may have been limited to funerals – moments of social transformation. In many regions, a limited number of men are still identified in burial as 'warriors'; in places like Lefkandi on Euboia, their burials also include objects from distant places, accessed by boat. While it is less clear that the concepts of 'sailor' and 'warrior' were combined in this period, elements of both remained significant.

In the eighth century, a growing population meant that there were more people travelling because they did not need to be present on their farmland quite as much. These travellers were able, once again, to change the pattern of warrior feasts focusing instead on interregional meeting places such as the sanctuary of Zeus at Olympia. Some sailors who need not have been warriors adapted an alphabet with their fellow-travellers from the Phoenician cities of the Levant. This was eventually used to write down oral epic and lyric poems, which reveal ideas about warfare that enhance those gleaned from the archaeological evidence.

It is possible that these connections brought new armour to the Greek world, and those who could were certainly equipping themselves with bronze cuirasses and helmets. It is perhaps this new equipment, or greater numbers, that allowed groups of warriors to sack cities such as Asine and to result in the deaths of over a hundred men at Paroikia, who were buried and perhaps died together. As elites were forced to consolidate their power, especially against local competitors, more people were incorporated into their war-bands, meaning that more people had the opportunity to acquire wealth and status through battle. More people could take the opportunity to travel the Mediterranean, seeking work as traders, mercenaries, or even settling in new homes overseas.

The ability to organise war-bands was of central importance to developing political institutions, which continued to rely on social gatherings locally as well as at the interregional sanctuaries. Even at the end of the seventh century, on Mytilene, such gatherings affected local politics. Elite leadership remained fluid and open to aggrandisers who wished to establish themselves at the pinnacle of society, but this process of aggrandisement was restricted by the transportation of social events from local funeral feasts to sanctuaries and methods of aggrandisement became increasingly connected to ideas and objects from overseas – as they had been throughout the EIA in those regions that maintained their overseas connections, such as Euboia.

The Greek world was changing in the eighth century and it continued to change into the seventh and beyond. These changes were rooted in the power structures and elite ideology which had been in place for many centuries. Certain factors, such as population increase and growing Mediterranean connectivity, were particularly strong features of the eighth century. Others, such as literacy, increase the evidential basis of our understanding of Greek history. Events in the eighth century were responses to changing conditions, but these did not result in completely new power structures; the period cannot be understood without knowing the forces in place when the century began.

Bibliography

Agelarakis, A.P. (2017) *Parian Polyandreia: The Late Geometric Funerary Legacy of Cremated Soldier's Bones on Socio-Political Affairs and Military Organizational Preparedness in Ancient Greece*. Oxford.

Ahlberg, G. (1971) *Fighting on Land and Sea in Greek Geometric Art*. Stockholm.

Antonaccio, C.M. (2002) 'Warriors, Traders, and Ancestors: The 'Heroes' of Lefkandi', in Højte (2002) 13–42.

Bachhuber, C. and Roberts, R.G. (eds) (2009) *Forces of Transformation: The End of the Bronze Age in the Mediterranean*. Oxford.

Baitinger, H. (2011) *Waffenweihungen in griechischen Heiligtümern*. Mainz.

Bakker, E.J. (ed) (2010) *A Companion to the Ancient Greek Language*. Chichester.

Bérard, C. (1970) *Eretria Foulles et Recherches III. L'Hérôon à la Porte de L'Ouest*. Bern.

Bettelli, M. (2001) 'A Supposed Mycenaean Spearhead from Eretria'. *SMEA*: 189–93.

Blandin, B. (2007) *Eretria XVII: Les Pratiques Funéraires D'Époque Géométrique à Érétrie Volume II – Catalogue, tableaux et planches*. Renens.

Boardman, J., Brown, M.A. and Powell, T.G.E. (eds) (1971) *The European Community in Later Prehistory. Studies in Honour of C.F.C. Hawkes*. London.

Bonfante, L. (1989) 'Nudity as Costume in Classical Art'. *AJA* 94 (4): 543–70.

Bonfante, L. and Karageorghis, V. (eds) (2001) *Italy and Cyprus in Antiquity: 1500–450 BC. Proceedings of an International Symposium Held at the Italian Academy for Advanced Studies in America at Columbia University, November 16–18, 2000*. New York.

Bouzek, J. (1988) 'An Alternative Reconstruction of the Kaloriziki Shield'. *RDAC*: 319–20.

Brock, R. and Hodkinson, S. (eds) (2000) *Alternatives to Athens. Varieties of Political Organization and Community in Ancient Greece*. Oxford.

Broodbank, C. (2013) *The Making of the Middle Sea. A History of the Mediterranean from the Beginning to the Emergence of the Classical World*. London.

Brouwers, J.J. (2007) 'From Horsemen to Hoplites: Some Remarks on Archaic Greek Warfare'. *BABesch* 82.2: 305–19.

Brouwers, J.J. (2013) *Henchmen of Ares. Warriors and Warfare in Early Greece*. Rotterdam.

Cairns, D.L. (ed) (2001) *Oxford Readings in Homer's Iliad*. Oxford.

Catling, H.M. (1965) Review of *Early Greek Armour and Weapons* by Antony Snodgrass. *Antiquity* 39: 150–52.

Catling, R.W.V. and Lemos, I. (1990) *Lefkandi II: The Protogeometric Building at Toumba. Part 1: The Pottery*. Oxford.

Coldstream, J.N. (2003) *Geometric Greece: 900–700 BC*. 2nd edn. London.

Coldstream, J.N. (2006) '"The Long, Pictureless Hiatus". Some Thoughts on Greek Figured Art between Mycenaean Pictorial and Attic Geometric', in Rystedt and Wells (2006) 159–63.

Courbin, P. (1957) 'Une tombe géométrique d'Argos'. *BCH* 81(1): 322–86.

Courbin, P. (1974) *Tombes Géométriques D'Argos, I*, Paris.

Crellin, R. (2020) *Change and Archaeology*. Abingdon.

Crielaard, J.P. (2006) 'Basileis at Sea: Elites and External Contacts in the Euboean Gulf Region from the End of the Bronze Age to the Beginning of the Iron Age', in Deger-Jalkotzy and Lemos (2006) 271–97.

Crielaard, J.P. (2016) 'Living Heroes: Metal Urn Cremations in Early Iron Age Greece, Cyprus and Italy', in Gallo (2016) 43–78.

Crouwel, J.H. (1981) *Chariots and Other Means of Land Transport in Bronze Age Greece*. Amsterdam.

Crouwel, J.H. (1992) *Chariots and Other Wheeled Vehicles in Iron Age Greece*. Amsterdam.

D'Onofrio, A.M. (2011) 'Athenian Burials with Weapons: The Athenian Warrior Graves Revisited', in Mazarakis Ainian (2011) 645–73.

Dahm, M.K. (2007) 'Not Twins at All. The Agora Oinochoe Reinterpreted'. *Hesperia* 76: 717–30.

Deger-Jalkotzy, S. (2006) 'Late Mycenaean Warrior Tombs', in Deger-Jalkotzy and Lemos (2006) 151–79.

Deger-Jalkotzy, S. (2008) 'Decline, Destruction, Aftermath', in Shelmerdine (2008) 387–415.

Deger-Jalkotzy, S. and Lemos, I.S. (eds) (2006) *Ancient Greece: From the Mycenaean Palaces to the Age of Homer*. Edinburgh.

Dezsö, T. (1998) *Oriental Influence in the Aegean and Eastern Mediterranean Helmet Traditions in the 9th–7th Centuries BC: The Patterns of Orientalization*. Oxford.

Dickinson, O.T.P.K. (2006) *The Aegean from Bronze Age to Iron Age: Continuity and Change Between the Twelfth and Eighth Centuries BC*. London.

Donlan, W. (1980) *The Aristocratic Ideal in Ancient Greece. Attitudes of Superiority from Homer to the End of the Fifth Century BC*. Lawrence, KS.

Douzougli, A. and Papadopoulos, J.K. (2010) 'Liatovouni: A Molossian Cemetery and Settlement in Epirus'. *JDAI* 125: 1–88.

Dué, C. and Ebbott, M. (2010) *Iliad 10 and the Poetics of Ambush: A Multitext Edition with Essays and Commentary*. Hellenic Studies 30. Washington, DC.

Ebbinghaus, S. (2005) 'Protector of the City, or the Art of Storage in Early Greece'. *JHS* 125: 51–72.

Eder, B. (2006) 'The World of Telemachus: Western Greece 1200–700 BC', in Deger-Jalkotzy and Lemos (2006) 549–80.

Emanuel, J.P. (2015) 'The Late Bronze-Early Iron Age Transition: Changes in Warriors and Warfare and the Earliest Recorded Naval Battles', in Lee, Whittaker and Wrightson (2015) 191–209.

Erickson, B.L. (2011) 'Public Feasts and Private Symposia in the Archaic and Classical Periods', in Glowacki, K. T. and Vogeikoff-Brogan (2011) 381–91.

Ervin, M. (1963) 'A Relief *Pithos* from Mykonos'. *AD* 18: 37–75.

Fisher, N. and Van Wees, H. (eds) (1998) *Archaic Greece: New Approaches and New Evidence*. London.

Fisher, N. and Van Wees, H. (eds) (2015) *'Aristocracy' in Antiquity: Redefining Greek and Roman Elites*. Swansea.

Foxhall, L. (1995) 'Bronze to Iron: Agricultural Systems and Political Structures in Late Bronze Age and Early Iron Age Greece'. *ABSA* 90: 239–50.

Foxhall, L. (2009) 'Gender', in Raaflaub and Van Wees (2009) 483–507.

Foxhall, L. (2013) 'Can We See the "Hoplite Revolution" on the Ground? Archaeological Landscapes, Material Culture, and Social Status in Early Greece', in *MoB* 194–221.

Frederiksen, R. (2011) *Greek City Walls of the Archaic Period 900–480 BC*. Oxford.

Frödin, O. and Persson, A.W. (1938) *Asine: Results of the Swedish Excavations 1922–1930*. Stockholm.

Galanakis, Y., Wilkinson, T., and Bennet, J. (eds) (2014) *ΑΘΥΡΜΑΤΑ: Critical Essays on the Archaeology of the Eastern Mediterranean in Honour of E. Susan Sherratt*. Oxford.

Gallo, F. (ed) (2016) *Omero: quaestiones disputatae*. Ambrosiana Graecolatina 5. Milan.

Georganas, I. (2018) '"Warrior Graves" vs. Warrior Graves in the Bronze Age Aegean', in Horn and Kristiansen (2018) 189–97.

Giannopoulos, T.G. (2008) *Die letzte Elite der mykenischen Welt. Achaia in mykenischer Zeit und das Phänomen der Kriegerbestattungen im 12.-11. Jahrhundert v.Chr.* Bonn.

Gimatzidis, S. and Weniger, B. (2020) 'Radiocarbon Dating the Greek Protogeometric and Geometric Periods: The Evidence of Sindos'. *PLoS ONE* 15.5: e0232906.

Glowacki, K. T. and Vogeikoff-Brogan, N. (eds) (2011) *Στέγα: The Archaeology of Houses and Households in Ancient Crete*. Hesperia Supplements 44. Princeton.

Greenhalgh, P.A.L. (1973) *Early Greek Warfare: Horsemen and Chariots in the Homeric and Archaic Ages*. Cambridge.

Griffo, P. (1946) 'Ultimi scavi e ultime scoperte in Agrigento'. *Quaderni di Archeologia Agrigento* 3: 30–31.

Härke, H. (1990) 'Warrior Graves'? The Background of the Anglo-Saxon Weapon Burial Rite'. *P&P* 126: 22–43.

Haggis, D.C., Mook, M.S., Margaret Scarry, C., Snyder, L.M., and West, W.C., III. (2004) 'Excavations at Azoria, 2002'. *Hesperia* 73:3: 339–400.

Hale, J.R. (2013) 'Not Patriots, Not Farmers, Not Amateurs: Greek Soldiers of Fortune and the Origins of Hoplite Warfare', in *MoB* 176–93.

Hall, J.M. (1995) 'How Argive was the "Argive" Heraion? The Political and Cultic Geography of the Argive Plain, 900–400 BC'. *AJA* 99: 577–613.

Hall, J.M. (2007) 'Polis, Community, and Ethnic Identity', in Shapiro (2007) 40–60.

Hall, J.M. (2014) *A History of the Archaic Greek World* ca. *1200–479 BCE*. 2nd edn. Oxford.

Hanson, V.D. (1995) *The Other Greeks: The Family Farm and the Agrarian Roots of Western Civilization*. New York.

Hanson, V.D. (2013) 'The Hoplite Narrative', in *MoB* 256–75.

Harrell, K. (2014) 'Man/Woman, Warrior/Maiden: The Lefkandi Toumba Female Burial Reconsidered', in Galanakis, Wilkinson and Bennet (2014) 99–104.

Hayden, B. (2009) 'Funerals as Feasts: Why Are They So Important?'. *CArchJ* 19: 29–52.

Hoffmann, H. (1972) *Early Cretan Armorers*. Mainz.

Højte, J.M. (ed) (2002) *Images of Ancestors*. Gylling.
Horn, C. and Kristiansen, K. (eds) (2018) *Warfare in Bronze Age Society*. Cambridge.
Hughes, B. and Robson, F. (eds) (2017) *Unconventional Warfare from Antiquity to the Present Day*. New York.
Hurwit, J.M. (1985) 'The Dipylon Shield Once More'. *ClAnt* 4.2: 121–26.
Hurwit, J.M. (2007) 'The Problem with Dexileos: Heroic and Other Nudities in Greek Art'. *AJA* 111.1: 35–60.
Jarva, E. (1995) *Archaiologia on Archaic Greek Body Armour*. Rovaniemi.
Jung, R. and Mehofer, M. (2013) 'Mycenaean Greece and Bronze Age Italy: Cooperation, Trade or War?' *ArchKorrBl* 43.2: 175–94.
Kagan, D. and Viggiano, G.F. (2013) 'The Hoplite Debate', in *MoB* 1–56.
Karageorghis, V. (ed) (1979) *Acts of the International Archaeological Symposium 'The Relations between Cyprus and Crete, ca. 2000–500 BC'*. Nicosia.
Kelder, J.M. (2012) 'Horseback Riding and Cavalry in Mycenaean Greece'. *Ancient West and East* 11: 1–18.
Knapp, A.B. and Van Dommelen, P. (eds) (2015) *The Cambridge Prehistory of the Bronze and Iron Age Mediterranean*. Cambridge.
Kramer-Hajos, M. (2008) *Beyond the Palace: Mycenaean East Lokris*. Oxford.
Kramer-Hajos, M. (2012) 'Sailor-Warriors and the End of the Bronze Age along the Euboean Gulf', in Mazarakis Ainian and Doulgeri-Intzesioglou (2012) 811–21.
Kramer-Hajos, M. (2016) *Mycenaean Greece and the Aegean World: Palace and Province in the Late Bronze Age*. Cambridge.
Krentz, P. (1985) 'The Nature of Hoplite Battle'. *ClAnt* 4: 50–61.
Krentz, P. (1994) 'Continuing the *Othismos* on *Othismos*'. *AHB* 8: 45–49.
Krentz, P. (2013) 'Hoplite Hell: How Hoplites Fought', in *MoB* 134–56.
Lagia, A. (2007) 'The Human Skeletal Remains', in Ruppenstein (2007) 273–81.
Langdon, S. (ed) (1997) *New Light on a Dark Age. Exploring the Culture of Geometric Greece*. Columbia, MO.
Langdon, S. (2008) *Art and Identity in Dark Age Greece, 1100–700 BCE*. Cambridge.
Latacz, J. (1977) *Kampfparänese, Kampfdarstellung und Kampfwirklichkeit in der Ilias, bei Kallinos und Tyrtaios*. Munich.
Lee, G., Whittaker, H., and Wrightson, G. (eds) (2015) *Ancient Warfare: Introducing Current Research, Volume I*. Newcastle.
Lemos, I. (2001) 'The Lefkandi Connection: Networking in the Aegean and the Eastern Mediterranean. In Memory of Mervyn Popham', in Bonfante and Karageorghis (2001) 215–26.
Lemos, I. (2020) 'Euboea', in Lemos and Kotsonas (2020) 787–813.
Lemos, I. and Kotsonas, A. (eds) (2020) *A Companion to the Archaeology of Early Greece and the Mediterranean*. Hoboken, NJ.

Liston, M.A. and Papadopoulos, J.K. (2004) 'The "Rich Athenian Lady" Was Pregnant: The Anthropology of a Geometric Tomb Reconsidered'. *Hesperia* 73.1: 7–38.

Liston, M.A. (2017) 'Human Skeletal Remains', in Papadopoulos and Smithson (2017) 503–60.

Lloyd, M. (2015) 'Death of a Swordsman, Death of a Sword: The Killing of Swords in the Early Iron Age Aegean *ca.* 1050 – *ca.* 690 BCE', in Lee, Whittaker and Wrightson (2015) 14–31.

Lloyd, M. (2017) 'Unorthodox Warfare? Variety and Change in Archaic Greek Warfare (ca. 700-ca. 480 BCE', in Hughes and Robson (2017) 231–52.

Lloyd, M. (2020a) 'Warrior Graves', in Ondine Pache (2020) 399–401.

Lloyd, M. (2020b) 'Weapons', in Lemos and Kotsonas (2020) 499–522.

Lorimer, H.L. (1947) 'The Hoplite Phalanx with Special Reference to the Poems of Archilochus and Tyrtaeus'. *ABSA* 42: 76–138.

Luraghi, N. (2006) 'Traders, Pirates, Warriors: The Proto-History of Greek Mercenary Soldiers in the Eastern Mediterranean'. *Phoenix* 60: 21–47.

Matthew, C. (2012) *A Storm of Spears: Understanding the Greek Hoplite at War*. Barnsley, UK.

Mazarakis Ainian, A. (ed) (2011) *The "Dark Ages" Revisited*. Volos.

Mazarakis Ainian, A. and Doulgeri-Intzesioglou, A. (eds) (2012) *Proceedings of the 3rd Archaeological Meeting of Thessaly and Central Greece 2006–2008. From Prehistory to the Contemporary Period, Volos 12–15 March 2009*. Volos.

Mohen, J.-P. (1987) 'Marmesse', *Trésors des Princes Celtes. Galeries nationale du Grand Palais 20 octobre 1987–15 février 1988*. Paris: 47–49.

Morgan, C. (1990) *Athletes and Oracles. The Transformation of Olympia and Delphi in the Eighth Century BC*. Cambridge.

Morgan, C. (2012) 'Setting Zagora in Context'. *MedArch* 25: 29–42.

Morris, I. (1987) *Burial and Ancient Society: The Rise of the Greek City-State*. Cambridge.

Morris, I. (1999) 'Archaeology and Gender Ideologies in Early Archaic Greece'. *TAPhA* 129: 305–17.

Morris, I. (2000) *Archaeology as Cultural History: Words and Things in Iron Age Greece*. Oxford.

Morris, I. (2001) 'The Use and Abuse of Homer', in Cairns (2001) 57–91.

Morris, I. (2009) 'The Eighth-Century Revolution', in Raaflaub and Van Wees (2009) 64–80.

Murray, O. (1993) *Early Greece*. 2nd edn. London.

Murray, S.C., Chorghay, I., and MacPherson, J. (2020) 'The Dipylon Mistress: Social and Economic Complexity, the Gendering of Craft Production, and Early Greek Ceramic Material Culture'. *AJA* 124.2: 215–44.

Nagy, G. (1997) 'The Shield of Achilles. Ends of the *Iliad* and Beginnings of the Polis', in Langdon (1997) 194–207.

Nakassis, D. (2020) 'The Economy', in Lemos and Kotsonas (2020) 271–91.
O' Brien, S. (2022) 'Warfare and "Social Complexity" from LH III to the Early Iron Age', in Shelton and Kvapil (2022), forthcoming.
Ondine Pache, C. (ed) (2020) *The Cambridge Guide to Homer*. Cambridge.
Papadopoulos, A. (2009) 'Warriors, Hunters and Ships in the Late Helladic IIIC Aegean: Changes in the Iconography of Warfare?', in Bachhuber and Roberts (2009) 69–77.
Papadopoulos, J.K. (2015) 'Greece in the Early Iron Age: Mobility, Commodities, Polities, and Literacy', in Knapp and Van Dommelen (2015) 178–95.
Papadopoulos, J.K. and Smithson, E.L. (eds) (2017) *The Athenian Agora Volume XXXVI: The Early Iron Age – The Cemeteries*. Princeton.
Popham, M.R. and Lemos, I.S. (1995) 'A Euboean Warrior Trader'. *OJA* 14.2: 151–57.
Popham, M.R. and Lemos, I.S. (1996) *Lefkandi III: The Toumba Cemetery. The Excavations of 1981 to 1994 (Plates)*. Oxford.
Popham, M.R., Sackett, L.H., and Thelemis, P.G. (1980) *Lefkandi I: The Iron Age Settlement*. London.
Popham, M.R., Touloupa, E., and Sackett, L.H. (1982) 'Further Excavation of the Toumba Cemetery at Lefkandi'. *ABSA* 77: 213–48.
Popham, M.R., Calligas, P.G., and Sackett, L.H. (1993) *Lefkandi II: The Protogeometric Building at Toumba, Part 2. The Excavation, Architecture, and Finds*. Oxford.
Raaflaub, K.A. and Van Wees, H. (eds) (2009) *A Companion to Archaic Greece*. Malden, MA.
Ridgway, D. (1997) 'Nestor's Cup and the Etruscans'. *OJA* 16: 325–44.
Rizza, G. (1979) 'Tombes de Chevaux', in Karageorghis (1979) 294–97.
Rombos, T. (1988) *The Iconography of Attic Late Geometric II Pottery*. Jonsered.
Ruppenstein, F. (ed) (2007) *Kerameikos XVIII: Die Submykenische Nekropole. Neufunde und Neubewertung*. Munich.
Rystedt, E. and Wells, B. (eds) (2006) *Pictorial Pursuits: Figurative painting on Mycenaean and Geometric pottery (Acta Instituti Atheniensis Regni Sueciae 4,53)*. Stockholm.
Scheidel, W. (2003) 'The Greek Demographic Expansion: Models and Comparisons'. *JHS* 123: 120–40.
Schepartz, L.A., Fox, S.C. and Bourbou, C. (eds) (2009) *New Directions in the Skeletal Biology of Greece*. Princeton.
Schwartz, A. (2013) 'Large Weapons, Small Greeks: The Practical Limitations of Hoplite Weapons and Equipment', in *MoB* 157–75.
Seaman, M.G. (2020) 'Early Greek Siege Warfare', in *NAGRW* 29–38.
Shapiro, H.A. (ed) (2007) *The Cambridge Companion to Archaic Greece*. Cambridge.
Shelmerdine, C.W. (ed) (2008) *The Cambridge Companion to the Aegean Bronze Age*. Cambridge.
Shelton, K. and Kvapil, L. (eds) (2022) *Brill's Companion to Warfare in the Bronze Age Aegean*. Warfare in the Ancient Mediterranean World 5. Leiden.

Sidnell, P. (2006) *Warhorse: Cavalry in Ancient Warfare*. London.

Silk, M. (2010) 'The Language of Greek Lyric Poetry', in Bakker (2010) 424–40.

Smith, S.K. (2009) 'Skeletal Evidence for Militarism in Mycenaean Athens', in Schepartz, Fox and Bourbou (2009) 99–109.

Snodgrass, A.M. (1964) *Early Greek Armour and Weapons: From the End of the Bronze Age to 600 BC*. Edinburgh.

Snodgrass, A.M. (1971) 'The First European Body-Armour', in Boardman, Brown and Powell (1971) 31–50.

Snodgrass, A.M. (1973) 'Bronze "Phalara" – A Review'. *HBA* 3.1: 41–50.

Snodgrass, A.M. (1974) 'An Historical Homeric Society?'. *JHS* 94: 114–25.

Snodgrass, A.M. (1980) *Archaic Greece: The Age of Experiment*. Berkeley, CA.

Snodgrass, A.M. (1987) *An Archaeology of Greece: The Present State and Future Scope of a Discipline*. Berkeley, CA.

Snodgrass, A.M. (1993) 'The "Hoplite Reform" Revisited'. *DHA* 19.1: 47–61.

Snodgrass, A.M. (1998) *Homer and the Artists. Text and Picture in Early Greece*. Cambridge.

Snodgrass, A.M. (2000) *The Dark Age of Greece. An Archaeological Survey of the Eleventh to Eighth Centuries BC*. 2nd edn. Edinburgh.

Snodgrass, A.M. (2013) 'Setting the Frame Chronologically', in *MoB* 85–94.

Stamatopoulou, M. and Yeroulanou, M. (eds) (2002) *Excavating Classical Culture. Recent Archaeological Discoveries in Greece*. Oxford.

Steele, P.M. (2020), 'Script and Literacy', in Lemos and Kotsonas (2020) 246–69.

Van de Moortel, A. and Zachou, E. (2004) '2004 Excavations at Mitrou, East Lokris'. *Aegean Archaeology* 7: 39–48.

Van Wees, H. (1992) *Status Warriors. War, Violence and Society in Homer and History*. Amsterdam and Leiden.

Van Wees, H. (1994a) 'The Homeric Way of War: The *Iliad* and the Hoplite Phalanx I'. *G&R* 41.1: 1–18.

Van Wees, H. (1994b) 'The Homeric Way of War: The *Iliad* and the Hoplite Phalanx II'. *G&R* 41.2: 131–55.

Van Wees, H. (1998) 'Greeks Bearing Arms: The State, the Leisure Class, and the Display of Weapons in Archaic Greece', in Fisher and Van Wees (1998) 333–78.

Van Wees, H. (2000a) 'The Development of the Hoplite Phalanx: Iconography and Reality in the Seventh Century', in *W&V* 125–66.

Van Wees, H. (2000b) 'Megara's Mafiosi: Timocracy and Violence in Theognis', in Brock and Hodkinson (2000) 52–67.

Van Wees, H. (2013) 'Farmers and Hoplites: Models of Historical Development', in *MoB* 222–55.

Van Wees, H. and Fisher, N. (2015) 'The Trouble with "Aristocracy"', in Fisher and Van Wees (2015) 1–57.

Viggiano, F.G. (2013) 'The Hoplite Revolution and the Rise of the Polis', in *MoB* 112–33.
Viggiano, F.G. and Van Wees, H. (2013) 'The Arms, Armor, and Iconography of Early Greek Hoplite Warfare', in *MoB* 57–73.
Viviers, D. (1994) 'La cité de Dattalla et l'expansion terriroriale de Lyktos en Crète centrale'. *BCH* 118.1: 229–59.
Vlachou, V. (ed) (2015) *Pots, Workshops and Early Iron Age Society: Function and Role of Ceramics in Early Greece*. Brussels.
Webster, T.B. (1955) 'Homer and Attic Geometric Vases'. *ABSA* 50: 38–50.
West, M.L. (2011) *The Making of the Iliad*. Oxford.
Whitley, J. (2002) 'Objects with Attitude: Biographical Facts and Fallacies in the Study of Late Bronze Age and Early Iron Age Warrior Graves'. *CArchJ* 12: 217–32.
Williams, D. (2015) 'Ship, Horse, Battle: Some Attic Geometric Fragments from the Sanctuary of Aphaia, Aigina, and Attic Geometric Gold Jewellery', in Vlachou (2015): 253–66.
Worley, L.J. (1994) *Hippeis: The Cavalry of Ancient Greece*. Boulder, CO.
Xydopoulos, I.K. (2017) 'Macedonians in Bottiaea. 'Warriors' and Identities in Late Iron Age and Archaic Macedonia', in Xydopoulos, Vlassopoulos and Tounta (2017) 72–98.
Xydopoulos, I.K., Vlassopoulos, K., and Tounta, E. (eds) (2017) *Violence and Community: Law, Space and Identity in the Ancient Eastern Mediterranean World*. Abingdon.
Zaphiropoulou, P.N. (2002) 'Recent Finds from Paros', in Stamatopoulou and Yeroulanou (2002) 281–84.
Zaphiropoulou, P.N. (2006) 'Geometric Battle Scenes on Vases from Paros', in Rystedt and Wells (2006) 271–77.

CHAPTER 3

The Anatolian Roots of Archaic Greek Warfare

Josho Brouwers

It is often claimed that the military practices of the people identified (by us) as Greeks were wholly – or to a large extent – their own invention, created more or less in isolation from outside influences. Supposed idiosyncrasies of Greek warfare are emphasised by some modern authors, who contrast the ways that the Greeks waged war with the modes of fighting used by other peoples – especially from ancient Southwest Asia and Egypt.[1] The most extreme example can be found in Victor Davis Hanson's *The Western Way of War*:

> We have put ourselves out of business, so to speak; for any potential adversary has now discovered the futility of an open, deliberate struggle on a Western-style battlefield against the firepower and discipline of Western infantry. Yet, ominously, the legacy of the Greeks' battle style lingers on, a narcotic that we cannot put away. [...] There is in all of us a repugnance, is there not, for hit-and-run tactics, for skirmishing and ambush?[2]

In this chapter, I argue that the developments in Greek warfare have to be placed in a wider cultural context. It is, therefore, offered as part of a wider strand in recent scholarship that emphasises the pluriform nature of Greek culture.[3] The Greeks of the Early Iron Age and especially the Archaic period did not operate in isolation. They maintained contacts with a variety of different peoples, and it is difficult – perhaps impossible? – to determine who influenced whom. As James Whitley has put it, '"Influence" is a term that encourages bad thinking'.[4] What I intend to demonstrate in this chapter is that 'Greek' warfare arose out of complex interactions between peoples in the Aegean and West Anatolia. It should be understood as a mode of warfare developed as part of an Aegean-Anatolian cultural *koine*, which was itself embedded within a larger Mediterranean world, rendering moot dichotomies such as West versus

1 Because the term 'Near East', like 'Far East', is Eurocentric, I have opted to use the more neutral term 'Southwest Asia' instead, and list Egypt (an important kingdom throughout the Bronze and Iron Ages) separately, as it is a part of Africa.

2 *WWW* 13; cf. Lynn (2003) 1–28.

3 See, for example, the papers collected in Dougherty and Kurke (2003).

4 Whitley (2001) 104 n. 7.

 | DOI:10.1163/9789004501751_004

East or Greek versus barbarian. Indeed, the case can even be made that 'Greek culture', as conceived by modern commentators, did not exist in the Archaic period: 'There were certainly "cultures" in late Archaic Greece, but a singular "Greek" (or Hellenic) culture would appear to exist only in the etic, descriptive shorthand of modern scholars'.[5]

Before turning to an examination of cultural exchange in and with Anatolia, it is necessary to briefly discuss the concepts of 'culture' and 'ethnicity', to lay some groundwork for what is to follow. In the first place, culture is not static or monolithic, but dynamic and pluralistic, constantly being negotiated and redefined, with many subcultures competing within a single culture. Particularly useful is a distinction between 'emic (internally perceived) and etic (externally observed) views of cultural homogeneity'.[6] Of course, etic and emic definitions may coincide, for example, in the case of 'elite culture' in Archaic Greece.[7] There is a tendency among modern commentators, as explained by Jonathan Hall, to look at, for example, perceived 'hoplite equipment' in ostensibly 'non-Greek' contexts and regard this as an example of Hellenization (an etic view), whereas this probably was not the case for the people involved from an 'emic' perspective.[8] Put differently, whoever used and/or deposited the equipment may or may not have regarded them as typical – let alone archetypal – of one particular culture or another.

Secondly, 'culture' cannot be equated simply with 'ethnicity'. Ethnicity itself can be defined on different levels: for example, the inhabitants of the cities in Ionia would have identified themselves individually as Ephesians, Smyrnaeans, and so on, as well as collectively as Ionians,[9] or as Hellenes. Like culture, ethnicity is a construct to express identity, but depends to a large extent on shared language, which is generally invisible when it comes to material culture.[10] Indeed, many cultures or subcultures cross perceived or observed ethnic boundaries. Material culture – that is, archaeologically identified assemblages – is not merely symbolic, but constituent to the identities of social groups. This includes physical remains, such as fortifications: walls are not merely *representative* of a community's strength, but visibly *contribute* to it. It also includes activities, some of which have left traces in the archaeological record, such as feasting and especially the consumption of alcohol, which are activities that are constituent to the identities of elites throughout the

5 Hall (2003) 30. See also, more broadly, Hall (2002).
6 Hall (2003) 23.
7 Hall (2003) 26; see also Antonaccio (2003) 62–63.
8 Hall (2003) 24; cf. Antonaccio (2003) 57–59.
9 On Ionians see esp. Crielaard (2009).
10 See esp. Derks and Roymans (2009) 1–9; Anson (2009).

ancient world: feasting is not merely a way to *express* one's wealth but *contributes actively* to the notion that one is rich. As regards the subject of this chapter, I would argue that the use of certain equipment or tactics by one group or another should not be regarded as the result of slavish emulation, but as a pathway toward expressing that group's own identity.[11]

In short, both culture and ethnicity are constructs: how they were defined depends largely on emic points of views, which to a large extent are replaced by etic points of view by modern scholars. Etic points of view cannot be entirely avoided: all knowledge about the past is, after all, constructed. However, in this chapter, I strive to avoid falling into this trap by examining the Anatolian evidence without seeking to pinpoint specific cultural or ethnic traits one way or another.

1 Historical Context

Anatolia formed an important bridge between the Aegean to the west and the wider Southwest Asian world to the east. Culturally, Anatolia has much in common with mainland Greece, including a historical 'Dark Age' between the collapse of complex polities in the second millennium and the rise of new complex polities in the first millennium, when historical evidence is limited.[12] Politically, it was by and large a patchwork of independent polities from the end of the Bronze Age down to the emergence of Persia in the mid-sixth century,[13] although various kingdoms would vie for control, including the Phrygian and the Lydian.[14] Regarding Lydian material culture in particular, Christopher Roosevelt has said that it

> is marked not by monolithic and unchanging assemblages that can be compared and contrasted with ease, but rather, by a pronounced eclecticism of Eastern and Western features and traditions that are sometimes imported intact, yet sometimes amalgamated locally into wholly indigenous compilations.[15]

11 I follow Smith (2003), who emphasises that material culture is not merely symbolic (e.g., 75–77 and 210–15). See also Antonaccio (2003), who emphasises how (material) culture is not merely instrumental, but reflexive.

12 Mellink (1991) 621.

13 For the Classical situation, Hagemajer Allen (2003) offers a starting point.

14 Mellink (1991) 625.

15 Roosevelt (2009) 7.

This statement applies equally to the material culture of the Phrygians and other West-Anatolian peoples (such as the Lykians and the Lydians), which can all be regarded as similarly 'eclectic'. The case that I try to make in this chapter is that the cultures of the Archaic Aegean would similarly have adopted aspects from both their eastern and (north)western neighbours. The notion of 'hybridity' in the context of this chapter – in which I argue that the material culture in Anatolia is impossible to disentangle – is useful when discussing acculturation.[16] Since material culture cannot be easily assigned to cultural ethnicities without resorting to circular reasoning, I will instead focus on geographic regions rather than named 'cultures'. It is the back-and-forth between different peoples in the Aegean and Anatolia – and beyond – that gave rise to what is conventionally conceived of as 'Greek' warfare.

The period under examination can be divided into two main chunks: (1) *ca.* 800 to 550, and (2) the second half of the sixth century. These divisions are not perfect, but the first period encompasses the age in which the Phrygian Kingdom flourished in Anatolia, the Neo-Assyrians achieved hegemony in Mesopotamia, and the Lydian and Neo-Babylonian Empires rose and fell. The period after approximately the mid-sixth century is dominated by the rise of the Achaemenid Persians, who would carve out the greatest empire the world had yet seen, and would subdue the Lydian Kingdom as well as the Greek cities on the western coast of Anatolia.

By the eighth century, there were Greek-speaking populations in Anatolia, with a large number of settlements on the coast and the islands. But Greeks were also active further afield. The 'Yamani' mentioned in Assyrian sources, some of whom served as mercenaries, consisted of Ionian Greeks, but probably also included other Aegean and Anatolian peoples, such as the Karians.[17] That the Assyrians used an umbrella designation for these peoples supports the idea that they were etically indistinguishable.

Over the course of the seventh century, the Lydian Kingdom rose to prominence. Its capital, Sardis, was frequented by Greeks, as evidenced by graffiti from Sardis.[18] The kingdom, centred more or less on the Hermus River, was large, with the Lydian Empire eventually, in the first half of the sixth century, growing to become 'more than twice the size of modern Greece'.[19] Ionians and Aeolians lived on the western borders of the Lydian kingdom and were

16 See Antonaccio (2003) 59–61; cf. Şale (2010) ('hybridization'). Cultural 'entanglement' (as opposed to 'world system' and 'Hellenization'): Dietler (1998).

17 Brown (1984).

18 For example, Cahill (2010) cat. no. 114, dated to *ca.* 575–525. Also available online on the website of the Sardis Expedition: http://sardisexpedition.org/en/artifacts/latw-114

19 Roosevelt (2009) 40.

eventually incorporated into it. Etically, Greeks and Lydians may have been quite similar.[20] Indeed, there is ample evidence for widespread bilingualism in Anatolia, a prerequisite for easy contact between Greeks and Lydians (and other peoples).[21] Furthermore, the people here adopted an alphabetic writing system, with evidence for the use of the alphabet among both Greeks and Phrygians dating to before the end of the eighth century.[22]

The Greeks in Anatolia may have been among the first of their people to fortify their settlements after the end of the Bronze Age. The development of fortifications in Greece after the fall of the Mycenaean palaces is a complex issue, as settlements in mainland Greece developed differently from those in Anatolia, and re-use and abandonment of sites is seldom taken into proper consideration.[23] However, the Greek cities in Anatolia perhaps not only built fortifications for defence, but also to compete with rival Lydian towns located further inland.[24] Among the first and largest were the circuit walls of Old Smyrna, the earliest of which were constructed in the tenth century. Eventually, however, Old Smyrna was besieged and captured by the Lydians under Alyattes in *ca.* 600. Excavations at Smyrna have uncovered the vast siege mound that the Lydians had cast up to defeat the city's monumental fortifications.[25] At other times, the tables were turned, and the Greeks managed to seize a Lydian town: the earliest fortification wall at Neandria in the Troad, dated to the seventh century, is thought to have been Lydian rather than Greek.[26]

Lydian rulers likely patronised the construction of stone temples in East Greek cities.[27] Ancient sources tell of dedications at Delphi – among other sanctuaries – made by the Phrygian King Midas and, more securely, the Lydian kings of the Mermnad dynasty from Gyges onwards.[28] The earliest true coins were struck in Lydia of electrum.[29] Coins spread from Anatolia westwards to Greece in the course of the sixth century. Similarly, the symposium is thought to have been adopted from Lydia, and the practice of reclining on couches while consuming alcohol can be shown, via depictions on painted vases,

20 Roosevelt (2009) 86–87.

21 For example, Hipponax frs. 92.1–2, 125; *Hom. hymn Aphr.* 113–16; cf. Strabo 13.4.12 and 13.4.17.

22 Mellink (1991) 621, 624–26 (Phrygian), 631. The Phrygians perhaps adopted the alphabet from Greeks.

23 For Archaic Greek fortifications, Frederiksen (2011), esp. 105–107; Hülden (2020); Seaman (2020).

24 Greenewalt and Rautbaum (2000).

25 On Old Smyrna, see, briefly: Lang (1996) 28–29; Wokalek (1973) 48–51.

26 Lang (1996) 33 and 37; Wokalek (1973) 43–44.

27 For further details, refer to Ratté (1993); see also Roosevelt (2009) 146–47.

28 For an overview of the evidence, see Kaplan (2006) 30–39.

29 See now the contributions in Wartenberg and Van Alfen (2020).

to have spread westwards across the Aegean over the course of the seventh century.[30] Finally, the Greek word *tyrannos* ('tyrant') is first used by Archilochos (fr. 19 West) to denote Gyges, the founder of the Lydian Mermnad dynasty. It is probably derived from the Lydian word *tûran*, meaning 'lord' or 'sole ruler'.[31]

2 Warfare in Anatolia

Having given a brief historical outline of Anatolia in the previous section, I will next examine a specific number of aspects of Anatolian warfare, namely specific pieces of military equipment (i.e., Karian inventions, linen corslets, and helmets), military organization, and finally depictions of warriors in Anatolian art.

2.1 *Military Equipment*

The idea that the hoplite panoply was something that set the Greeks apart from other peoples appears to be wholly an invention of modern scholarship. Indeed, Herodotos credits the Karians with inventing a number of elements most commonly considered to be Greek (Hdt. 1.171), namely: (a) a method for fastening crests to helmets;[32] (b) the idea for putting blazons on shields; and finally (c) attaching handles to shields to replace the earlier method of carrying shields using leather straps (cf. descriptions in the Homeric epics).

But Anthony Snodgrass argued in 1964 that, at least where the Karians were involved, Herodotos was mistaken. Instead, he found the idea that 'the Carians habitually used these devices' was 'quite acceptable'.[33] Rather than having invented the items that Herodotos credits them with, Snodgrass suggests that the Karians, 'as the first great mercenary force of historical times', were the ones who made those items famous, concluding that

> the pattern is emerging [...] of the Greeks themselves going and seeking out these innovations from their source, however far that might be. The nearer barbarians to the East and North seem to have contributed comparatively little, and in many cases were soon to be attracted into the Greek orbit. There is every sign that Caria conformed to this pattern.[34]

30 On the Eastern qualities of the symposium, Morris (1996) 32–33; cf. Hammer (2004).
31 Dale (2011) 21.
32 Cf. Alkaios fr. 388 West.
33 Snodgrass (1964) 107.
34 Snodgrass (1964) 117–18.

In this way, Snodgrass reinforced the essential Hellenocentric notion that the initiative is always with the Greeks. But his argument that, 'It is difficult to see what part Caria could have played in its [i.e., the crest's] diffusion',[35] is weak. The evidence all dates to before 700, and it is not hard to imagine that the Karians adopted crests from the Assyrians before the Greeks did, or even did so at the same time. We know, as Snodgrass is also aware,[36] that Ionian and Karian mercenaries operated side-by-side in the eastern Mediterranean, and sometimes no doubt alongside Lydian mercenaries, too. After all, Herodotos tells us that Gyges sent Lydian mercenaries to Egypt (Hdt. 2.152). Greeks and Anatolian peoples maintained relations with each other as well as with peoples further afield. The argument should not focus so much on what the Greeks brought to the table – and to argue away anything that seems to diminish the 'Greek achievement' – but rather to acknowledge the complex web of interactions that gave rise to certain practices and objects. Seen in this light, the Karians do not have to be pushed to the side-lines, but may instead be regarded as equal partners to the Greeks, and neither can be considered in isolation.

If the evidence for helmet plumes, blazons, and shield handles does not definitively point in an easterly direction, there is another piece of military equipment that almost certainly does: the linen corslet, referred to in Greek texts as the *linothorax*.[37] This type of body-armour is known exclusively from texts and figurative art, as it was made of perishable materials and no examples have survived archaeologically. The idea that the corslet depicted in vase-paintings was not – largely – made of linen, but of leather is current among reenactors and amateur historians, but that need not distract us here. The evidence for leather body-armour from the ancient world is so slim as to be negligible.[38] The earliest textual reference to a linen corslet is found in Homer's *Iliad*. Here, linen corslets are associated with Trojan Warriors (*Il.* 2.830), but the lesser Aias is also said to wear one (*Il.* 2.259). Eero Jarva has argued that some Protoattic vase fragments of the early seventh century depict linen corslets, based largely on what look like *pteryges* cut along the bottom edge.[39] If this interpretation is correct, this means that the earliest attestations of linen corslets in Greek art are more or less contemporary with the Homeric descriptions.

Linen corslets are not shown in other examples of Greek art before the sixth century, but are mentioned in early literary sources, most notably by Alkaios

35 Snodgrass (1964) 115.
36 Snodgrass (1964) 117 and esp. n. 78.
37 But note Nosch (2014) 37.
38 For a full and up-to-date treatment of the linen corslet, including a complete review of the ancient evidence, refer to Aldrete *et al.* (2013).
39 Jarva (1995) Figure 19.

from Mytilene, Lesbos (fr. 140 Voigt), whose *floruit* is usually placed around 600.[40] Ian Morris has emphasised that the poet's description of a stockpile of weapons and armour is not only a reference to Homer, but also 'evokes the east'.[41] Herodotos mentions Assyrians with linen breastplates (Hdt. 7.63). Linen corslets were also used in the Persian Empire from at least the early fifth century onwards, as shown by a seal from Persepolis dating to the sixteenth year of Xerxes' reign.[42] In this context, it is worth pointing out that some ostensibly 'Greek' places in the Eastern Aegean and Western Anatolia, such as the cities on the islands of Lesbos, seem, at least in the Archaic period, to have had closer ties with the rest of Anatolia than with mainland Greece.[43] It seems likely that linen corslets were first used in Anatolia and then spread westwards to mainland Greece.

Finally, there is a particular type of helmet that seems characteristic of Anatolia. Archaeologists have investigated a debris layer in Sardis that is associated with the Persian capture of the city. In this layer, excavators unearthed a skeleton of a man in his early twenties, as well as fragments of a bronze-decorated iron helmet.[44] It is not clear if the skeleton and the helmet belong together, nor is it clear if the helmet is Lydian or Persian (or something else altogether, as both armies made use of mercenaries). The helmet uses a radial construction with triangular plates similar to the much later Roman *Spangenhelm*, and the find has pushed up the date for helmets constructed like this by six centuries.[45] The only other pre-Roman helmet that most resembles the one from Sardis comes from excavations at Old Smyrna and has been dated to the first half of the sixth century.[46] Iron helmets composed of multiple pieces were perhaps invented in Anatolia.

This type of helmet, composed of different pieces, seems closely related to the so-called Ionian helmet. Among the earliest evidence are fragments of a krater from Smyrna, dated 640–630, features warriors that are perhaps equipped with this type of helmet.[47] The typical form covers most of the head and neck, leaving the face open. Many examples have a raised plate protecting

40 For an in-depth discussion, including the use of leather instead of linen, see Jarva (1995) 33–44.
41 Morris (2000) 177; also *infra*, n. 24.
42 Schmidt *et al.* (1957) 30 and plate 9.
43 Dale (2011) 21–23 and 23 n. 43.
44 For a description and photos, including reconstruction of the helmet, see the website of the Sardis excavations, and especially: https://sardisexpedition.org/en/artifacts/latw-211.
45 Greenewalt and Heywood (1992) 10–11; Greenewalt (1992) 264–65.
46 Greenewalt and Heywood (1992) 12–13 and 23 n. 19.
47 Akurgal (1983) plate 109a.

the forehead; a pronounced raised ridge runs from the front of the helmet to a little beyond the area of the crown. Cook suggests that this ridge indicated the helmet may (originally) have been made entirely out of leather. The examples known from Greek art, however, seem to have hinged cheek pieces, which means that those helmets must have been made entirely from metal.[48] This helmet type is known only from figurative art, especially from small aryballoi fashioned in the shape of a warrior's head, which were most probably made in workshops somewhere in the Eastern Aegean or Western Anatolia.[49] Examples of those kinds of vases have also been unearthed in Lydia,[50] where they must have held some significance. The unusual structure of the helmet makes it seem unlikely that it is derived from any other types of Greek helmets.[51] It must have been assembled from several pieces of metal rather than a single sheet, and thus might be related to the aforementioned helmets from Sardis and Old Smyrna, or otherwise fits in an Anatolian tradition where helmets were made out of multiple pieces. Hinged cheek pieces are perhaps also an Eastern Aegean or West Anatolian invention; I know of no earlier examples than the Ionian helmet and the mid-sixth-century helmet unearthed at Sardis.

2.2 *Military Organization*

Contrary to their reputation of being the 'most cowardly of men' (Hdt. 4.142), the Greeks in Anatolia fought not only frequently amongst themselves, like their compatriots on the Greek mainland, but also engaged in war against the armed forces of a state – the Lydian Kingdom – that was larger than the average Greek city-state. We do not know exactly how the Lydian army was organised. The evidence is slim, but what we do have, such as the siege mound at Smyrna and the evidence from Greek written sources (e.g., Hdt. 1.15–22), suggests that the Lydians were a military force to be reckoned with.

Mimnermos of Smyrna wrote exhortation poetry to encourage his countrymen to fight against the Lydians, probably in the age of Gyges (fr. 13a West) or his successor, Ardys. In one fragment (fr. 14 West), Mimnermos tells of a warrior who attacked, on his own, the Lydian cavalry's *pukinas phalaggas* ('thick ranks'). The word used for the cavalry, derived from *hippomachos*, along with the context, suggests strongly that these men were 'true cavalry', that is, they fought from horseback. Use of the word *phalanx*, in the plural, seems similar to

48 Cook (1981) 123 n. 99.
49 Brief overview in Biers (1984) 5 n. 5.
50 Example from modern Ikiztepe: Özgen and Özturk (1996) 135.
51 *Contra* Connolly (1998) 60.

how Homer uses the term, and might indicate simply that there were numerous cavalry on the field, rather than the cavalry being deployed in some kind of massed formation.[52] The adjective, however, can be taken to imply that they were packed unusually close together, which does suggest that the Lydian horse were massed in some way.

It seems likely that the Greeks of the Archaic period first came into contact with organised armies in Anatolia. We also know that Greeks from Anatolia and the islands fought as mercenaries in the armies of the great kingdoms and empires of Southwest Asia and Egypt, which all fielded large and well-organised armies. Aside from the Sumerian 'Stela of the Vultures' – which predates the Chigi Vase by some two thousand years – we have, for example, Bronze-Age wooden models recovered from tombs in Egypt that depict soldiers deployed in ranks and files.[53] The Parian poet Archilochos, whose *floruit* is dated to the mid-seventh century, is usually thought to have served as a mercenary. He also offers the earliest use of the term *stratēgos*, that is, 'army-leader' (cognate with *stratos*) or simply 'general' (fr. 114.1 West). Elsewhere, he refers to a general as *archos* (fr. 113.1 West). A mercenary force would have required some form of strict hierarchy, especially when not all of the combatants might have been known to each other and their ties were as likely to be based on expected payment as on bonds of blood or friendship.[54] We know from the early sixth-century inscription from the temple of Abu Simbel in Egypt, which was Ionian in lettering but Doric in dialect,[55] that the 'non-native speakers' (i.e., foreign mercenaries, including Greeks) were led by a single man, Potasimto, a known general,[56] with a subordinate commander by the name of Psammetichos (undoubtedly named in honour of the Pharaoh Psamtik),[57] son of Theokles, in command of the Greek contingent. This contingent probably included Anatolian mercenaries, such as Karians and possibly Lydians.

---

52 For a recent treatment of the topic, see Echeverría (2012).

53 See, for example, Ferrill (1985) 49.

54 See also Hale (2013), who argues that massed formations were developed neither by the 'middle-class' farmer-hoplites hypothesised by writers like Hanson and other proponents of that view, nor by the rich, aristocratic warriors favoured by Hans van Wees and others, but rather by the professional Greek mercenaries who served in the armies of kingdoms and empires of the ancient Southwest Asia and Egypt. See also the chapters by Lloyd and Van Wees in this volume.

55 Jeffery (1961) 355.

56 Vittmann (2003) 201.

57 Haider (1996) 110–11.

2.3 *Depictions of Warriors in Anatolian Art*

Excavations at Gordion, once the capital of the kingdom of the Phrygians, within a structure known as 'Megaron 3' (possibly the central building of the citadel and originally used for feasting),[58] have revealed a square piece of ivory inlay that depicts a rider with a large, round shield, helmet with cheek pieces, and a long lance. Similar horsemen are also shown on other pieces of ivory inlay from Gordion.[59]

The date for these inlays is a matter of some contention, with some suggesting the building was destroyed around 800 and others favouring a date around 700.[60] The former date is based on carbon-dating. The second date is based on the idea that the mounted warrior was influenced by Greek art. From the late eighth century, warriors with shields are often depicted riding horses or otherwise associated with horses. The shield in question is always the characteristic 'Argive' shield, equipped with a double grip (*porpax* and *antilabē*) and commonly associated with hoplites. Building on the work of Detienne,[61] I suggested in an earlier article that the introduction of the double-grip shield is closely associated with horse-riding.[62] Key evidence for this includes the Early Protocorinthian Evelyn aryballos of the late eighth century, which depicts a warrior on foot (with shield) following a young man on horseback (without shield). The exact same scene is encountered on an Early Ripe Corinthian aryballos, produced about a century later, where the two figures are provided with labels: the warrior is called a *hippobatas* and the young man a *hippostrophos*.[63] Peter Greenhalgh referred to this as a 'knight and squire' motif, and argued persuasively that the *hippobatas* was mounted infantry, who would ride to battle and dismount to fight on foot, his squire taking the reins of his horse and leading it away from the fighting.[64] The motif of the 'knight and squire' is commonly encountered in vase-paintings and other Greek art throughout the Aegean in the seventh and sixth centuries.[65]

Horse-riding in Greece was already known in Mycenaean times,[66] though apparently rare. Riders only become common in Greek art from the late eighth

58 Kealhofer (2005) 15; DeVries (1980) 35–36.
59 DeVries (1980) 34 and 47 figs. 7 (horseman) and 8 (battle-scene).
60 Kealhofer (2005) 27 and 69–70; Summerer (2008).
61 Detienne (1968).
62 Brouwers (2007).
63 Greenhalgh (1973) 58 fig. 36.
64 Greenhalgh (1973) 59–61.
65 Greenhalgh (1973); Brouwers (2007, 2013). Both works are illustrated throughout; my 2007 article lists the evidence and provides illustrations for key pieces.
66 For example, Greenhalgh (1973) 44 Figure 28.

century onwards,[67] and when we look for possible sources of influence – either for the practice of riding itself or for the very act of depicting horsemen in art – we must look east. The Assyrians are widely considered the first to field true cavalry in the Early Iron Age. Archer argues persuasively that 'the people of the Near East themselves were the ones pioneering this development [i.e., true cavalry], not the people of the steppes', adding that 'Over the course of the lifetime of the Neo-Assyrian Empire (934–612 BCE), a clear line of development in military technology can be seen to take place'.[68] In the *Iliad*, the Trojans and other Anatolian peoples are commonly associated with horses and horse-breeding.[69] In the Archaic period, the Phrygians and Lydians gained fierce reputations on account of the quality of their cavalry.[70] I would suggest that the ivory inlay from Gordion should be placed in this same tradition and be considered a precursor to the later *hippobatai*, or perhaps simply an example of a Phrygian *hippobatas*.

Located further inland, a chamber tomb from Tatarlı (Phrygia), near modern Dinar and dated to the second half of the sixth century, has yielded interesting painted decoration on one of the wooden walls.[71] The painting is divided into three friezes: the top shows seated felines (presumably lions), the middle frieze depicts heavily-armed warriors, and the bottom frieze shows what seems like a procession that includes three heavy chariots. These chariots are similar to chariots from Assyrian reliefs, with their high boxes and large wheels with eight spokes;[72] they are not at all similar to the light chariots with four-spoked wheels depicted in Greek art. The warriors of the middle frieze wear (metal?) helmets with crests, greaves, and carry large, round shields with rims that are almost certainly of the *porpax-antilabē* variety. The warriors are grouped in pairs, one facing the other. We only see the front sides of the shields, so the men facing right are carrying the shields on the right arm instead of the left.[73] They

67 See esp. Greenhalgh (1973); Brouwers (2007); see also Snodgrass (1971).

68 Archer (2010) 60.

69 For example, *Il.* 2.230 (Trojans as 'breakers of horses'); *Il.* 3.185–86 and 10.431 (Phrygians); *Il.* 3.189 (Amazons).

70 For example Mimnermos fr. 14 West; Hdt. 1.27. For the later roles of cavalry in the Greek world, partly shaped by the mounted strength of the Greeks' Anatolian neighbours, see Konijnendijk's chapter in this volume.

71 Bingöl (1997) 39–42.

72 Cf., for example, the Lachisch relief of Sennacherib: Healey (1991) 47.

73 I would argue strongly against attaching value-statements to how the painter composed the scene, *contra* Bingöl (1997) 42, 'Vergeleicht man die Schilde auf der berühmten Chigi-Kanne, die bereits *richtig* dargestellt sind, so wird der *Atavismus* des provinziellen Malers klar' (original emphasis). Implicit in this argument is that the painter of this tomb strives for – but somehow fails to achieve – something akin to the Chigi Vase, which is an

carry a long spear each in their shield-hands, and brandish curved swords in their free hands. The positions of the legs, which recall the typical Archaic way of depicting figures while running, with knees bent, suggests that the figures are not fighting, but instead engaged in a ritual war-dance,[74] like the Pyrrhic.

Excavation of the Kızıbel burial mound at Elmalı, close to the southern coast in Lykia and a little west of modern Antalya, have yielded impressive wall-paintings dated to the second half of the sixth century,[75] probably the final quarter. The tomb measures 2.4m by 2.0m and is 2.3m at the highest point. The chamber has a gabled roof. The paintings have suffered damage over the past two millennia, but many scenes can still be recognised. The south wall depicts Medousa and includes the birth of Chrysaor and a wingless Pegasos. The north wall depicts several scenes, including a galley at sea with many figures aboard; another scene depicts a figure seated on a throne and a procession of men, women, and horses. On the east wall, there is a frieze with heavily-armed warriors, a chariot, and horses. Below is arguably the main scene: a heavily-armed warrior on a chariot, presumably representing the deceased, saying farewell to his family. He wears a cuirass with a rosette on his chest and carries a short spear. His helmet seems Corinthian, with a large crest, but has hinged cheek pieces similar to an Ionian helmet. The charioteer wears a bell-shaped cuirass and a regular Corinthian helmet with crest. The same wall also features a hunting scene involving three 'Skythian' warriors.

Stylistically, these wall-paintings have drawn comparisons with Greek art. However, the tombs themselves, located far from the coast – where most of the Greeks in Anatolia are thought to have settled – were probably used for the entombment of natives rather than Greeks. We also cannot be sure that the paintings were made by Greeks as work-for-hire. Again, it seems better not to try and untangle the supposed different cultures at work here – an etic observation – and, instead, to consider these tombs as arising organically from the interplay between different peoples in a region where there was a relatively free flow of information.[76] If these graves must be classified, it seems safer to do so along socio-political lines, and from an emic point of view. In that regard, these decorated Anatolian tombs are similar to painted graves such as those from Southern Italy, especially at Paestum, or from Etruscan *necropoleis*. The crucial point to make here is that such decorated tombs are part and parcel

---

assumption of the author's that seems to me unfounded. On evidence of chariot use in Greece, see Lloyd in this volume.

74 Bingöl (1997) 41.

75 Mellink and Angel (1970) 251–53; Bingöl (1997) 43–48, with illustrations, considers the wall-paintings 'Lydian'.

76 Cf. Roosevelt (2009) 75.

of an elite subculture common throughout the ancient Mediterranean;[77] they are not unique to Greece or even the Aegean.

Similar to the wall-paintings in these tombs is a Lydian alabastron made from silver and dated to the sixth century. It was looted from a burial mound in the regions of Uşak and Manisa in western Turkey. It features, among other things, a battle-scene between heavily-armoured men, equipped with Corinthian helmets, cuirasses, tunics, greaves, *porpax-antilabē* shields, and thrusting spears.[78] There is hide or fabric suspended from the bottom of some of these warriors' shields, perhaps to offer further protection for the legs. These are similar to the 'aprons' later depicted attached to some shields in Attic vase-paintings of the fifth century.[79] The different scenes, divided among a number of friezes, bear comparison to earlier Levantine silver bowls of the mid-seventh century, such as the famous Amathus Bowl and the Arjan Bowl.[80] It has been said that such scenes inspired Homer's description of the new shield of Achilles in book 18 of the *Iliad*.[81] Indeed, the presumably Phoenician-made Amathus Bowl contains a few warriors with ostensibly 'Greek' equipment, who we should probably regard as Aegean-Anatolian mercenaries.

3 Conclusions

The purpose of this article is to question the notion that the military practices of the Greeks had sprung into being without any outside influences. As stated in the introduction, this article fits into a recent scholarly line of enquiry that emphasises the pluriform character of Greek culture. As this article has shown, the Greeks of the Early Iron Age and the Archaic period were part of an Aegean-Anatolian cultural *koine*, itself embedded within the wider Mediterranean world.

A necessarily brief overview of interactions in Anatolia revealed that it is next to impossible to disentangle different peoples in an ethnic or cultural sense. Speakers of Greek, Lydian, and other languages lived side by side in cities in Anatolia, and it is often difficult if not impossible for us to determine what culture (if any) these people identified themselves as being. Many communities in Anatolia appear to have been bilingual. Lydian kings patronised

77 For example, Crielaard (2009) 62 refers to Ionians as 'members of an international elite'.
78 Özgen and Özturk (1996) 125 fig. 154.
79 For example, Boardman (1975) 165 Figure 279.
80 Amathus Bowl: Myres (1933). Arjan Bowl: Majidzadeh (1992).
81 Comparisons between the Amathus Bowl and the shield of Achilles – among other elements – are important in fixing the date of Homer's world in Crielaard (1995a).

the construction of stone temples in 'Eastern Greek' cities, and we know that, for example, the Phrygian king Midas sent dedications to Delphi and other sanctuaries. In short, the Aegean-Anatolian world was highly interconnected.

An examination of Anatolian warfare seems to confirm that elements long held to be unique to the ancient Greeks more likely arose out of the complex interplay between different peoples. Herodotos credits the Karians with inventing a number of elements of the Greek hoplite panoply, a notion once dismissed by Snodgrass, but which I would suggest is not entirely without merit. We know, for example, that Ionians and Karians together served as mercenaries in the armies of kingdoms and empires of the Southwest Asia and Egypt. Furthermore, items such as linen corslets are almost certainly an invention whose origin must be traced to Anatolia if not further afield. A particular type of helmet, composed of different pieces and unearthed at Sardis, may even be closely related to the so-called Ionian helmet, a popular type of headgear often associated with Anatolian Greeks, but its origins may be sought further east.

When we turn to an examination of tactics and military organization, it again seems likely that Anatolia played a crucial role here. Unlike their compatriots in mainland Greece, the Anatolian Greeks would have fought not just against each other, but also against larger, better organised forces, such as the armies of the Lydian Kingdom. We do not know how the Lydians organised their armed forces, but the scant evidence that we do have, including the construction of a massive siege mound at Smyrna, suggests that the Lydian war machine was fairly sophisticated. By contrast, the siege techniques of the mainland Greeks remained primitive until well into the Classical period. Other evidence, including the poetry of Mimnermos of Smyrna, provides valuable insights into the possible organization of the Lydian forces. It is probably not a coincidence that terms like *stratēgos* ('army leader' or 'general') first pop up in Greek texts from the islands, nor that the earliest known Greek mercenaries in Egypt came from Ionia.

Finally, art gives us important glimpses at what warfare may have been like in Anatolia. At Gordion, we find evidence for the practice of what might be mounted infantry: men on horseback with round shields and spears. I suggest that the mounted infantry known from Greek art of the seventh and sixth centuries has an Anatolian, perhaps even specifically Phrygian, origin. Wall-paintings in tombs at a number of sites in Anatolia provide further insights with regards to the use of the Argive – that is, hoplite – shield, helmets, chariots, and horses. Sometimes, these wall-paintings depict elements for which there are no known Greek counterparts, like the sickle-swords from the tomb at Tatarlı.

In short, there is sufficient evidence to suggest that the Greek way of war did not develop in isolation, but arose out of the complex interplay of different peoples within the larger Aegean-Anatolian world. For certain elements, like the Ionian helmet or the practice of fielding mounted infantry, it seems clear that we cannot distinguish, for example, 'Greek' from 'Lydian', and to try to do so is counterproductive. More research is necessary, but this chapter can at least stimulate further discussion.[82]

Bibliography

Akurgal, E. (1983) *Alt-Smyrna I. Wohnschichten und Athenatempel*. Ankara.

Aldrete, G.S., Bartell, S., and Aldrete, A. (2013) *Reconstructing Ancient Linen Body Armour: Unraveling the Linothorax Mystery*. Baltimore, MD.

Anson, E.M. (2009) 'Greek Ethnicity and the Greek Language'. *Glotta* 85: 5–30.

Antonaccio, C.M. (2003) 'Hybridity and the Cultures within Greek Culture' in Dougherty and Kurke (2003) 57–74.

Archer, R. (2010) 'Chariotry to Cavalry: Developments in the Early First Millennium BC', in Fagan and Trundle (2010) 57–79.

Biers, W.R. (1984/1985) 'A Helmeted Ionian'. *JWAG* 42/43: 2–5.

Bilde, P.G. and Petersen, J.H. (eds) (2008) *Meetings of Cultures in the Black Sea Region: Between Conflict and Coexistence*. Aarhus.

Bingöl, O. (1997) *Malerei und Mosaik der Antike in der Tuerkei*. Mainz.

Boardman, J. (1975) *Athenian Red Figure Vases: The Archaic Period*. London.

Boardman, J., Brown, M.A., and Powell, T.G.E. (eds) (1971) *The European Community in Later Prehistory. Studies in Honour of C.F.C. Hawkes*. London.

Brouwers, J.J. (2007) 'From Horsemen to Hoplites: Some Remarks on Archaic Greek Warfare'. *BABesch* 82.2: 305–19.

Brouwers, J.J. (2013) *Henchmen of Ares. Warriors and Warfare in Early Greece*. Rotterdam.

Brown, R. (1984) 'Greeks in Assyria: Some Overlooked Evidence'. *CW* 77.5: 300–303.

Cahill, N.D. (ed) (2010) *The Lydians and Their World*. Istanbul.

Connolly, P. (1998) *Greece and Rome at War*. 2nd edn. London.

Cook, R.M. (1981) *Clazomenian Sarcophagi*. Mainz.

82 My thanks to Cezary Kucewicz and Roel Konijnendijk for organising the London colloquium on Greek warfare and for inviting me as a speaker. The present chapter is a revised and much-expanded form of my paper, which has benefited greatly from the insightful comments received at the colloquium, and the editors of this volume. My thanks also to Sean Manning, who read an earlier draft and supplied a number of useful references.

Crielaard, J.P. (1995a) 'Homer, History and Archaeology: Some Remarks on the Date of the Homeric World', in Crielaard (1995b) 201–88.

Crielaard, J.P. (ed) (1995b) *Homeric Questions. Essays in Philology, Ancient History, and Archaeology, Including the Papers of a Conference Organized by the Netherlands Institute at Athens, 1993*. Amsterdam.

Crielaard, J.P. (2009) 'The Ionians in the Archaic period. Shifting Identities in a Changing World', in Derks and Roymans (2009) 37–84.

Cusick, J. (ed) (1998) *Studies in Culture: Interaction, Culture Change and Archaeology*. Carbondale, IL.

Dale, A. (2011) 'Alcaeus on the Career of Myrsilos: Greeks, Lydians and Luwians at the East Aegean-West Anatolian Interface'. *JHS* 131: 15–24.

Derks, T. and Roymans, N. (eds) (2009) *Ethnic Constructs in Antiquity: The Role of Power and Tradition*. Amsterdam.

Des Courtils, J. and Moretti, J.-C. (eds) (1993) *Les grand ateliers d'architecture dans le monde égéen du Vie siècle av. J.C. Actes du colloque d'Istanbul 23–25 mai 1991*. Istanbul.

Detienne, M. (1968) 'La phalange: problèmes et controverses', in Vernant (1968) 119–42.

DeVries, K. (ed) (1980) *From Athens to Gordion*. Philadelphia.

Dietler, M. (1998) 'Consumption, Agency, and Cultural Entanglement: Theoretical Implications of a Mediterranean Colonial Encounter', in Cusick (1998) 288–315.

Dougherty, C. and Kurke, L. (eds) (2003) *The Cultures within Ancient Greek Culture: Contact, Conflict, Collaboration*. Cambridge.

Droß-Krüpe, K. (ed) (2014) *Textile Trade and Distribution in Antiquity*. Wiesbaden.

Echeverría, F. (2012) 'Hoplite and Phalanx in Archaic and Classical Greece: A Reassessment'. *CPh* 107.4: 291–318.

Fagan, G.G. and Trundle, M. (eds) (2010) *New Perspectives on Ancient Warfare*. Leiden.

Ferrill, A. (1985) *The Origins of War: From the Stone Age to Alexander the Great*. London.

Frederiksen, R. (2011) *Greek City Walls of the Archaic Period 900–480 BC*. Oxford.

Greenewalt, C.H. (1992) 'When a Mighty Empire Was Destroyed: The Common Man at the Fall of Sardis, ca. 546 BC'. *PAPHS* 136.2: 247–71.

Greenewalt, C.H. and Heywood, A.M. (1992) 'A Helmet of the Sixth Century B.C. from Sardis'. *BASO* 285: 1–31.

Greenewalt, C.H. and Rautman, M.L. (2000) 'The Sardis Campaigns of 1996, 1997, and 1998'. *AJA* 104: 643–81.

Greenhalgh, P.A.L. (1973) *Early Greek Warfare: Horsemen and Chariots in the Homeric and Archaic Ages*. Cambridge.

Hagemajer Allen, K. (2003) 'Becoming the "Other": Attitudes and Practices at Attic Cemeteries', in Dougherty and Kurke (2003) 207–36.

Haider, P.W. (1996) 'Griechen im Vorderen Orient und in Ägypten bis ca. 590 v. Chr.', in Ulf (1996) 59–115.

Hale, J.R. (2013) 'Not Patriots, Not Farmers, Not Amateurs: Greek Soldiers of Fortune and the Origins of Hoplite Warfare', in *MoB* 176–93.

Hall, J.M. (2002) *Hellenicity: Between Ethnicity and Culture*. Chicago.

Hall, J.M. (2003) '"Culture" or "Cultures"? Hellenism in the Late Sixth Century', in Dougherty and Kurke (2003) 23–34.

Hammer, D. (2004) 'Ideology, the Symposium, and Archaic Politics'. *AJPh* 125.4: 479–512.

Healey, M. (1991) *The Ancient Assyrians*. Oxford.

Hülden, O. (2020) *Das griechische Befestigungswesen der archaischen Zeit: Entwicklungen – Formen – Funktionen*. Vienna.

Jarva, E. (1995) *Archaiologia on Archaic Greek Body Armour*. Rovaniemi.

Jeffery, L.H. (1961) *The Local Scripts of Archaic Greece. A Study of the Origin of the Greek Alphabet and its Development from the Eighth to the Fifth Centuries BC*. Oxford.

Kaplan, P. (2006) 'Dedications to Greek Sanctuaries by Foreign Kings in the Eighth through Sixth Centuries BCE'. *Historia* 55.2: 129–52.

Kealhofer, L. (2005) *The Archaeology of Midas and the Phrygians: Recent Work at Gordion*. Philadelphia.

Lang, F. (1996) *Archaische Siedlungen in Griechenland: Struktur und Entwicklung*. Berlin.

Lynn, J.A. (2003) *Battle: A History of Combat and Culture from Ancient Greece to Modern America*. Philadelphia.

Majidzadeh, Y. (1992) 'The Arjan Bowl'. *Iran* 30: 131–44.

Mellink, M.J. (1991) 'The Native Kingdoms of Anatolia', in *CAH* 3.2: 619–65.

Mellink, M.J. and Angel, J.L. (1970) 'Excavations at Karataş-Semayük and Elmalı, Lycia, 1969'. *AJA* 74.3: 245–59.

Morris, I. (1996) 'The Strong Principle of Equality and the Archaic Origins of Greek Democracy', in Ober and Hedrick (1996) 19–48.

Morris, I. (2000) *Archaeology as Cultural History: Words and Things in Iron Age Greece*. Oxford.

Myres, J.L. (1933) 'The Amathus Bowl: A Long-Lost Masterpiece of Oriental Engraving'. *JHS* 53: 25–39.

Nosch, M.-L. (2014) 'Linen Textiles and Flax in Classical Greece: Provenance and Trade', in Droß-Krüpe (2014) 17–42.

Ober, J. and Hedrick, C. (eds) (1996) *Dēmokratia: A Conversation on Democracies, Ancient and Modern*. Princeton.

Özgen, I. and Öztürk, J. (1996) *The Lydian Treasure: Heritage Recovered*. Ankara.

Ratté, C. (1993) 'Lydian Contributions to Archaic East Greek Architecture', in Des Courtils and Moretti (1993) 1–12.

Roosevelt, C.H. (2009) *The Archaeology of Lydia, from Gyges to Alexander*. Cambridge.

Şale, T. (2010) 'An Archaic Ivory Figurine from a *Tumulus* near Elmalı: Cultural Hybridization and a New Anatolian Style'. *Hesperia* 79.1: 53–78.

Schmidt, E.F. (1957) *Persepolis II: Contents of the Treasury and Other Discoveries*. Chicago.
Seaman, M.G. (2020) 'Early Greek Siege Warfare', in *NAGRW* 29–38.
Smith, A.T. (2003) *The Political Landscape: Constellations of Authority in Early Complex Polities*. Berkeley, CA.
Snodgrass, A.M. (1964) 'Carian Armourers: The Growth of a Tradition'. *JHS* 84: 107–18.
Snodgrass, A.M. (1971) 'The First European Body-Armour', in Boardman, Brown and Powell (1971) 31–50.
Summerer, L. (2008) 'Indigenous Responses to Encounters with the Greeks in Northern Anatolia: The Reception of Architectural Terracottas in the Iron Age Settlements of the Halys Basin', in Bilde and Petersen (2008) 263–86.
Ulf, C. (ed) (1996) *Wege zur Genese griechischer Identität: Die Bedeutung der früharchaischen Zeit*. Berlin.
Vernant, J.-P. (ed) (1968) *Problèmes de la guerre en Grèce ancienne*. Paris.
Vittmann, G. (2003) *Ägypten und die Fremden im ersten vorchristlichen Jahrtausend*. Mainz.
Wartenberg, U. and Van Alfen, P. (eds) (2020) *White Gold. Studies in Early Electrum Coinage*. New York.
Whitley, J. (2001) *The Archaeology of Ancient Greece*. Cambridge.
Wokalek, A. (1973) *Griechische Stadtbefestigungen. Studien zur Geschichte der Frühgriechischen Befestigungsanlagen*. Bonn.

CHAPTER 4

The War Dead in Archaic Sparta

Cezary Kucewicz

The most famous epigram commemorating casualties of war in antiquity was inscribed on a memorial set up for the Spartans who fell at Thermopylai. According to Herodotos, the short epigram was composed by Simonides and praised the obedience of men who sacrificed their lives for the Spartan polis: 'Stranger, tell the people of Lakedaimon, that we lie here obedient to their commands' (7.228).[1] Although we cannot be sure who buried the fallen warriors at Thermopylai, the inscription testifies not only to the famed Spartan imperative of fighting until death but also to the Spartan practice of burying the war dead in *polyandria* on or near the battlefield.[2] Apart from Thermopylai, the custom is attested across a wide array of literary sources and archaeological evidence from the Classical era, including actual tombs of the Lakedaimonian soldiers who died in the expedition to Athens in 403 and were buried in the Kerameikos cemetery. Interring corpses on the battlefield was a quick and cost-effective way to dispose of the dead, especially when compared to the 'ancestral custom' of the Classical Athenians who cremated their fallen and brought back the remains for public burial at home.[3] Furthermore, burying the dead in foreign soil had an additional political and symbolic significance, providing a constant reminder for local communities of past Spartan invasions

1 My own translation. All other translations are taken from the Loeb Classical Library unless otherwise indicated. For more on the epigrams on the Graeco-Persian Wars, see Higbie (2010).

2 Herodotos (7.228) tells us that the dead from Thermopylai 'were buried where they fell'; he does not indicate who buried them but it is unlikely that it was the Persians, who intentionally piled the corpses in a heap for their naval forces to see (8.24–25). It seems more likely that the Lakedaimonian and Thespian dead were buried later by local or medising Greeks. The commemorative monuments were commissioned by the Delphic Amphictyony; in addition to them, the Spartans also inscribed the names of the fallen on a monument back in Sparta (Paus. 3.14.1). For more on the Thermopylai war dead, see *GSW*4 168–73; Low (2006) 99–101.

3 The *polyandrion* in the Kerameikos contained bodies of 23 Spartans who died during the expedition to Athens led by King Pausanias (Xen. *Hell.* 2.4.31–34). For more on the tomb, see *GSW*4 133–34, 208; Hodkinson (2000) 252, 257–59; Stroszeck (2006); Christesen (2021) 15–22. The practical and logistical difficulties of the Athenian custom of repatriating the war dead are discussed by Rees (2018).

 | DOI:10.1163/9789004501751_005

while also supporting any claims that the Spartans might have had to disputed territories in the Peloponnese.[4]

Despite the strength of the evidence for the Spartan custom of battlefield burials, an anecdote recorded by Plutarch in his collection of Lakonian sayings espouses a seemingly different tradition concerning the treatment of the war dead in Sparta. The well-known *exemplum*, which has served as an archetypal illustration of the toughness of Spartan women, relates the story of a mother emboldening her son leaving for war.[5] Speaking words of encouragement while handing a shield to her son, the mother tells him to return 'either this or on this' (Plut. *Mor*. 241f); or, in other words: 'come back home alive with your shield or be brought back dead upon it by your comrades'. The anecdote highlights the ideal of death in battle, a source of pride for the entire family of the fallen warrior, as opposed to cowardice, symbolised most evocatively by throwing away one's shield in combat, which brought shame (*atimia*) on the warrior and his relatives. It also appears to indicate that fallen Spartan warriors – with or without their shields – were brought back home to their families. A similar notion of repatriating the war dead features in other Lakonian apophthegms, as well as a few epigrams from the Hellenistic era.[6] But since such evidence contradicts the Spartan custom of battlefield burials, most scholars dismiss it as fictitious and not representative of the actual practices of the ancient Spartans. Accordingly, any mention of warriors buried at home in the apophthegms have been seen as products of post-Classical times, 'redolent of Hellenistic and later moral and philosophical ideals, not the ethics of the classical Spartan polis'.[7]

4 E.g., Plut. *Mor*. 233c; *Vit. Ages*. 31.8. See also Low (2006).

5 Hammond (1979/80) provides an exhaustive study of the history of the *exemplum* in the ancient and later sources.

6 The image of a fallen warrior carried back home on his shield features in an epigram for Thrasyboulos quoted by Plutarch (*Mor*. 235a), whose death in the conflict against the Argives may also be commemorated in an epigram from the Hellenistic period (*Greek Anthology* 7.229). Other examples in Plutarch include references to mothers burying their sons (*Mor*. 241a; 241c). The Hellenistic epigrams portray a mother proudly burying her eight sons; and six Spartans who fell fighting in Messenia and had their ashes brought back to Sparta by their brother (*Greek Anthology* 7.434–35). Such epigrams were most likely composed to commemorate famous Spartans of the past, such as Thrasyboulos, who fell during the battle of Thyrea (Gow and Page 1965, 261–62). Aelian (*VH* 12.21) also relates that Spartan mothers inspected the wounds on their sons' bodies.

7 Hodkinson (2006) 144. On the fictitious nature of Lakonian apophthegms, see *GSW*4 243 n. 427; Hodkinson (2000) 253–54; *contra* Missoni (1986); Nafissi (1991) 290–309, who argued that the apophthegms reflect a standard practice of home burials for the bravest warriors from the Archaic era onward. In a controversial article, Robertson (1983) used the Spartan mother *exemplum* as indicative of a wider Panhellenic convention of repatriating the war dead; after heavy criticism from Pritchett (*GSW*4 94 n. 1), he later revised his claim and

Other evidence, however, indicates that bringing the war dead back home might have been practised in pre-Classical Sparta. In particular, a fragment of a martial elegy composed by Tyrtaios asserts that dead warriors were buried in individual tombs in Sparta (Fr. 12.29). Since Tyrtaios' *floruit* is commonly dated to the latter half of the seventh century, most researchers assume that the burial practices for the war dead must have undergone a major change sometime during the sixth century.[8] The date or specific circumstances behind this change has, nonetheless, been left largely unexplored; it is simply asserted that at some point the Spartans moved toward a more uniform and egalitarian way of treating their war dead which reflected the equal standing of their citizen 'peers' (*homoioi*). The main reason for the lack of investigation into the change in military burial customs stems from the scanty nature of the evidence for the early Spartan polis. Accounts for early Sparta come mostly from later periods and tend to credit the legendary lawgiver Lycurgus with creating the socio-political institutions known from the Classical era, in an effort to legitimise and trace the image of an austere and militaristic society back to an immemorial past. As such, our understanding of the institutional development of Sparta has been marred by the problems of the so-called 'Spartan mirage'.[9]

An examination of the treatment of the war dead in early Sparta, as this paper will demonstrate, allows us to approach the issue of Spartan socio-political history from a different perspective. Since changes to burial practices and the commemoration of the war dead tend to herald wider institutional development within the military forces of other Greek poleis, an investigation into the origins of Spartan battlefield burials may offer a similar glimpse into the nature and organisation of Archaic Spartan armies, overlooked in the previous studies of the subject.[10] Such an approach, drawing from a number of recent

---

conceded that the Spartans followed a 'distinctive' custom of burying the war dead on the battlefield, with occasional exceptions (1992, 166–75). Van Wees (2018a, 223) suggested that the custom of bringing the war dead back to be cremated at home may indicate a new practice in the Hellenistic period. Since the Spartans practised inhumation from the Protogeometric to Classical periods, and introduced cremation only in the Hellenistic period, the prominence of cremations in the anecdotes about the Spartan war dead of the past is a clear anachronism.

8 Cartledge (1987) 337; Hodkinson (2000) 251, (2006) 145; Van Wees (2006a) 133, (2018a) 225; Rawlings (2007) 196; Christesen (2018) 314.

9 For a good summary of the challenges of the 'Spartan mirage' and the invention of tradition in studying early Spartan history, see Hodkinson (1997).

10 The only recent study on the Spartan war dead is Low (2006), which focused primarily on the Classical period and the different forms of commemorating the war dead in Sparta and Lakonia. *GSW*4 discussed several Spartan *polyandria* in his detailed survey of Greek

historical studies which emphasise the significance and multifaceted nature of the 'cultural work of the dead', as well as how the funeral treatment and commemoration of soldiers constitute 'a fundamental element in the definition of political communities', holds the potential to provide some much-needed insight into the military development of the early Spartan polis – a topic traditionally marred by controversy and scholarly disagreement.[11] The aim of this chapter is, therefore, to provide a comprehensive survey of evidence for the early Spartan custom of war burials and to establish its place within the social and military history of Archaic Sparta.

1 Tyrtaios and Early Sparta

The earliest literary evidence for Spartan burial practices comes from the poetry of Tyrtaios, whose martial elegies have been commonly associated with the Second Messenian War (*ca.* 640–600). His poems, composed to exhort the Spartans in their fight to put down the rebellion in Messenia, provide the only contemporary source for the early Spartan army, its ethos, fighting techniques, and relation to the wider community. The surviving fragments are unfortunately few and often pre-determined by the agendas of the later authors who chose to preserve them, but they do offer a precious glimpse of late seventh-century Sparta.[12] The main themes of what remains of Tyrtaios' elegies concern the importance of fighting on behalf of the polis; bravery and solidarity in combat are praised as paramount to military success, while death in battle is hailed as beautiful and bringing 'imperishable glory' (*kleos esthlon*) to a fallen warrior and his family.[13] The physical embodiment of the latter, as

---

war dead burials (94–259), devoting one chapter exclusively to Spartan conventions (241–46). Other studies include Missoni (1986); Nafissi (1991) 290–309.

11 Bérard (2020) 6. On 'the cultural work of the dead', see Laqueur (2015). For recent studies on the commemoration of the war dead, in ancient Greece and beyond, see Low *et al.* (2012); Giangiulio *et al.* (2019).

12 The major surviving fragments of Tyrtaios (Frs 10–12), recorded by Lycurgus and Stobaeus, emphasise the ideal of patriotism and depict a warfare dominated by heavy-armed warriors who, according to some interpretations, fought in a phalanx formation. Such depictions might have been deliberately chosen for their relevance to later audiences. Other fragments of Tyrtaios' poetry preserved in papyri imply that missile combat and light-armed troops had a more prominent role in Archaic warfare (e.g., Frs 19, 23–23a). For more, see Van Wees (2000) 149, and Lloyd, this volume.

13 For more on military themes in Tyrtaios, see *GSW*4 37–44; Wheeler (1991) 129–32; Van Wees (2000) 149–54, *GWMR* 172–74; Rawlings (2007) 54–55; Schwartz (2009) 115–23. On

Tyrtaios remarks, consists of the individual tomb, pointed at and remembered by the whole community:

> And if he falls among the front ranks, pierced many times through his breast and bossed shield and corslet from the front, he loses his own dear life but brings glory to his city, to his people, and to his father. Young and old alike mourn him, all the city is distressed by the painful loss, and his tomb (*tumbos*) and children are pointed out among the people, and his children's children and his line after them. Never do his name and good fame (*kleos esthlon*) perish, but even though he is beneath the earth he is immortal, whoever it is that furious Ares slays as he displays his prowess by standing fast and fighting for land and children.
>
> Fr. 12.23–34

According to Tyrtaios, the main rewards for any Spartan perishing in the fight against the Messenians were a prominent tomb and immortal fame for him and his descendants in the collective memory of the polis. Such inducements clearly emulated the language of heroic death familiar from the Homeric epics, but they also testify to the practices of burial and commemoration of the Spartan war dead in late-seventh century.[14] Tyrtaios' mention of the tombs 'pointed out among the people' indicates that the casualties of the Messenian wars were repatriated and buried individually at home; their funerals were conspicuous affairs, as the loss was felt by the entire community, 'young and old alike'.[15] Based on the fragment, it would appear that the Spartans brought back the bodies of their dead and provided them with public honours and funerals at home. But was this really the case? And can it tell us anything about the nature of early Spartan armies?

---

the ideal of 'beautiful death' (*kalos thanatos*) in Archaic and Classical Sparta, see Loraux (1995) 63–74.

14 The epic language of Tyrtaios Fr. 12.27–34 and its parallels to Homeric poems are highlighted by Fuqua (1981), who argued that the appeal of Tyrtaios' promises to the fallen Spartans was enhanced by the contemporary religious practice of hero cults, suggesting 'the possibility of heroization for the individual' (226). For an archaeological study on the veneration of the communal dead in early Sparta, see Pavlides (2011). The military and political significance of Tyrtaios Fr. 12 has been studied by Luginbill (2002).

15 The prominence of the funerals for elite Spartans in Tyrtaios' times is further conferred by two ivory fibula plaques found at the Artemis Ortheia temple. The plaques, dated to the third quarter of the seventh century, depict scenes of *prothesis*; the nature of their decoration and material suggests that they were prestige objects commissioned by wealthy families. For more, see Nafissi (1991) 330–31; Hodkinson (2000) 238.

The custom of repatriation of the casualties of war from the drawn-out conflict in Messenia would have certainly required a considerable effort, both on the level of logistics and army administration. And although, apart from Tyrtaios, any literary or archaeological evidence for such a process is virtually non-existent for Archaic Sparta, helpful insights can be drawn from a comparison to equivalent practices in Athens.[16] In Classical Athens, the custom of bringing back the war dead (*patrios nomos*) was governed by a set of established procedures, which ensured that the cremated remains of all fallen warriors were returned to the city (Thuc. 2.34). The costly and complicated process was supported by the administrative and financial infrastructure of the polis, as casualties of major battles could have measured in the hundreds and required transportation over considerable distances across the Mediterranean. The degree of complexity involved in the custom of *patrios nomos*, which occupied a symbolic place in the Athenian self-image of civic equality, reveals the high level of institutionalisation of the citizen army in Classical Athens.[17] The treatment of the war dead prior to the reforms of Kleisthenes, however, was most likely not overseen by the state, as burial of the men who fell fighting on behalf of the polis was a private matter, facilitated by the family and companions of the deceased, reflecting the less formal organisation of Archaic citizen militias in Athens and elsewhere. Because of this pattern, some scholars assumed that the repatriation of the Spartan war dead from Messenia implied by Tyrtaios must have followed similar principles, being a private, and not public, affair.[18] This, in turn, provides important information about the organisation of Spartan armies of the seventh century; again, comparison with Athens may prove particularly useful.

16 According to Kurtz and Boardman (1971) 191, a well-known mid-sixth-century Lakonian cup by the Hunt Painter depicted 'warriors returning home bearing the bodies of the dead'. The cup portrayed pairs of warriors carrying the corpses of their fallen comrades over their shoulders; the vivid depiction of wounds, with blood pouring from the bodies of the dead, indicates that the image shows a scene in the immediate aftermath of battle. Since 'no two men could have carried a corpse any great distance over Greek roads', as suggested by *GSW*4 104, it is impossible to determine whether the dead were brought back to Sparta or buried on the battlefield. For more on the cup, see Hodkinson (2000) 268 n. 48; Saunders (2008) 85–86.

17 Seminal studies on the Athenian custom of *patrios nomos* are Loraux (1981); Clairmont (1983). Arrington (2015) and Kucewicz (2021) supplement and challenge some of their assumptions.

18 Rawlings (2007) 196. Cf. Hodkinson (2000, 238, 251), who suggested that their funeral rites were public; and Van Wees (2018a, 222), who concludes that it is not clear whether the burials were private or public.

The Archaic Athenian army has been the subject of some controversy, but recent studies suggest that the legislation of Solon at the beginning of the sixth century was crucial in defining the formal criteria for military service, imposing an obligation to serve in the militia on the wealthiest citizens in Athens.[19] The army, as a result, was relatively small, comprising around 15–20% of the adult male population, as the business of war was firmly in the hands of the Athenian elites. The social status of those who fought and died in early Athenian armies is best represented in their funerary monuments, which beginning from *ca.* 600 marked their graves all over Attika. The latter consisted predominantly of impressive statues of naked youths, known as *kouroi*, or painted stone relief *stelai* portraying the deceased. Since military attributes feature heavily in a number of such monuments, it has been assumed that some of them marked the graves of men who fell in combat.[20] This assumption is further confirmed by the epitaphs inscribed on their bases.[21] And while such monuments testify to the elite status of the Archaic Athenian militia, they also imply that the war dead were brought back home to be buried by their families. Their funerals and grave markers provided a medium for the families of the deceased to come to terms with the loss, but also to glorify their sons by conspicuous monuments, located in highly visible areas in their respective communities.[22] The small size of the militias meant that the number of casualties was usually low; considering that most conflicts in the sixth century took place within or close to Attika, transporting the corpses back would not have been a logistical problem. The treatment and commemoration of the war dead in Archaic Athens indicates, therefore, the elite status of the citizens serving in the militia, as well as the relatively low level of institutionalisation of the latter. Full public control of

19 The obligation to serve in the Athenian citizen militia, most likely introduced by Solon, was imposed on the top three property classes: the *pentakosiomedimnoi*, the *hippeis*, and the *zeugitai*. All three classes, as convincingly demonstrated by Van Wees (2006b), were composed of leisured elites of Archaic Athens. *Contra* Pritchard (2018, 39–48), who dismissed the military dimension of the Solonian *tele* but offered no plausible alternative for army mobilisation before the reforms of Kleisthenes. Non-elite Athenians could have also served in the citizen militias on a voluntary basis, predominantly as light-armed warriors or retainers; their post-mortem fate, however, is unclear. Considering the cost and practical difficulties of repatriation, one might assume that non-elite fallen were buried on the battlefield (in archaeologically invisible ways). For more on Archaic Athenian army, see Van Wees (2013a, 2018c); Valdés Guía (2019a, 2019b); Kucewicz (2021) 135–76.

20 See, for instance, Humphreys (1980) 104–105; Sourvinou-Inwood (1995) 221–97; Kucewicz (2021) 101–17. For potential seventh-century graves of Athenian war dead, see Doronzio (2018) 115, 144.

21 E.g., *CEG* 13 (Tettichos), 19 (Xenokles), 27 (Kroisos). All are discussed by Tentori Montalto (2017).

22 As argued by Arrington (2015) 19–32; Kucewicz (2021) 101–17.

the military forces has usually been associated with the democratic reforms of 508/7, which, among other changes, removed control over the war dead from the private to the public domain.[23]

Can this evidence tell us anything about the procedures and organisation of Spartan armies of the seventh century? A simple analogy is ruled out – the conspicuous burials of the elites in Archaic Athens find at first glance no equivalent in the Spartan archaeological record. With the possible exception of the famous Lakonian hero reliefs, the purpose of which and relation to burials is uncertain, funerary monuments in Archaic Sparta and the wider Peloponnese are rare.[24] The prominent tombs alluded to by Tyrtaios seem, therefore, to be archaeologically invisible; this means that any indications of the social standing of the men buried in such tombs may be seemingly impossible to uncover, despite the rich tradition of elite lifestyle and luxury in seventh-century Sparta.[25] Displays of material goods were limited instead to sanctuaries – such as the temple of Artemis Ortheia, the Amyklaion, and the Menelaion – which from the mid-seventh century begin to fill with ivories, bronzes and other luxurious items.[26] But even though the Athenian patterns of social differentiation may not easily apply to Spartan burial practices, the Spartan elites may have employed different ways to commemorate their deceased, including the dead from the Messenian war.

The main body of evidence which indicates that the graves of some Spartans may have received special distinction consists of a series of terracotta relief amphoras, commonly viewed as funerary in purpose. Dated to the period from *ca.* 625 to 550, the vases are large in size and decorated on only one side, which has led scholars to suspect that they served as grave markers, thus providing a rare exception to the general scarcity of funerary material from Archaic Sparta.[27] The central themes of the iconography adorning the

23 Kucewicz (2021) 119–31. For more on the introduction of democracy and its effect on the Athenian practice of war, see Pritchard (2010a, 2018).

24 Kokkorou-Alevras (2011). For more on Lakonian hero reliefs, see n. 81 below.

25 On elite lifestyle and values in seventh-century Sparta, see Van Wees (1999) 2–6, (2018b) 249–51, where he outlines the dining culture of Archaic Spartan *andreia*. On luxury and conspicuous consumption in Archaic Lakonian art, see Hodkinson (1998); Powell (1998).

26 Artemis Ortheia: Dawkins (1929); Amyklaion: Kalligas (1992); Demakopoulou (2009); Menelaion: H. Catling (1976–77); R. Catling (1986).

27 The main study on the Lakonian terracotta relief amphoras is Christou (1964). For more recent discussions, see Hodkinson (2000) 240–43; Förtsch (2001) 99–104. In his forthcoming study on Lakonian grave markers, Christesen (forthcoming) puts some doubt on the scholarly consensus that the terracotta relief amphoras had a strictly funerary purpose. The common interpretation stems largely from Christou's find of one of the amphoras next to a cluster of graves which he dated to the seventh century; the graves, however,

amphoras revolve around the subject of war: depictions of heavy-armed warriors, engaged in a variety of combat scenes, are interspersed with images of chariots and light-armed troops. Such iconography recalls the heroic fighting scenes from the Homeric poems, where warfare consists of both hand-to-hand and missile combat, the dead are fought over and recovered in the heat of the battle, and the leading warriors use chariots to travel around the battlefield. The 'heroic' overtones of the images, however, also find some parallels with the martial elegies of Tyrtaios, whose tone and language were heavily inspired by the Homeric epic tradition, and whose war depictions similarly combined close- and long-range combat.[28] The combination of the elaborate character and martial iconography of the amphoras suggests they may have served as grave markers for tombs of wealthy Spartans who died in combat in the late seventh and early sixth centuries.[29] The appearance and continuity of the style could represent, as Stephen Hodkinson suggested, 'an attempt by the wealthy to establish prominent markers for their family graves by utilising an iconography in keeping with the new Tyrtaian ideology'.[30] This form of display, as he concluded, was nonetheless relatively limited, especially when compared to other Greek poleis, where imposing funerary monuments were inscribed with

---

have been more recently dated to the Byzantine period, which in turn falsified Christou's claim concerning the primary usage of the Archaic amphora (see Christesen (2018), Appendix 7, 43–44). Since most of the amphoras were found in the vicinity of sanctuaries, it could be argued that they served primarily as monumental mixing bowls. The find spots, however, are not necessarily incompatible with the potential funerary character of the amphoras, as argued by Hodkinson (2000) 240. The fact that all of the known examples were decorated on one side only suggests that the amphoras were likely intended to be lain on their side, which gives enough support to see them as funerary objects used to mark the graves of the wealthiest Spartans. Fitzhardinge's claim (1980, 52) that the amphoras would have been buried in the grave mounds (i.e., not as markers above) is less convincing.

28 Tone and language: Fuqua (1981); Gerber (1997) 104–107. War depictions: see nn. 13–14 above and Lloyd, this volume.

29 The overall dominance of martial themes on the amphoras is noteworthy when compared to Lakonian pottery of the Archaic period, generally characterised by a lack of interest in military subjects: Powell (1998, 128), counted only 14 vases showing warriors. The amphoras, furthermore, depicted scenes of fighting over the dead (e.g., Christou 1964, 272 fig. 84) highlighting the importance of retrieving individual fallen during combat. Chariots are also prominent, leading Nafissi (1991, 335) to argue that the amphoras intentionally evoked a heroic ideal in order to reflect the hierarchical structure of the army. The military imagery of the amphoras is arguably less exceptional within the wider context of Archaic Spartan art, as suggested to me by Paul Christesen. A case in point are the terracotta metopes from a cult building excavated near the Eurotas bridge, which depict pairs of hoplites and horsemen. Steinhauer (1982) dated the metopes to *ca.* 570.

30 Hodkinson (2000) 242–43.

epitaphs commemorating the deceased. In the absence of the latter in Sparta, the association of the terracotta amphoras with Tyrtaios' tombs of the fallen warriors remains tentative. Recent archaeological finds, however, indicate that grave markers might not have been the sole distinguishing mark of Spartan elite burials.

The latest surveys of the archaeological evidence from Sparta challenge a number of long-held preconceptions about Spartan funerary practices.[31] Firstly, they show that burials were located both within (intracommunal) and outside (extracommunal) the inhabited areas of the polis, revising the previous scholarly consensus that the Spartans buried their dead only within the city. The unique nature of Spartan burial customs within the wider Greek world has also been called into question, as similarities between other contemporary poleis in the Peloponnese, such as Argos and Corinth, have been highlighted; most noteworthy among them was the general absence of any major investment in marking tombs in the Archaic and Classical eras. The results also brought a significant number of new graves to attention, of which around thirty-one date to the Archaic era.[32] Due to the paucity of grave goods, the main criterion which differentiated burials in Archaic Sparta was their location. The dead were buried either in dense clusters located in peripheral areas on the outside of the city; or in small plots within or in close proximity to the residential and commercial portions of the urban centre. The contrast between burials in cemeteries on the margins of the city (e.g., the Olive Oil Cemetery), intensively used from the Archaic times onward and largely invisible to the inhabitants of Sparta, and individual graves in the city centre, located along the sides of the main roads and across hillsides in the heart of the polis, suggests that such differentiation was likely not random. And while current archaeological evidence does not provide any definite answers as to the identity of the people buried in each location, some preliminary theories have been put forward.[33]

31 Christesen (2018) provides the most accessible overview of recently published graves in Sparta from the Protogeometric to the Hellenistic period. I apply his terminology of 'intracommunal' and 'extracommunal' burials throughout this chapter; Sparta had no defensive wall in the Archaic and Classical periods making the usual designation of intramural and extramural burials not applicable.

32 Christesen (2018) 321–25, Appendix 3 (11–17). Any information about the age and sex of the buried is usually missing; Christesen listed at least eleven adult and three child burials.

33 In her study of Archaic funerary statuary in the Peloponnese, Kokkorou-Alevras (2011) suggested that the few surviving statues of lions might have marked the graves of the war dead in the region, similar to the Classical lion monuments commemorating the collective burials and/or cenotaphs for casualties of war. The difficulties in establishing exact

In his review of Spartan burials Paul Christesen hypothesised that the contrasting spatial contexts for burials which begin to emerge in Archaic Sparta could have been motivated by social factors. The areas along roadsides within the city, located in the vicinity of spaces used on an everyday basis, were undoubtedly desirable places for burials and therefore sought out by the wealthy families in Sparta, who could afford to purchase plots of land in the city centre. One such area, for instance, was the main north-south road of ancient Sparta, the Aphetaïs road, where at least twelve graves were discovered and dated to the Archaic period; Pausanias (3.12.8) tells us that the royal graves of the Eurypontids were located at the south end of the road in Limnai, confirming its special status within the city. Up to nine Archaic graves were also found on the edges of the Gerokomeiou hill, leading scholars to suspect that another major road must have ran along the slope of the hill, parallel to the Aphetaïs road.[34] The peripheral cemeteries, by contrast, removed from the civic areas and the eyes of the Spartans, must have been considerably less prestigious and thus cheaper if not altogether free to use for funerary purposes; they may have been accordingly meant for the less wealthy or socially disadvantaged residents of the polis. As a result, the burial locations, as Christesen concluded, could have served different functions: 'If burial in a formal, communal cemetery located on the edges of Sparta spoke to inclusion within the community, then burial within the urban fabric of Sparta may have indicated that the individuals in question enjoyed elevated social standing'.[35] Although such a hypothesis may still need further validation from archaeological finds across the city, the Archaic graves alongside the central roads of Sparta provide a good candidate for the tombs of the fallen warriors mentioned by Tyrtaios, which, being 'pointed out among the people', must have been located in highly visible areas of the polis.[36]

Associating the intracommunal graves in Archaic Sparta with the warrior tombs of Tyrtaios does not, however, necessarily help us to decide whether such burials were private or public. The areas where the graves were located made them highly conspicuous within the Spartan community, especially if further marked by the aforementioned terracotta amphoras, but they could have been deliberately chosen as special parts of the city reserved only for

find-spots, combined with the relatively low number of surviving statues, makes the theory speculative and difficult to apply to Archaic Sparta.

34 For spatial distribution of intracommunal burials in Sparta, see Christesen (2018) 332–35.

35 Christesen (2018) 355.

36 Christesen provides a short discussion of the fragment of Tyrtaios in his survey study (Appendix 8, 47), but does not set it within the context of his spatial differentiation hypothesis.

the war dead, interred and distinguished by the polis upon their death. The notion of public home burials for individual casualties of war, however, would presuppose the existence of relatively high levels of administration within the armies of seventh-century Sparta, as each casualty would have to be accounted for and provided with a state-funded funeral. And although the latter possibility cannot be ruled out, the evidence from other contemporary Greek poleis, along with ancient traditions stressing the private and decentralised character of warfare in early Greece, makes the notion unlikely.[37] It is more plausible to assume that the Spartans of Tyrtaios' age followed the private repatriation customs of the Archaic Athenians – an assumption further supported by the fact that no surviving literary sources mention the existence of any public cemeteries for the war dead in Sparta.[38] We should consequently suspect that the burial areas within the centre of the city were used by any wealthy citizens who wished and could afford to commemorate their deceased, military or not, in a prominent spot; the alternative, burial in extracommunal cemeteries, would have certainly carried far less appeal.[39] The tombs of the war dead from the fragment of Tyrtaios would, accordingly, simply confirm that the Spartan elite buried their dead in prime city locations. His poem, as Hodkinson suggested, might have been an 'attempt to manipulate such visible private funeral rites for state purposes' but it does not appear to entail that the war dead from the Messenian conflict were buried at public expense in Sparta.[40] Their burials were most likely private.

This interpretation can, in turn, shed important light on the nature of early Spartan armies. For if we assume, following Tyrtaios, that the fallen from the Messenian wars were returned to Sparta and buried by their families in the most prestigious locations of the polis, it would lead us to the conclusion that Spartan citizen armies in the seventh century must have been dominated by the elite. The latter, inspired by the epic songs of martial glory which featured

37 E.g., Thuc. 1.5–6. For more on the decentralised character of early Greek warfare, see Raaflaub (1997) 51–53; Gabrielsen (2007) 250–53; Pritchard (2010a) 7–15.

38 Pausanias (3.14.1) relates that the monuments for Brasidas, the regent Pausanias, Leonidas and the fallen from Thermopylai were located in the same area near the theatre, south of the Akropolis, but makes no indication that the area was part of a public cemetery for the war dead.

39 The lesser status of extra-communal cemeteries is confirmed by their high exploitation from the Archaic to Hellenistic period. The Olive Oil Cemetery, as Christesen (2018, 326) reported, 'was intensively used; earlier graves were destroyed to make way for later ones and piles of bones, especially skulls, from earlier burials were collected outside newer ones. New graves were in some cases built directly on top of older ones; the cover stones of older graves sometimes served as the floors of new graves'.

40 Hodkinson (2000) 238.

heavily in the elite culture of the period, comprised the majority within the citizen militias; they fought exclusively as heavy-armed warriors, being able to afford the equipment, and set their actions against the heroes of the epics. Non-elite Spartans, on the other hand, could and no doubt did play a role in the army, but they served primarily as the light-armed *gymnetes* operating 'beneath the shields' of their socially superior comrades, as depicted by Tyrtaios (Fr. 11.35–8; 23a.12–14); their burials, we should expect, matched their social standing and took place either on the battlefield or in the citizen cemeteries away from the main civic areas.[41] While the practicalities of the private repatriation system for the war dead are not explicit in our sources, we could imagine that the initial retrieval and handling of the corpses would have been the responsibility of the friends and attendants of the deceased, who then transferred the bodies on carts to their kin for burial in Sparta.[42] Taking the example of Archaic Athens, where the treatment of the war dead was governed by a similar procedure, we can presume that the elite character of the early Spartan militias meant that the armed forces were relatively small, especially

41 Since Tyrtaios explicitly addresses the light-armed in his poems, we may conclude that the latter did include Spartan citizens not able to afford the full hoplite equipment, as suggested by Van Wees (2018b, 250). Nafissi (2009, 128) made a similar link between Tyrtaios' *gymnetes* and 'small landholders or peasants' who 'were to a greater or lesser extent in socio-economic difficulties'. Although we cannot be sure how non-elite Spartan war dead were buried in the Archaic era, it seems plausible to suspect that their internment varied depending on circumstances and private arrangements. Our knowledge about the *perioikoi*, their burials and early service in the Spartan armies, remains highly speculative; the lack of public involvement in the early war burials suggests that the treatment of non-citizen fallen would have been similarly private, displaying diversity with respect to where and how individuals were buried. For a recent overview on the *perioikoi* in Sparta, including their military service, see Ducat (2018). For the light-armed in Archaic Greece, see Kucewicz, Lloyd and Konijnendijk in this volume.

42 According to a much later tradition enshrined in Diodoros (8.27.2) and Polyainos (1.17), before an unspecified battle with the Messenians the Spartans inscribed their names on wooden sticks (*skytalides*) which they fastened to their wrists, so that 'if they died, they would not be unidentified by their kinsmen'. Although such practice seemingly highlights the importance of individual recognition of casualties in Archaic Spartan armies, potentially as part of a system of private repatriation by kinsmen, its historical value has to be doubted; there is no Archaic or Classical evidence for the use of 'dog tags' by any Greek, let alone Spartan, citizen armies. For more on the Spartan *skytalides* and identifying the war dead, see Vaughn (1991), esp. 55–57. Missoni's suggestion (1986, 73–81) that the Spartans used Dipylon shields to transport their fallen home before the introduction of the smaller Argive shields (*ca.* 700) is too implausible. Considering the physical effort required to carry individual dead over long distances, carting the bodies would have made far more practical sense.

when compared to the armies of the Classical era.[43] The number of casualties after any military engagement was, as a result, limited, making the transferral of corpses easier to manage, even considering the logistical difficulties of crossing the Taygetos.

Finally, the lack of any public control over the war dead indicates that the level of institutionalisation of the Spartan army in the seventh century was low and no different from other Greek poleis of the era.[44] As the boundaries between the private and public aspects of citizen militias remained blurry for most early Greek poleis, the elite's desire to bury and commemorate their war dead in socially meaningful ways provided one example of their dominance over the business of war-making. Sparta was no exception. The 'imperishable glory' bestowed upon the dead warrior through his burial reflected an ideology of individualism, reminiscent of the magnificent funerals for the Homeric heroes;[45] this stood against the later Spartan ideals of egalitarianism of the *homoioi*, imbedded in the social institutions designed to suppress individuality in favour of the community and central to the military identity of Classical Spartan armies. Common burial on the battlefield was an important expression of the latter, revealing the wider ideological and institutional change from elite to communal, and private to public principles in the Spartan practice of war.

---

43 The social class of leisured landowners, which usually comprised the heavy-armed warriors in the armies of Archaic Greek poleis, might have been somewhat larger in Sparta because of their control of agricultural land in Messenia and the system of helotage, established after the First Messenian War. Despite this, the difficulties and sheer length of the later conflict in Messenia suggest that the numerical strength of the Spartan army was a far cry from its Classical counterparts. Indeed, Tyrtaios' poems are often seen by scholars as a rallying call for Spartan citizens to join the conflict (e.g., Hodkinson 2006, 116). According to Luginbill (2002), the specific purpose of Tyrtaios Fr. 12 might have been to encourage more Spartans to enrol in the militia in 'a response to some manpower crisis in the Spartan army' (413).

44 In a recent paper on commemorating pre-Classical battles in Greece, Bergmann (2019) concluded that the majority of Archaic war burials consisted of public battlefield *polyandria*, while also arguing that the custom of private repatriation for the war dead by the Greek poleis 'is not supported by the evidence' (117). This formed part of her larger claim that post-battle activities, such as burial, dedications and commemoration, were controlled and regulated by the Greek poleis since the eighth century. War burials, according to her, are the earliest example of this, 'because those first graves are attested archaeologically and archaeology can go back to a time where (reliable) literary sources cannot' (123). The only archaeological example of an Archaic *polyandrion* which Bergmann discussed are the Parian *polyandria* (see n. 78 below); her other evidence for early war burials rely on literary accounts only and consist almost entirely of Pausanias, whose testimony for early Greek (and Spartan) history is problematic (see below).

45 E.g., *Il.* 23.127–257, 777–804; *Od.* 24.35–94.

The gradual process of institutionalisation of citizen armies which spread throughout the Greek world over the sixth century meant that the elite's control over the military affairs of their poleis was curbed and transferred to the wider community of citizens. As new public offices and positions appeared within the army structures, alongside laws which regulated and enhanced conscription procedures, the armies grew in size, increasingly relying on non-elite members to swell the ranks of their heavy infantries. In a similar way, the bodies of the war dead were removed from the private hands of the elite families and claimed by the state, which took over the burial ceremonies and commemoration of those who fell fighting on its behalf. These phenomena, however, have been notoriously difficult to trace in Archaic Sparta, which, after successfully consolidating its hold over Messenia, continued to rely on the leisure class of wealthy landowners for its military force. The institutionalisation of the Spartan army did, nonetheless, occur at some point in the sixth century, paving the way for the highly organised armies of the Classical era. The change in the customs concerning the war dead, no longer brought back to Sparta but instead buried on or near the battlefield, was an important early sign of this process which deserves closer investigation.

2 Sixth-Century Revolution?

The history of early Sparta was defined, according to Classical authors, by a long period of turmoil and bad governance (Hdt. 1.65; Thuc. 1.18). The radical change which ended the state of *kakonomia* was widely credited by ancient sources to the constitution of Lycurgus, who transformed the political, social and military structure of the Spartan polis. Although the Lycurgan reforms remain an insoluble chronological puzzle, modern scholars believe that the main turning-point(s) in Archaic Spartan history occurred during the sixth century. The major catalyst for what Moses Finley famously referred to as the 'sixth-century revolution' has been traditionally seen in the Second Messenian War, which prompted the Spartans to impose new laws and customs designed to maintain their hold over Messenia and its helot population; their result was a unique social system, characterised by a culture of austerity and a state of permanent military readiness among the Spartiate community and its reformed citizen army.[46] Finley suggested that this revolution was a piecemeal process but different dates have been subsequently offered for the key institutional developments in sixth-century Sparta. Some placed it during or in the

46 Finley (1971) 161–77.

immediate aftermath of the Second Messenian War; others associated it with the activities of the ephor Chilon in the 550s; a few pushed it back to the end of the sixth century and the reign of Kleomenes I.[47] The move from private repatriation to common battlefield burial of the war dead likely featured as part of such reforms, which significantly impacted the military organisation of Archaic Sparta; the ongoing scholarly debates mean, however, that its place within the sixth-century revolution remains unclear.

One way to approach the historical circumstances of the switch from private to public treatment of the Spartan casualties of war would be to set it within the larger context of burial practices in late Archaic Sparta.[48] Apart from the fragment of Tyrtaios, early testimonies about Spartan funerary customs – military or not – are lacking. Although the burial ceremonies for Spartan kings were discussed at some length by Herodotos (6.58), the first account which dealt with the funerary customs for non-royal Spartans at more length was the Aristotelian *Polity of the Lakedaimonians*. According to a brief excerpt of it preserved in Herakleides Lembos (373.13 Dilts), the graves of the Lakedaimonians were 'modest and the same for all' (Fr. 611.13). The work undoubtedly included more details on the subject which have not survived, but the lost sections almost certainly influenced the accounts of later authors writing about Archaic and Classical Sparta. By far the most important of them was Plutarch, whose description of the funerary reforms of Lycurgus is our main literary source for Spartan burial customs.[49] In addition to highlighting the principle of uniformity and the modest character of burials reflected in the Aristotelian *Lak. Pol.*, the passages from Plutarch's *Instituta Laconica* and *Life of Lycurgus* discuss a number of other measures introduced by the legendary lawgiver:

> Lycurgus did away with all superstitious fear connected with burials, granting the right to bury the dead within the city, and to have the memorials (*mnemeia*) near the shrines. He also abolished the pollutions associated with death and burial. He permitted the people to bury nothing with their dead, but only to enfold the body in a red robe and olive leaves, and all to treat their dead alike. He also did away with the inscriptions on

---

47 Second Messenian War: Cartledge (1977) 27, (2002) 113–37; Rahe (2016). Mid-sixth century: Nafissi (1991) 99–150, (2009); Trundle (2001). Late-sixth century: Van Wees (2018a, 2018b). For more references, see Van Wees (2018a) 234 nn. 134–35.

48 Main studies on Spartan burial customs are: Cartledge (1987) 331–43, (2012); Nafissi (1991) 277–341; Toher (1991); Richer (1994); Hodkinson (2000) 237–70; Cavanagh *et al.* (2011); Christesen (2018).

49 On Spartan burial customs and Plutarch, see Toher (1991) 169–73; Hodkinson (2000) 243–56, 260–62; Christesen (2018) Appendix 8, 46–51.

> memorials (*mnemeia*), except of those who had met their end in war (*en polemo*), and also did away with mourning and lamentation. (*Mor.* 238d)
>
> Furthermore, Lycurgus made most excellent regulations in the matter of their burials. To begin with, he did away with all superstitious terror by allowing them to bury their dead within the city, and to have memorials of them near the sacred places, thus making the youth familiar with such sights and accustomed to them, so that they were not confounded by them, and had no horror of death as polluting those who touched a corpse or walked among graves. In the second place, he permitted nothing to be buried with the dead; they simply covered the body with a scarlet robe and olive leaves when they laid it away. To inscribe the name of the dead upon the tomb was not allowed, unless it were that of a man who had fallen in war (*en polemo*), or that of a woman who had died in sacred office. He set apart only a short time for mourning, eleven days; on the twelfth, they were to sacrifice to Demeter and cease their sorrowing. (*Vit. Lyc.* 27.1–2)

According to Plutarch, the defining features of Spartan funerary practice were twofold: a) burying the dead was allowed within the city, which eradicated any superstition concerning the proximity of graves to living spaces, while also allowing the youth to familiarise themselves with the sight of death and corpses; b) the burials were simple, no grave goods or inscriptions commemorating the dead were allowed, with the period of mourning also restricted;[50] the only exception was made for men who died in war and sacred women.[51] The regulations of Lycurgus were, therefore, built around the wider Spartan ideals of simplicity, disdain for wealth and egalitarianism of the community of *homoioi*. Special emphasis was placed on the areas of war and sacrifice for

50 Plutarch's account implies that every Spartan was buried with a scarlet robe (*phoinikis*) and olive leaves; the Spartan dead at Kerameikos from 403 were tightly wrapped in cloth, usually identified with the *phoinikis*: Stroszeck (2006) 104. Aelian, however, claimed that olive branches were buried with all who fought bravely, but scarlet robes were reserved only for the 'outstandingly brave ... given special honours at burial' (*VH* 6.6). Despite the communal nature of Spartan burials for the war dead in the Classical period, special honours could have been awarded to some warriors after the battle (see below), making Aelian's claim potentially plausible. For more on his statement and its implications, see Missoni (1986) 70–72; Nafissi (1991) 291; Van Wees (2018a) 221–22; cf. Hodkinson (2000) 247–49.

51 The identity of 'sacred women' (*hierai*) has been a subject of controversy; some scholars amend the text to read 'women who died in childbirth', following post-Classical inscriptions commemorating women who died in labour. For more, see Hodkinson (2000) 260–62; Brulé and Piolot (2004); Dillon (2007); Papapostolou (2011).

the polis as only those who fell fighting for the community were distinguished with an inscribed memorial. Furthermore, the presence of the dead in the city was intentionally normalised and prepared the Spartan citizens for the gruesome realities of the battlefield and handling the corpses of their comrades. Combined with the excerpt from Aristotle's *Lak. Pol.*, the image of Spartan burial customs projected in the literary sources is dominated by strong egalitarian and militaristic principles, credited back to the era of Lycurgus. But can we trace any evidence of such regulations in the archaeological record for early Sparta?

Plutarch's claim that there was no ban on burials within the city is, as we saw earlier, confirmed by archaeological evidence, which indicates that during the Archaic era graves began to appear around the very centre of Sparta. The practice of intracommunal burials was, however, common from the Protogeometric period onward and so its potential origins cannot be ascribed to any specific Archaic reform.[52] By contrast, a potential prohibition on grave goods may match the noticeable decline in the number of pots found in graves from the Geometric to Archaic periods. The uncertainty regarding precise dates for Archaic burials, nonetheless, means that any definite changes are difficult to place in time; pottery grave goods are absent from all Classical burials, but rare examples dated to the sixth century imply that the disappearance of the latter was likely a gradual process. Finally, it has been recently suggested that the disappearance of mourners from Spartan art and poetry around 600 may reflect regulations on lamenting imposed around the same time in Sparta.[53] The latter would have matched the rationale behind funerary legislation introduced by other Greek poleis in the sixth century, which also address the issues surrounding the concept of pollution (*miasma*), mentioned by Plutarch as part of Lycurgus' measures.[54] Considering the many difficulties of using Spartan artistic evidence for contemporary practices – in particular black-figure pottery often destined for foreign markets – such suggestions remain speculative and cannot offer us any firm dates for changes in the burial customs in sixth century Sparta.[55]

---

52 On Protogeometric and Geometric burials in Sparta, see Christesen (2018) 318–21.

53 Van Wees (2018a, 220–23) associates the ban on lamentation (and possibly other aspects of funerals) with other sumptuary laws restricting the spending on houses, introduced around 600.

54 Traditions of early funerary codes are associated with Athens, Ioulis, Delphi, Mytilene, Sikyon and Gortyn. For full texts and discussion, see Toher (1991); Blok (2006).

55 On Lakonian pottery and its distribution in the Mediterranean, see Hodkinson (1998) 97–102; Pipili (2018). The sixth-century dating of funerary reforms in Sparta has been suggested by Toher (1991) 170; Papapostolou (2011) 495; Van Wees (2018a) 220–23.

The most concrete and securely dated piece of evidence which validates the account of Plutarch are the so-called ΕΝ ΠΟΛΕΜΟΙ *stelai*. These monuments consisted of an undecorated stone pillar inscribed with the name of the deceased and the words 'in war', or *en polemōi*, which reflects the exact phrase used by Plutarch, therefore confirming his assertion that the Spartan war dead were commemorated with special inscriptions. Around thirty such *stelai* have been discovered, both within Sparta and the wider region of Lakonia, dating from the mid-fifth until the first century.[56] Their immediate purpose, however, remains a subject of controversy: were they actual graves for Spartans who died in war or were they cenotaphs for warriors buried elsewhere? Or were they graves for men who died of war wounds after returning home? The main reason behind the uncertainty stems from the fact that Plutarch does not provide any definite information on what the Spartans did with the dead after a battle.[57] Although his testimony about the ΕΝ ΠΟΛΕΜΟΙ monuments sheds an important, if somewhat controversial, light on Spartan funerary practices, their relevance to the wider military customs is unclear. Moreover, since the earliest surviving *stelai* come from the mid-fifth century, their relation to any Archaic reforms is similarly uncertain. As such, the latter appear to follow the rest of Plutarch's statements in *Instituta Laconica* and *Life of Lycurgus* which, as Christesen recently concluded, altogether provide 'a generally accurate picture of Spartan burial practice in the Classical and early Hellenistic periods'; for the customs of the Archaic era we simply have to look elsewhere.[58]

The earliest Spartan war dead for which we have literary testimony are the fallen from the First Messenian War (mid-eighth century). According to Pausanias, after an indecisive battle the Spartans and Messenians each sent heralds to arrange a truce for the recovery of the dead and 'when this was agreed on both sides, they proceeded at once to bury them' (4.8.13). Despite its neatness, the account of Pausanias, our main and often only source for early

56 Low (2006) offers the most detailed study of the ΕΝ ΠΟΛΕΜΟΙ *stelai*, including a full catalogue with dates, locations, and a few photographs. See also Hodkinson (2000) 249–56; Christesen (2018) Appendix 9, 52–55.

57 Plutarch's only comment on the burial for rank-and-file warriors comes from his *Life of Agesilaus*, where he states that the custom for those who died abroad was 'to leave their bodies there after burial' (40); he contrasts this with the rites afforded to the kings, whose bodies were returned to Sparta. Considering the Spartan practice of collective battlefield burial related by numerous Classical authors, it seems most plausible to view the ΕΝ ΠΟΛΕΜΟΙ *stelai* as memorials for fallen warriors buried elsewhere, as convincingly argued by Hodkinson (2000) 249–56. Cf. Missoni (1986), esp. 68–72; Nafissi (1991) 294–301, who suggested that the *stelai* adorned actual graves for distinguished Spartans who fell in war.

58 Christesen (2018) Appendix 8, 51.

Spartan wars, cannot be taken at face value; his narratives, drawn primarily from Hellenistic sources, are heavily inspired by later conventions of hoplite battles – featuring separate units of hoplites, cavalry, and light-armed troops, as well as truces for the recovery of the dead – which had little in common with the realities of Archaic warfare.[59] Pausanias' reference to the battlefield burial of the Spartans is, accordingly, to be doubted, especially when compared against the evidence of Tyrtaios and the practice of private repatriation of the dead implied in his poems.[60] The fate of the Spartan war dead from the battle of Hysiai (*ca.* 669/8) is similarly unclear; Pausanias claims to have seen the *polyandria* of the victorious Argives at nearby Kenchreai but does not comment on the burial of the Spartan casualties (2.24.8).[61] The first physical Spartan *polyandrion* which Pausanias does report in his work comes from the battle of Thyrea (*ca.* 546), known more famously as the Battle of the Champions.[62] Most of our knowledge about the events of the battle derives from Herodotos (1.82), who relates in detail the initial arranged combat between 300 Lakedaimonians and Argives, as well as the subsequent full-scale fight which the Spartans won despite heavy losses. Pausanias adds little to the story, apart from first stating that the tumuli (*taphoi*) for the dead from the first encounter were piled up by the sole surviving Spartan and two Argives; later, as he exited the site of battle, he also mentioned common graves (*polyandria*), though he does not make it clear whether they contained the bodies of those who fell in the general melee (2.38.5–6). Nevertheless, scholars have seen the Thyrean *polyandria* as the first instance of Spartans burying their dead on or near the battlefield, which ever since the battle became the standard practice of their armies.[63]

---

59 Pausanias' source for the story came from a third century rhetorician Myron of Priene, whom Pausanias himself convicts of 'indifference to truth and to probability' (4.6.4). On the traditions surrounding early Messenian history, see Pearson (1962); Luraghi (2008). On memorials (including *polyandria*) and their role in shaping communal memories of war in Pausanias, see Roy (2019).

60 *Contra GSW*4 159, who accepted the authenticity of Pausanias' story and concluded that the tradition behind his account, 'whatever its source, clearly points to burial on the battlefield'. In his later discussion of the Spartan burial practices for the war dead he used the passage to claim that the Spartans buried their dead on the battlefield since the First Messenian War (243); the evidence of Tyrtaios Fr. 12.29 did not feature in his study.

61 On Pausanias and the battle of Hysiai, see Franchi (2012).

62 On the Thyrean *polyandria*, see *GSW*4 160–61. Although Pausanias provides the only testimony for the graves, some attempts have been made to connect the burials to a fifth-century inscription on a cenotaph (*SEG* 13.266) found on Mount Zavista; Low (2006) 104 n. 25, provides key references, concluding that the connection is implausible.

63 Hodkinson (2000) 251, (2006) 145; Christesen (2018) 314.

The few references about Spartan military activities in the late Archaic period indicate that the new custom of battlefield burials indeed ended the elite tradition of bringing the bodies of the war dead back home. Herodotos tells us that Archias, who died during the Spartan siege of Samos (*ca.* 525), was given a public funeral by the Samians (3.55); likewise, Anchimolios, who led the task force against the Peisistratids, being a Spartan 'citizen of repute', died in the fighting at Phaleron (*ca.* 512) and was buried in Attika (5.63). The war dead from Selinous, commemorated with an epitaph recorded by Plutarch (*Vit. Lyc.* 20; *Mor.* 217f), may likely have been the remnants of the Spartans accompanying Dorieus' ill-fated expedition to Sicily in the late-sixth century (Hdt. 5.46).[64] Finally, after the battle of Plataia in 479 the Lakedaimonians buried their dead on the spot, making three graves: one for 'the holy ones (*irees*), among whom were Poseidonios and Amompharetos and Philokyon, and Kallikrates', the second for 'the rest of the Spartans', and the third for the helots (Hdt. 9.85). These examples clearly imply that from the mid-sixth century the Spartan war dead were buried abroad. It has been suggested, however, that the early practice might have been to bury prominent individuals on the battlefield and take the rest of the fallen home. Since Pausanias does not explicitly say what happened to the casualties of the full-scale battle at Thyrea, it is possible, as Van Wees argued, that the battlefield *polyandrion* contained only the bodies of the 300 'champions'; the rest of the dead were taken home. Such distinction of battlefield burial, attested also by the examples of individual Spartans buried abroad, mirrored the customs of the Homeric epics, where 'the great Greek heroes expected burial in a tumulus near the battlefield, but the cremated remains of the anonymous Greek multitudes ... were taken home for burial by relatives and friends'.[65] Common battlefield burials for all, as Van

64 Since the epitaph explicitly mentions extinguishing tyranny, it most likely refers to Euryleon's deposition of Peithagoras related by Herodotos (5.46). Pausanias (3.16.4) reports a shrine (*heroon*) to some (Athenodoros?) members of the expedition to Sicily; the exact text of the passage has been disputed by scholars, with some suggesting that the monument commemorated the dead from Anthene, a city on the Argive-Lakonian border. Pausanias, however, gives no indication whether the shrine was also a grave, which seems unlikely. For a longer discussion, see *GSW*4 161–63, who concluded that the *heroon* in Sparta was a cenotaph for the dead buried at Selinous. Cf. Van Wees (2018a) 222–23.

65 Van Wees (2006a) 132. The case of Anchimolios is particularly relevant in this context, since Archias was not buried by the Spartans but given a public funeral by the Samians. Herodotos reports that Anchimolios' tomb was located in Alopeke near the temple of Herakles at Kynosarges; its proximity to Phaleron suggests that he might have been buried on or near the battlefield, but Herodotos tells us nothing about who buried him or what happened to the rest of the Spartans, who were forced back to their ships after suffering heavy casualties. Jacoby (1944, 44 n. 28), suggested that an Athenian friend might

Wees concluded, were accordingly introduced later as part of a move toward greater equality within Spartan armies, perhaps not until *ca.* 500 or even the battle of Plataia.[66]

Such interpretation, however, poses several problems, especially regarding the evidence of Tyrtaios and the pre-550 burials for the war dead at Sparta. For if the mark of distinction was to bring the bodies of the elite Spartan warriors home and provide them with conspicuous tombs honoured by the entire community, it seems unlikely that a reversal of the practice to burial on the battlefield for the chosen few and repatriation for the multitudes would have been seen as a special honour for the former.[67] But since the few mentions of individual Spartans buried on the field of battle clearly imply that the bodies of distinguished citizens were not returned home, it is more plausible to assume that *all* Spartan fallen were interred on the battlefield from *ca.* 550. This reading, furthermore, does not preclude the possibility of distinction in burial for outstanding individual warriors. The Spartan *polyandria* at Plataia, as we saw, were divided into three graves, among which one was reserved for the so-called 'holy ones'; the identity of the latter has been the subject of an ongoing scholarly debate, but since the names listed by Herodotos match the list of individual Spartans who displayed most valour in the battle, we should assume that the 'special honours' afforded to them consisted of their separate burial (9.71–2).[68] In a similar way, the tomb of the Lakedaimonians in Athens from 403 suggests that selected individuals were buried with special care. The excavations revealed that the central chamber of the tomb contained three bodies, accompanied by an inscription naming two Spartans, Chairon and

have buried Anchimolius privately; Missoni (1986, 62), proposed that the Peisistratids facilitated Anchimolios' burial. Pritchett (*GSW*4 163–64) argued that the Spartan fallen were buried *en masse* by the Athenians.

66 Van Wees (2018a) 222–23.

67 The ultimate distinction in burial was granted by the Spartans to their kings, exceptionally honoured in death as heroes (Xen. *Lak. Pol.* 15.9). Herodotos (6.58) provides a detailed description of the customs accompanying royal funerals in Sparta, adding that an image (*eidolon*) is made for any king who died in war, to be carried around the city. The latter practice, however, almost certainly applied only when the body of the king could not be recovered (e.g., Leonidas); the standard, as related by Xenophon (*Hell.* 5.3.19) and Plutarch (*Vit. Ages.* 40.3), was to return the body of any kings who died abroad back to Sparta preserved in honey (or wax). The origins of the custom (inevitably ascribed to Lycurgus) may be, therefore, indicative of the elite funerals for the war dead enshrined in Tyrtaios, characterised by their grand nature and the repatriation of the body. As such, the royal funerals of the Classical period may have drawn their ideology from the elite traditions of the past. For more on burial ceremonies for Spartan kings, see Cartledge (1987) 331–43; Nafissi (1991) 284–90; Hodkinson (2000) 262–63.

68 For the Lakedaimonian tombs at Plataia, see *GSW*4 174–75; Hodkinson (2000) 257–58; Christesen (2021).

Thibrachos, followed by a single letter identified as either M or L. The first two, as we learn from Xenophon's account of the campaign, served as polemarchs, while the latter may have referred to an Olympic victor Lakrates (*Hell.* 2.4.33). Their separate burial implies, therefore, not only that distinction in common battlefield burials was practised by the Spartans, but also that military rank and social standing played a factor in their post-mortem treatment during the Classical and conceivably also late Archaic periods.[69]

Another piece of evidence which suggests that common battlefield burial for all was practised by the Spartan armies in the second half of the sixth century comes from the so-called Oath of Plataia. It consists of an inscription from a fourth century *stele* from Acharnai which relates the Athenian Ephebic Oath (lines 1–20) and a number of pre-battle pledges made by the Athenians before a battle against the barbarians (lines 21–51). Due to the close resemblance to the oaths sworn by the allied Greek forces during the Graeco-Persian Wars recorded in literary sources, the latter lines are commonly known as the Oath of Plataia.[70] The late date of the *stele* has led most scholars to dismiss it as forgery, but some studies have shown that there are sufficient grounds to believe that the Oath of Plataia reflects a genuine document from the late Archaic period.[71] Despite its Athenian context, the origins of the oath might also hold important information on the military history of early Sparta, as demonstrated by Van Wees, who traced some of its components back to the official vows taken by the members of the Spartan 'sworn bands'.[72] Among pledges of commitment and obedience, the oath importantly included a vow to bury the dead on the battlefield:

> I shall fight while I am alive,
> and I shall not regard being alive as more important than being free.
> And I shall not leave my *taxiarchos* or my *enomotarches*, whether he is alive or dead.
> And I shall not go away unless the leaders [*hegemones*] lead us away,

69 For more on the Spartan *polyandrion* in Kerameikos, see n. 3 above.

70 Herodotos (7.132) relates that the Greek forces swore an oath to punish any poleis which freely surrendered to the Persians but does not provide any chronological context. The oath is first cited by Lycurgus (1.80–81), who claims that the Greeks swore it before the battle of Plataia; and later Diodoros (11.29.3), who dates it to the 480 meeting at the Isthmus. Both Lycurgus and Diodoros mention the vow to bury the dead.

71 The majority opinion is represented by: Robert (1938) 307–16; Meiggs (1972) 504–507; *GSW*4 116–17; RO 440–48; Cartledge (2013). Scholars who argued that the Oath refers to a genuine agreement sworn by the allied Greek forces before the battle of Plataia (or Marathon) include: Siewert (1972); Robertson (1983) 82; Van Wees (2006a); Krentz (2007).

72 Van Wees (2006a).

and I shall do whatever the generals order.
And the dead among my fellow-fighters I shall bury on the spot,
and unburied I shall leave no-one.[73]

The Spartan origin of the oath is immediately discernible from its reference to an enomotarch, who commanded the smallest unit within the army, known as the 'sworn band' (*enomotia*), structured around the famous institution of dining messes (*syssitia*) and unique to the military organisation of the Spartan polis.[74] The mention of the enomotarch and a taxiarch also implies, as Van Wees observed, a relatively basic unit structure, especially when compared to the Classical division of the Lakedaimonian armies into *morai, lochoi,* and *pentekostys*. This limited organisation indicates that the text of the oath must have dated back to a time when such divisions were not yet in place, leading him to posit that most of its pledges were modelled on a late Archaic oath of the sworn bands, reformulated to suit the purposes of the Greek coalition army during the Graeco-Persian Wars.[75] The vows to follow the orders of the commanders (*hegemones*) would have been the most likely additions in the context of the latter, affirming the position of overall command held by the Spartans, though similar formulations might have already featured in the vows sworn by the member poleis of the Peloponnesian League, formed in the second half of the sixth century under Spartan leadership. The specific pledge to bury the dead on the spot and leave none unburied, however, most likely belonged to the set of original oaths which bound the 'fellow-fighters' of the *enomotia*.[76] The appearance of such military oaths in Archaic Sparta might be, accordingly, associated with the emergence of communal messes which, according to Herodotos, played a key role within Spartan military system (1.65). Recent studies date the introduction of the latter to the last quarter of the sixth

73 Lines 23–31, trans. Van Wees (2006a) 126.

74 The dining messes, referred to as *syssitia, phiditia,* or *andreia*, consisted of around fifteen members (Plut. *Vit. Lyc.* 12), forming a subdivision of an *enomotia* comprised of either thirty-two or thirty-six men. For more on *enomotiai*, their place within the Spartan army organisation and relation to the *syssitia*, see Lazenby (1985) 9–18, 67–71. Outside Sparta, tactical division into *enomotiai* featured only in the army of the Ten Thousand, led by Klearchos and later Cheirisophos, both Spartans (Xen. *An.* 3.4.21; 4.3.26).

75 Van Wees (2006a) 134–35.

76 If the provision about burial referred exclusively to the later Panhellenic oath, the term 'fellow-fighters' (*symmachesamenoi*) would have been most likely substituted for the more appropriate 'allies' (*symmachoi*). The central pledge of obedience to the *hegemones*, as Van Wees concluded (2006a, 135), 'interrupts the natural flow of the clauses which precede and follow it, and is therefore best treated as an insertion into a previously shorter and more coherent oath' which included the remaining pledges.

century, but it does not necessarily follow that the practice of burying the dead on the battlefield was first established alongside the *syssitia*.[77] The burial of the fallen from Thyrea (*ca.* 546) suggests instead that the custom was in place by the mid-sixth century, providing us with a *terminus post quem*; but its origins may conceivably reach back further.

The key developments related to the treatment of the war dead in Archaic Sparta occurred in the first half of the sixth century, after the Second Messenian War and before the conquest of the Thyreatis. Considering the lack of any reliable archaeological evidence for *polyandria* for the casualties of war in the early Greek world, any attempt to trace the institutional circumstances behind the introduction of the custom in Sparta requires us to turn to other sources which may point us in the right direction.[78] Removing the war dead from the hands of their kin, and therefore denying the latter an opportunity for a conspicuous funeral at home, must have been felt by Spartan elites attempting to cope with the passing of family members. The process of transferral of control over the bodies of fallen warriors, which Reine-Marie Bérard referred to as 'la dématérialisation du cadavre', has been commonly associated with a number of cultural and ideological developments in other poleis, reflecting both the alternative ways of families trying to make peace with their loss, and the state's efforts to install a different discourse to glorify the dead on a communal level.[79] We should suspect that similar changes marked Sparta's adoption of the practice of foreign military burials; and even though the evidence base is comparably sparse, there are some indications that the change occurred in the second quarter of the sixth century.

---

77 Van Wees (2018b).

78 The only pre-Classical *polyandria* in the wider Greek world are the graves from Akragas in Sicily and Paroika on Paros; none are known from mainland Greece. For more on the Akragas and Paroika *polyandria*, see Kurtz and Boardman (1971) 257; Morgan (2001) 32–33; Zaphiropoulou (2002); Agelarakis (2017).

79 Bérard (2020) 33. The immediate example of such developments on a state level is the Classical institution of public funerals at Athens, which apart from a ceremonial burial consisted of a funeral oration and games (Thuc. 2.34). Its introduction coincided with the emergence of new artistic discourses associated with the war dead, such as the red-figure warrior loutrophoroi and the white-ground lekythoi. Their iconography, as shown by Hannah (2010) and Arrington (2015) 208–17, 239–74, provided Athenian families with alternative (and private) ways to commemorate their fallen sons, in addition to the civic ideology of *patrios nomos*. In Sparta, the resistance of the elites to the sixth-century changes in funerary legislation has been linked to new forms of family cults, which Nafissi (1991, 309–27) studied in relation to the Aigeid family. For more on the 'difficult harmony' of public and private commemoration of the war dead in Greek poleis, see Marchiandi and Mari (2016); Bérard (2020) 33–38.

Putting the spotlight back on the material evidence, the first notable development is the disappearance of the terracotta amphoras around the 550s. These, as we observed earlier, likely marked the graves of the Spartan elite war dead, brought back home for intracommunal burial in the most prestigious locations in the city. Although the chronological time span for the latest examples of the amphoras has been disputed, they do provide a rough estimate for when the conspicuous burials of the war dead might have ceased, especially if we accept their link to the tombs of the latter.[80] The disappearance of the terracotta amphoras, furthermore, coincided with the increased prominence of hero cults in Archaic Sparta, as demonstrated by the first instances of the so-called 'hero reliefs' commonly dated to the mid-sixth century.[81] The early stone and terracotta reliefs, which typically show a seated couple often identified with mythical figures, have been associated with the gradual rise of communal values within the Spartan polis, which extended to religious, political, and potentially also military spheres. The context for this phenomenon, as Nicolette Pavlides suggested, overlapped with the decline of commemoration of the individual through burial; hero cults, according to her, 'express this shift in importance from ties to the family and the individual to those persons and

80 I follow the chronological range suggested by Hodkinson (2000) 240; others, such as Christou (1964), dated the disappearance of the amphoras to the early sixth century. As far as I am aware, there are no examples dated after 550, which, if my argument here is accepted, can offer a potential *terminus ante quem* for the practice of home burials for the war dead. It has to be noted that the cessation of production of relief amphoras in the sixth century was a wider phenomenon across the Greek world, not limited to Lakonia only; this trend might be seen as reflecting the general decrease in funerary display among the Greek poleis of late Archaic period, but its relation to specific socio-political developments remains problematic and likely varied from one polis to another. Unfortunately, precise dating of the Archaic intracommunal burials in Sparta is not possible, thus preventing us from associating any topographical changes in the latter with the termination of the custom of home burials for the war dead. It is notable, however, that the intracommunal burials of the Classical period are no longer clustered around the Aphetaïs road, which may indicate that the road held a special funerary status in Sparta, potentially related to the burials of the individual war dead. Considering the limited number of excavations, especially toward the southern end of the road where the graves of the Eurypontid kings were located, any such hypotheses remain, of course, speculative.

81 Although the first hero reliefs appear in Lakonia in the mid-sixth century, other hero cults (e.g., Menelaos and Helen) are attested in the seventh century. The link between the terracotta amphoras and the hero reliefs in Sparta has been noticed by Nafissi (1991) 334–35, who commented on the replacement of the amphoras' warrior themes by family ideals implicit in the hero reliefs; Nafissi loosely linked the implementation of the new 'ideologia della morte', visible especially in the processes of heroisation in Sparta, to Chilon's ephorate and the battle of Thyrea (338–41). For more on Lakonian hero reliefs, see Hibler (1993); Salapata (1993); Pavlides (2011) 554–58.

areas of society that were important for the entire polis', thus reflecting 'the new civic identity developing in Sparta in the Archaic period'.[82] The specific relation of the hero reliefs to the developments in Spartan burial practices continues to be a matter of debate, but the emergence of the monuments point to a change in the public discourses regarding the heroic dead; the shift from the private to communal might, therefore, be seen as indicative of similar shifts occurring in the discourses surrounding the war dead. In addition, the archaeological evidence of hero cults and the veneration of heroic dead can be linked to Sparta's wider political activities in the period and, in particular, its claim for Peloponnesian hegemony, of which the notion of bringing back the bones of mythical heroes to be re-buried in Sparta played a significant part.

The beginning of Spartan hegemony in the Peloponnese is closely related to the series of conflicts with the Arkadian city of Tegea (*ca.* 560–550). Herodotos relates that after being second best in the earlier wars the Lakedaimonians gained the upper hand during the reign of Anaxandridas and Ariston. The reason for their success, as he declares, was an intervention from Delphi instructing the Spartans to recover the bones of Agamemnon's son Orestes from Tegea. Following the oracle, a certain Lichas, one of the Spartan officials called 'Benefactors', found the giant bones of the mythical hero buried in a blacksmith's forge and returned them to Sparta; ever since then, Herodotos concludes, the Lakedaimonians easily won all their battles against the Tegeans and put most of the Peloponnese under their control (Hdt. 1.66–8). The story has been interpreted by scholars as part of the 'philachaian policy' adopted by Sparta to emphasise its pre-Dorian 'Achaian' heritage by providing a link to the famous kings of the past.[83] Indeed, Orestes was likely not the only member of the Atreid dynasty that the Spartans 'repatriated' during the mid-sixth century; Pausanias reports that the tomb of Orestes' son Tisamenos was also located in Sparta, after his body was removed from Helike by command of the Delphic oracle (7.1.8).[84] Although no dates are offered, based on its similarity to the Orestes episode, the return of Tisamenos' bones can be also dated to the 550s and Sparta's efforts to extend its influence into northern Peloponnese. Such

82 Pavlides (2011) 562.

83 The new policy may be also reflected in the poem by Stesichoros, who contrary to Homer claimed that Agamemnon's palace was located in Sparta and not Mycenae (216 Campbell). The 'philachaian policy' is usually ascribed to the activities of the ephor Chilon in the mid-550s: Huxley (1962) 69; Cartledge (2002) 120. On the politics of repatriating bones of mythical heroes (including Orestes), see Huxley (1979); McCauley (1999); Blomart (2000); Hall (2007) 333–38; Coppola (2008); Fragkaki (2016).

84 Leahy (1955) placed the event between the Tegean episode and the Spartan intervention in Sikyon (*ca.* 555).

manipulation of stories about mythical heroes, whose burials were ascribed with political significance, provided the Spartans with mythological precedents to assert their claim to hegemony in the Peloponnese, leading to the creation of the so-called Peloponnesian League. The Orestes and Tisamenos episodes, in addition, also reflect an important shift in Spartan foreign policy from aggression and 'Helotisation' to more diplomatic measures.[85] The location of the burials of the war dead, from both the mythical past and the present, likely played a key role in the practical and ideological foundations of the League.

The Peloponnesian League which formed in the wake of Sparta's victory against Tegea was the first 'hegemonic, multi-state, political organisation' in the Greek world;[86] it united its members in a predominantly defensive alliance with Sparta, designed to offer protection from non-member states in times of war. The success of the League meant that Sparta became the leading power not only in the Peloponnese but also in the wider Greek world, as reported by Herodotos in his story about the Lydian king Croesus seeking alliance with Sparta as the strongest polis in Greece (1.56, 69). Although the exact structure of the Peloponnesian League in the Archaic period remains uncertain, it must have relied from its very beginning on the military leadership and strength of the Lakedaimonians.[87] Despite the series of losses against Tegea in the first half of the sixth century and a potential defeat at Orchomenos (*FGrH* 115 F69), Sparta was successful in all other conflicts during the reign of Leon and Hegesikles (Hdt. 1.65; *ca.* 575–560), leading some scholars to posit that a major military reorganisation occurred during or immediately after the Second Messenian War. But considering that some of the most characteristic institutions of the army (*enomotiai, syssitia*), as defined by Herodotos, did not emerge until the end of the sixth century, we should suspect that the transformation of the Spartan militia occurred later and was most likely a gradual process. The onset of the latter has been often associated with the ephorate of Chilon and the creation of the Peloponnesian League, reflecting Sparta's ambition

85 Cartledge (2002) 120.

86 Hodkinson (1997) 92.

87 The Peloponnesian League most likely changed in structure and character either in the last years of the sixth century or sometime during the fifth century, leading some scholars to refer to the earlier organisation as the 'Spartan alliance'. Based on Herodotos (1.68.6), who informs us that most of the Peloponnese had been subjected to Sparta in the aftermath of its victory against Tegea, it is nonetheless clear that Sparta's hegemony was established in the mid-sixth century. For longer discussions, see de Ste. Croix (1972) 101–24; Cawkwell (1993); Yates (2005); Bolmarcich (2008).

to impose and maintain its hegemony in the newly formed alliance.[88] Any contemporary evidence for concrete public measures aimed to improve the organisation and efficiency of the citizen militia has been, nonetheless, largely lacking. Seen in this light, the switch from the elite practice of private repatriation for the war dead to the public custom of battlefield burial for all can be interpreted as the earliest indication of the process of institutionalisation of Spartan armies in the 550s.[89]

The move to bury all military casualties on the battlefield could have been dictated by practical considerations, providing a quick way of internment during campaigns in the Peloponnese and beyond. In Athens, the transferral of military burials from private to public hands coincided with the general growth of the citizen army toward the end of the sixth century, suggesting that the switch to battlefield *polyandria* may have been necessitated by the need to process greater numbers of corpses in more efficient ways.[90] A similar process

88 The connection between Chilon and the Peloponnesian League is specifically made in the Rylands Papyrus 18 (*FGrH* 105.1), which indicates that Chilon and Anaxandridas began the process of suppression of tyrannies which the League became famous for.

89 Another measure aimed to control the post-battle procedures would have concerned the practice of dedications of war spoils in sanctuaries. The first collective dedications of captured armour in Olympia appear around the mid-sixth century; the inscriptions name the citizens of the donor poleis (in the nominative plural), indicating increased levels of public control over post-battle activities exercised by Greek armies from the late sixth century onward. Sparta, however, is notable for its absence among the dedicants. According to a tradition enshrined in Plutarch, the Spartans did not dedicate captured arms and armour 'because they are taken from cowards' (Mor. 224B, 224F; cf. Ael. *VH* 6.6). Archaic dedications of weapons, especially of spearheads, javelin heads, and arrowheads, did nonetheless feature in large numbers at the sanctuaries of Apollo Maleatas and Apollo Tyritas in the Thyreatis/Kynouria (Pavlides 2018); swords, inscribed helmets, and bronze sheets from shield decorations were also found at the Amyklaion, most of which date from the first half of the sixth century: *SEG* 11.690–1; Jeffery (1961) 190; Kalligas (1992) 34; Demakopoulou (2009) 103. The lack of inscribed weapon dedications (esp. shields and helmets) from the Spartans post-550, collective or private, might suggest that captured weapons and armour, like the ones collected by the Spartan Orthryades following the battle of the Champions (Hdt. 1.82.5), were in public control by the second half of the sixth century; instead of dedicating 'raw' spoils, the Spartan custom might have been to convert them to other forms of tithes and victory monuments (e.g., Xen. *Hell.* 4.3.21; Paus. 3.11.3; 5.10.4; 10.9.7–8). Such arguments, however, have to remain *ex silentio*. For more on Archaic weapon dedications, with references to Sparta, see *GSW*3 240–95; Jackson (1991); Baitinger (2011); Pavlides (2018); Schröder (2020) 24–53. For earlier weapon dedications, see Lloyd, this volume.

90 Significantly, the earliest public burials of the Athenian war dead consisted of battlefield *polyandria* (Chalkis, Lemnos, Marathon, Plataia); the repatriation customs of *patrios nomos* were introduced after the Graeco-Persian Wars. While the question of why the Spartans did not adopt a similar custom of public repatriation falls outside the scope

may have been at work in Sparta, as the conquests in the Peloponnese in the early sixth century provided new lands and opportunities for social advancement, allowing more men to join the elite of leisured landowners and, by extension, the citizen militia. The main indications for population growth have been usually seen in the unsuccessful attempts to subjugate Tegea and distribute its lands among the citizens bereft of allotments in Messenia.[91] More convincing evidence, however, is provided by landscape studies, which indicate a widespread shift from an early Archaic pattern of large and isolated settlements to small and medium-size farmsteads in Lakonia from the mid-sixth century.[92] Located mostly in previously unsettled areas, the appearance of new rural sites points to an increase in small-scale farming, which suggests that more Spartans were able to become landowners eligible for service in the army. The influx of new manpower in the militias, some of whom might not have been members of the traditional elites, led to reforms designed to facilitate the growing number of warriors (and casualties) and potentially also the changing social identity of the armed forces. Public battlefield burials, therefore, might have been introduced not only for purely pragmatic reasons, but also as an ideological pronouncement of the egalitarianism of those who fought and sacrificed their

of this study, the immediate answer may be that the Spartan polis never developed the substantial administrative and financial apparatus required for mass repatriation of the war dead.

91 Rahe (2016) 118 stressed the difference between the nature of Spartan control over Messenia after the revolt. Following the First Messenian War, as he suggested, 'the Lacedaemonian aristocrats, operating as raiders, extracted one-half of their harvest every year … after the Second Messenian War, ritualised aristocratic raiding gave way to direct rule and systematic exploitation, and equal allotments of land were assigned to individual Spartan hoplites' (144). This reading can be further supported by statements from Plutarch (*Mor.* 194b) and Aelian (*VH* 13.42), who date the Spartan subjugation of Messenia to 600. Van Wees (2013b, 239) argued that social stability in Archaic Sparta was achieved by repeated conquests of new lands in the Peloponnese, 'which enabled large numbers of Spartiates to become gentlemen farmers, and staved off demands for a redistribution of land'. According to him, however, the onset of Spartan egalitarianism occurred later than the 550s and went hand in hand with the failed territorial expansions toward the end of the sixth century, after which Sparta had to turn 'to a different way of avoiding internal conflict by creating a rigid distinction between leisure-class landowning citizens and a subject labor force, and inhibiting competition for wealth by imposing on all citizens a strictly egalitarian material culture'. He consequently dated the practice of common battlefield burials to the end of the sixth century as part of a widespread austerity programme of socio-political reforms (2018a, 222–23). While his theory provides a number of convincing points, the evidence for the burials of the Spartan war dead discussed in this paper suggests that the influx of egalitarian policies in Sparta would have begun already in the mid-sixth century.

92 R. Catling (2002); Foxhall (2013) esp. 212.

lives on behalf of the polis. Conversely, the custom of burying the Spartan war dead on foreign soil may have also had a significant symbolic importance, with the *polyandria* serving as physical reminders of the Spartan military might and, especially in the context of the Peloponnesian League, its political claim over the neighbouring regions. The common tombs from Thyrea can accordingly be seen as the first attested instance of the new practice, employed to further Sparta's territorial interests in the Argolid; within a couple decades the custom was further formalised in the oaths of the *enomotiai*, sworn perhaps also in modified version by Sparta's Peloponnesian allies on any campaigns of the League, and eventually the Panhellenic forces before the battle of Plataia.

To sum up, the combination of literary and archaeological sources indicates that the custom of burial on the battlefield was introduced in Sparta in the 550s. Its origins can be related to the military victories over Tegea and later Argos, which led to the formation of the Peloponnesian League. Sparta's hegemonic position within the latter set off a programme of military reforms designed to institutionalise the citizen army, of which the transferral of the bodies of the war dead into public domain might have been among the first measures.[93] Although the introduction of public regulations concerning the treatment of casualties of war in Sparta was a novelty in the Greek world, it need not have required a complex system of army administration; battlefield burials provided considerably less logistical problems compared to the custom of repatriation practised in Classical Athens. The abandonment of the previous tradition of the elites, who brought back the bodies of the fallen for home burials, necessitated, however, the creation of alternative forms of commemoration in Sparta, which redefined the notions of heroic dead and emphasised the importance of communal values in the burial of the warrior dead.[94] Finally, contrary to

93 Other innovations were probably introduced at the same time but identifying them remains a matter for further study. The most likely reforms concern the issues of army organisation, in particular measures enhancing mobilisation, unit division and the replacement of the old kinship army structure of three Dorian tribes (Tyrtaios Fr. 19; Paus. 4.7.8).

94 In this context, the annual festival of Parparonia which celebrated the Spartan victory at Thyrea may be interpreted as a deliberate public initiative to bestow special honours on the first Spartan fallen buried *en masse* on the battlefield, commemorated in accordance to the new communal discourse of heroic death. The festival of Gymnopaidiai also featured hymns in honour of the casualties of the battle of the Champions: see Wade-Gery (1949) 80; Franchi (2018). The memory of individual Spartans who died in the battle of Thyrea, enshrined in the epigrams of the Hellenistic era, could suggest that an alternative tradition of private commemoration might have been once also associated with the battle. For more on the epigrams and the Parparonia, see Robertson (1992) 179–207.

scholars who posit a radical transformation of the Spartan polis in the mid-sixth century, the process of army institutionalisation in Sparta was probably a gradual affair. While the practice of battlefield burials heralded the customs of later eras, the key components of the Classical Spartan militias, including the army division into *enomotiai* based on the system of common messes, emerged only toward the end of the sixth century, as part of other socio-economic reforms which revolved around the ideals of military excellence and egalitarianism of the Spartan community of the *homoioi*. The "sixth-century revolution" did not happen all at once; but its beginnings in the military sphere can be found in the change in how and where the Spartans buried their war dead.

3 Conclusion

The early history of the Spartan war dead can be divided into two separate phases, each governed by different principles and each revealing important insights about the nature of Archaic Spartan armies. The earlier was characterised by the conspicuous burials of the Spartan elites, whose prominence in and domination over the seventh-century citizen militias was reflected in the custom of private repatriation of the dead from the battlefields. This practice, enshrined in Tyrtaios' martial elegies, centred on the importance of home funerals and tombs for the fallen warriors, which bestowed 'imperishable glory' on the deceased and their families in the collective memory of the polis. Following Sparta's military successes in the mid-sixth century and the establishment of the Peloponnesian League, the practice of bringing back the bodies of the war dead was succeeded by a system of common battlefield burials for all. The state's appropriation of the bodies of fallen warriors was arguably among the first steps in the process of army institutionalisation, which reflected the growth and professionalisation of the Spartan militias of the late Archaic period. The impact of the reform, confirmed by the rise of new community-wide discourses on heroic death, highlighted the gradual shift toward the ideology of military excellence and equality of the *homoioi*, which eventually led to the socio-political structures of Classical Sparta. Despite the egalitarianism of the new practice, a lasting vestige of the elite ideology of the past remained in the celebrated poems of Tyrtaios and in the tombs of the war dead scattered around the prestigious areas of the city. And so, even though the famous Hellenistic saying of the Spartan mother requesting her son to return with his shield or upon it was a literary device and had little in common with the military realities of the Classical era, its allusion to the practice of individual repatriation of the war dead may have contained the proverbial grain of truth:[95] a reference to the elite custom once practised in Archaic Sparta.[96]

Bibliography

Agelarakis, A.P. (2017) *Parian Polyandreia: The Late Geometric Funerary Legacy of Cremated Soldiers' Bones on Socio-Political Affairs and Military Organizational Preparedness in Ancient Greece*. Oxford.

Arrington, N.T. (2015) *Ashes, Images, and Memories: The Presence of the War Dead in Fifth-Century Athens*. Oxford.

Baitinger, H. (2011) *Waffenweihungen in griechischen Heiligtümern*. Mainz.

Baumbach, M., Petrovic, A., and Petrovic, I. (eds) (2010) *Archaic and Classical Greek Epigram*. Cambridge.

Bekker-Nielsen, T. and Hannestad, L. (eds) (2001) *War as a Cultural and Social Force*. Copenhagen.

Bérard, R.-M. (2020) 'La politique du cadaver: Traitements funéraires et usages civiques des morts à la guerre en Grèce archaïque et classique'. *Annales HSS* 75–1: 3–38.

Bergmann, B. (2019) 'Beyond Victory and Defeat: Commemorating Battles Prior to Persian Wars', in Giangiulio, Franchi and Proietti (2019) 111–29.

Blok, J.H. (2006) 'Solon's Funerary Laws: Questions of Authenticity and Function', in Blok and Lardinois (2006) 197–247.

Blok, J.H. and Lardinois, A.P.M.H. (eds) (2006) *Solon of Athens: New Historical and Philological Approaches*. Leiden.

Blomart, A. (2000) 'Les manières grecques de déplacer les héros: Modalités religieuses et motivations politiques', in Pirenne-Delforge and Suárez de la Torre (2000) 351–64.

Bolmarcich, S. (2008) 'The Date of the 'Oath of the Peloponnesian League'. *Historia* 57.1: 65–79.

95 Missoni (1986) 81, also suggested that the Spartan mother apophthegm originated 'in un'epoca molto antica', but his attempt to associate the custom with Dipylon shields is less convincing (see n. 42 above).

96 I am indebted to a number of people who offered their comments and critical insights at various stages of writing this paper; they include Paul Christesen, Caroline Musgrove, Robin Osborne, Nick Sekunda and the co-editors of this volume. The remaining deficiencies and mistakes are entirely my own. The work on this paper has been funded from a postdoctoral project awarded by the National Science Centre, Poland (project number 2018/28/C/HS3/00418).

Brouma, V. and Heydon, K. (eds) (2018) *Conflict in the Peloponnese: Social, Military and Intellectual. Proceedings of the 2nd CSPS PG and Early Career Conference, University of Nottingham 22–24 March 2013*. CSPS Online Publication 4.

Brulé, P. and Piolot, L. (2004) 'Women's Way of Death: Fatal Childbirth or Hierai?', in Figueira (2004) 151–78.

Cartledge, P. (1977) 'Hoplites and Heroes: Sparta's Contribution to the Technique of Ancient Warfare'. *JHS* 97: 11–27.

Cartledge, P. (1987) *Agesilaos and the Crisis of Sparta*. London.

Cartledge, P. (2002) *Sparta and Lakonia: A Regional History 1300 to 362 BC*. 2nd edn. Abingdon.

Cartledge, P. (2012) 'Spartan Ways of Death', in Cataldi, Bianco and Cuniberti (2012) 21–38.

Cartledge, P. (2013) *After Thermopylae: The Oath of Plataea and the End of the Graeco-Persian Wars*. Oxford.

Cataldi, S., Bianco, E., and Cuniberti, G. (eds) (2012) *Salvare le poleis, costruire la concordia, progettare la pace*. Alessandria.

Catling, H.W. (1976–77) 'Excavations at the Menelaion, Sparta 1973–1976'. *AR* 23: 24–42.

Catling, R.W.V. (1986) 'Excavations at the Menelaion: 1985'. *Lakonika Spoudai* 10: 205–16.

Catling, R.W.V. (2002) 'The Survey Area from the Early Iron Age to the Classical Period (c. 1050–c. 300 BC)', in Cavanagh, Crouwel, Catling and Shipley (2002) 151–256.

Cavanagh, H., Cavanagh, W., and Roy, J. (eds) (2011) *Honouring the Dead in the Peloponnese: Proceedings of the Conference Held at Sparta 23–25 April 2009*. CSPS Online Publication 2.

Cavanagh, W., Crouwel, J., Catling, R.W.V., and Shipley, G. (eds) (2002) *The Laconia Survey I: Methodology and Interpretation*. London.

Cavanagh, W., Gallou, C., and Georgiadis, M. (eds) (2009) *Sparta and Laconia from Prehistory to Premodern*. London.

Cawkwell, G.L. (1993) 'Sparta and her Allies in the Sixth Century'. *CQ* 43.2: 364–76.

Christesen, P. (2018) 'The Typology and Topography of Spartan Burials from the Protogeometric to the Hellenistic Period: Rethinking Spartan Exceptionalism and the Ostensible Cessation of Adult Intramural Burials in the Greek World'. *BSA* 113: 307–63.

Christesen, P. (2021) 'Herodotus 9.85 and Spartiate Burial Customs'. *C&M* 69: 1–73.

Christesen, P. (forthcoming) 'Grave Markers in Laconia from the Protogeometric through the Hellenistic Periods'.

Christou, C. (1964) 'Ο ΝΕΟΣ ΑΜΦΟΡΕΥΣ ΤΗΣ ΣΠΑΡΤΗΣ: ΟΙ ΑΛΛΟΙ ΜΕΤ' ΑΝΑΓΛΥΦΩΝ ΑΜΦΟΡΕΙΣ ΤΟΥ ΛΑΚΩΝΙΚΟΥ ΕΡΓΑΣΤΗΡΙΟΥ'. *AD* 19 A: 164–265.

Clairmont, C.W. (1983) *Patrios Nomos: Public Burial in Athens During the Fifth and Fourth Centuries B.C.: The Archaeological, Epigraphic-Literary, and Historical Evidence*. Oxford.

Coppola, A. (2008) *L'eroe ritrovato: Il mito del corpo nella Grecia classica*. Venice.
Dawkins, R.M. (ed) (1929) *The Sanctuary of Artemis Orthia at Sparta: Excavated and Described by Members of the British School at Athens*. London.
Demakopoulou, K. (2009) 'Το μυκηναϊκό ιερό στο Αμυκλαίο: μια νέα προσέγγιση', in Cavanagh, Gallou and Georgiadis (2009) 95–104.
Dillon, M. (2007) 'Were Spartan Women Who Died in Childbirth Honoured with Grave Inscriptions?' *Hermes* 135.2: 149–65.
Doronzio, A. (2018) *Athen im 7. Jahrhundert v. Chr.: Räume und Funde der frühen Polis*. Berlin.
Ducat, J. (2018) 'The *Perioikoi*', in Powell (2018) 596–614.
Duplouy, A. and Brock, R.W. (eds) (2018) *Defining Citizenship in Archaic Greece*. Oxford.
Figueira, T.J. (ed) (2004) *Spartan Society*. Swansea.
Finley, M.I. (1971) *The Use and Abuse of History: From the Myths of the Greeks to Lévi-Strauss, the Past Alive and the Present Illuminated*. London.
Fisher, N. and Van Wees, H. (eds) (1998) *Archaic Greece: New Approaches and New Evidence*. Swansea.
Fitzhardinge, L.F. (1980) *The Spartans*. London.
Flower, M. and Toher, M. (eds) (1991) *Georgica: Greek Studies in the Honour of George Cawkwell*. London.
Förtsch, R. (2001) *Kunstverwendung und Kunstlegitimation im archaischen und frühklassischen Sparta*. Mainz.
Foxhall, L. (2013) 'Can We See the "Hoplite Revolution" on the Ground?' Archaeological Landscapes, Material Culture, and Social Status in Early Greece, in *MoB* 194–221.
Fragkaki, M. (2016) 'The "Repatriation" of Orestes and Theseus'. *Antesteria* 5: 285–302.
Franchi, E. (2012) 'La battaglia di Isie e l'identità argiva: un caso di invenzione della tradizione?', in Franchi and Proietti (2012) 43–66.
Franchi, E. (2018) 'Commemorating the War Dead in Ancient Sparta: The Gymnopaidai and the Battle of Hysiai', in Brouma and Heydon (2018) 24–39.
Franchi, E. and Proietti, G. (eds) (2012) *Forme della memoria e dinamiche identitarie nell'antichità greco-romana*. Trento.
Fuqua, C. (1981) 'Tyrtaeus and the Cult of Heroes'. *GRBS* 22: 215–26.
Gabrielsen, V. (2007) 'Warfare and the State', in *CHGRW* 248–72.
Gerber, D.E. (1997) *A Companion to the Greek Lyric Poets*. Leiden.
Giangiulio, M., Franchi, E., and Proietti, G. (eds) (2019) *Commemorating War and War Dead: Ancient and Modern*. Stuttgart.
Gow, A.S.F. and Page, D.L. (eds) (1965) *The Greek Anthology: Hellenistic Epigrams*. Cambridge.
Hägg, R. (ed) (1999) *Ancient Greek Hero Cult: Proceedings of the Fifth International Seminar on Ancient Greek Cult*. Stockholm.
Hall, J.M. (2007) 'Politics and Greek Myth', in Woodard (2007) 331–54.

Hammond, M. (1979/80) 'A Famous 'Exemplum' of Spartan Toughness'. *CJ* 75.2: 97–109.

Hannah, P. (2010) 'The Warrior *Loutrophoroi* of Fifth-Century Athens', in Pritchard (2010b) 266–303.

Hibler, D. (1993) 'The Hero-Reliefs of Lakonia: Changes in Form and Function', in Palagia and Coulson (1993) 199–204.

Higbie, C. (2010) 'Epigrams on the Persian Wars: Monuments, Memory and Politics', in Baumbach, Petrovic and Petrovic (2010) 183–201.

Hodkinson, S. (1997) 'The Development of Spartan Society and Institutions in the Archaic Period', in Mitchell and Rhodes (1997) 83–102.

Hodkinson, S. (1998) 'Lakonian Artistic Production and the Problem of Spartan Austerity', in Fisher and Van Wees (1998) 93–117.

Hodkinson, S. (2000) *Property and Wealth in Classical Sparta*. Swansea.

Hodkinson, S. (2006) 'Was Classical Sparta a Military Society?' in Hodkinson and Powell (2006) 111–62.

Hodkinson, S. and Powell, A. (eds) (1999) *Sparta: New Perspectives*. London.

Hodkinson, S. and Powell, A. (eds) (2006) *Sparta & War*. Swansea.

Humphreys, S.C. (1980) 'Family Tombs and Tomb Cult in Ancient Athens: Tradition or Traditionalism?'. *JHS* 100: 96–126.

Huxley, G.L. (1962) *Early Sparta*. London.

Huxley, G.L. (1979) 'Bones for Orestes'. *GRBS* 20.2: 145–48.

Jackson, A.H. (1991) 'Hoplites and the Gods: The Dedication of Captured Arms and Armour', in *HCGBE* 228–49.

Jacoby, F. (1944) 'Patrios Nomos: State Burial in Athens and the Public Cemetery in the Kerameikos'. *JHS* 64: 37–66.

Jeffery, L.H. (1961) *The Local Scripts of Archaic Greece: A Study of the Origin of the Greek Alphabet and its Development from the Eighth to the Fifth Centuries B.C.* Oxford.

Kalligas, P.G. (1992) 'From the Amyklaion', in Sanders (1992) 31–48.

Kokkorou-Alevras, G. (2011) 'Funerary Statuary of the Archaic Period in the Peloponnese', in Cavanagh, Cavanagh and Roy (2011) 269–88.

Krentz. P. (2007) 'The Oath of Marathon, not Plataia?'. *Hesperia* 76.4: 731–42.

Kucewicz, C. (2021) *The Treatment of the War Dead in Archaic Athens: An Ancestral Custom*. London.

Kurtz, D. (ed) (2008) *Essays in Classical Archaeology for Eleni Hatzivassiliou 1977–2007*. Oxford.

Kurtz, D. and Boardman, J. (1971) *Greek Burial Customs*. Ithaca.

Laqueur, T.W. (2015) *The Work of the Dead: A Cultural History of Mortal Remains*. Princeton.

Lazenby, J.F. (1985) *The Spartan Army*. Mechanicsburg.

Leahy, D.M. (1955) 'The Bones of Tisamenus'. *Historia* 4.1: 26–38.

Loraux, N. (1981) *L'Invention d'Athènes. Histoire de l'oraison funèbre dans la « cité classique »*. Paris.

Loraux, N. (1995) *The Experiences of Tiresias: The Feminine and the Greek Man*. Princeton.
Low, P. (2006) 'Commemorating the Spartan War-Dead', in Hodkinson and Powell (2006) 85–109.
Low, P., Oliver, G., and Rhodes, P.J. (eds) (2012) *Cultures of Commemoration: War Memorials, Ancient and Modern*. Oxford.
Luginbill, R.D. (2002) 'Tyrtaeus 12 West: Come Join the Spartan Army'. *CQ* 52.2: 405–14.
Luraghi, N. (2008) *The Ancient Messenians: Constructions of Ethnicity and Memory*. Cambridge.
Luther, A., Meier, M., and Thommen, L. (eds) (2006) *Das Frühe Sparta*. Stuttgart.
Marchiandi, D. and Mari, M. (2016) 'I funerali per i caduti in guerra. La difficile armonia di pubblico e privato nell'Atene del V secolo a.C.'. *MediterrAnt* 19.1–2: 177–201.
McCauley, B. (1999) 'Heroes and Power: The Politics of Bone Transferral', in Hägg (1999) 85–98.
Meiggs, R. (1972) *The Athenian Empire*. Oxford.
Missoni, R. (1986) 'Idealità e prassi degli spartani circa i caduti in Guerra', in *Decima Miscellanea Greca e Romana*. Rome: 61–81.
Mitchell, L.G. and Rhodes, P.J. (eds) (1997) *The Development of the* Polis *in Archaic Greece*. Abingdon.
Morgan, C. (2001) 'Symbolic and Pragmatic Aspects of Warfare in the Greek World of the 8th to 6th Centuries BC', in Bekker-Nielsen and Hannestad (2001) 20–44.
Nafissi, M. (1991) *La nascita del kosmos: studi sulla storia e la società di Sparta*. Naples.
Nafissi, M. (2009) 'Sparta', in Raaflaub and Van Wees (2009) 117–37.
Palagia, O. and Coulson, W. (eds) (1993) *Sculpture from Arcadia and Lakonia*. Oxford.
Papapostolou, M. (2011) '"Honourable Death": The Honours Paid in Ancient Sparta to Dead War-Heroes and Mothers Dying in Child-Birth', in Cavanagh, Cavanagh and Roy (2011) 493–500.
Pavlides, N. (2011) 'Worshipping Heroes: Civic Identity and the Veneration of the Communal Dead in Archaic Sparta', in Cavanagh, Cavanagh and Roy (2011) 551–76.
Pavlides, N. (2018) 'The Sanctuaries of Apollo Maleatas and Apollo Tyritas in Laconia: Religion in Spartan-Perioikic Relations'. *ABSA* 113: 279–305.
Pearson, L. (1962) 'The Pseudo-History of Messenia and its Authors'. *Historia* 11: 397–426.
Pipili, M. (2018) 'Laconian Pottery', in Powell (2018) 124–53.
Pirenne-Delforge, V. and Suárez de la Torre, E. (eds) (2000) *Héros et héroïnes dans les mythes et les cultes grec: Actes du Colloque organisé à L'Université de Valladolid du 26 au 29 mai 1999*. Liège.
Powell, A. (1998) 'Sixth-Century Lakonian Vase-Painting: Continuities and Discontinuities with the 'Lykourgan' Ethos', in Fisher and Van Wees (1998) 119–46.
Powell, A. (ed) (2018) *A Companion to Sparta*. Hoboken, NJ.
Pritchard, D.M. (2010a) 'The Symbiosis Between Democracy and War: The Case of Ancient Athens', in Pritchard (2010b) 1–62.

Pritchard, D.M. (ed) (2010b) *War, Democracy and Culture in Classical Athens*. Cambridge.
Pritchard, D.M. (2018) *Athenian Democracy at War*. Cambridge.
Raaflaub, K.A. (1997) 'Soldiers, Citizens, and the Evolution of the Early Greek *Polis*', in Mitchell and Rhodes (1997) 49–59.
Raaflaub, K.A. and Van Wees, H. (eds) (2009) *A Companion to Archaic Greece*. Chichester.
Rahe, P.A. (2016) *The Spartan Regime: Its Character, Origins, and Grand Strategy*. New Haven, CT.
Rawlings, L. (2007) *The Ancient Greeks at War*. Manchester.
Rees, O. (2018) 'Picking Over the Bones: The Practicalities of Processing the Athenian War Dead'. *JAH* 6.2: 167–84.
Richer, N. (1994) 'Aspects des funérailles à Sparte', *Cahiers du Centre Gustave Glotz* 5: 51–96.
Robert, L. (1938) *Études épigraphiques et philologiques*. Paris.
Robertson, N. (1983) 'The Collective Burial of Fallen Soldiers at Athens, Sparta and Elsewhere'. *EMC* 27: 78–92.
Robertson, N. (1992) *Festivals and Legends: The Formation of Greek Cities in the Light of Public Ritual*. Toronto.
Roy, J. (2019) 'Memorials of War in Pausanias', in Giangiulio, Franchi and Proietti (2019) 147–56.
Salapata, G. (1993) 'The Lakonian Hero Reliefs in the Light of the Terracotta Plaques', in Palagia and Coulson (1993) 189–97.
Sanders, J.M. (ed) (1992) *PHILOLAKON: Lakonian Studies in Honour of Hector Catling*. London.
Saunders, D. (2008) 'Dead Warriors and Their Wounds on Athenian Black-Figure Vases', in Kurtz (2008) 85–94.
Schröder, J. (2020) *Die Polis als Sieger: Kriegsdenkmäler im archaisch-klassischen Griechenland*. Berlin.
Schwartz, A. (2009) *Reinstating the Hoplite: Arms, Armour and Phalanx Fighting in Archaic and Classical Greece*. Stuttgart.
Siewert, P. (1972) *Der Eid von Plataiai* (Vestigia 16). Munich.
Sourvinou-Inwood, C. (1995) *'Reading' Greek Death: To the End of the Classical Period*. Oxford.
Stamatopoulou, M. and Yeroulanou, M. (eds) (2002) *Excavating Classical Culture: Recent Archaeological Discoveries in Greece*. Oxford.
Ste Croix, G.E.M. de (1972) *The Origins of the Peloponnesian War*. London.
Steinhauer, G. (1982) 'Πήλινες Μετώπες από τη Σπάρτη'. *ASAA* 60: 329–41.
Stroszeck, J. (2006) 'Lakonisch-rotfigurige Keramik aus den Lakedaimoniergräbern am Kerameikos von Athen (403 v. Chr.)'. *AA* 2: 101–20.

Tentori Montalto, M. (2017) *Essere primi per il valore: gli epigrammi funerari greci su pietra per i caduti in guerra (VII–V sec. a.C.). Quaderni della rivista di cultura classica e medioevale 16*. Pisa.

Toher, M. (1991) 'Greek Funerary Legislation and the Two Spartan Funerals', in Flower and Toher (1991) 159–75.

Trundle, M. (2001) 'The Spartan Revolution: Hoplite Warfare in the Late Archaic Period'. *War & Society* 19.2: 1–17.

Valdés Guía, M. (2019a) 'War in Archaic Athens: *Polis*, Elites and Military Power'. *Historia* 68.2: 126–49.

Valdés Guía, M. (2019b) 'The Social and Cultural Background of Hoplite Development in Archaic Athens: Peasants, Debts, *Zeugitai* and *Hoplethes*'. *Historia* 68.4: 388–412.

Van Wees, H. (1999) 'Tyrtaeus' *Eunomia*: Nothing to Do with the Great Rhetra', in Hodkinson and Powell (1999) 1–41.

Van Wees, H. (2000) 'The Development of the Hoplite Phalanx: Iconography and Reality in the Seventh Century', in *W&V* 125–66.

Van Wees, H. (2006a) '"The Oath of the Sworn Bands": The Acharnae Stela, the Oath of Plataea and Archaic Spartan Warfare', in Luther, Meier and Thommen (2006) 125–64.

Van Wees, H. (2006b) 'Mass and Elite in Solon's Athens: The Property Classes Revisited', in Blok and Lardinois (2006) 351–89.

Van Wees, H. (2013a) *Ships and Silver, Taxes and Tribute: A Fiscal History of Archaic Athens*. London.

Van Wees, H. (2013b) 'Farmers and Hoplites: Models of Historical Development', in *MoB* 222–55.

Van Wees, H. (2018a) 'Luxury, Austerity and Equality in Sparta', in Powell (2018) 202–35.

Van Wees, H. (2018b) 'The Common Messes', in Powell (2018) 236–68.

Van Wees, H. (2018c) 'Citizens and Soldiers in Archaic Athens', in Duplouy and Brock (2018) 103–43.

Vaughn, P. (1991) 'The Identification and Retrieval of the Hoplite Battle-Dead', in *HCGBE* 38–62.

Wade-Gery, T. (1949) 'A Note on the Origin of the Spartan Gymnopaidiai'. *CQ* 43: 79–81.

Wheeler, E.L. (1991) 'The General as Hoplite', in *HCGBE* 121–70.

Woodard, R.D. (ed) (2007) *The Cambridge Companion to Greek Mythology*. Cambridge.

Yates, D.C. (2005) 'The Archaic Treaties between the Spartans and Their Allies'. *CQ* 55.1: 65–76.

Zaphiropoulou, P.N. (2002) 'Recent Finds from Paros', in Stamatopoulou and Yeroulanou (2002) 281–84.

CHAPTER 5

Women, Diversity, and War off the Battlefield in Classical Greece

Jennifer Martinez Morales

Throughout history women have been key agents in the evolution of conflicts. From the First World War to the Falklands War women were crucial to the war effort, both at home and in the arms industry. Yet, despite all their hard work they have received less recognition for their efforts. This is also the case in the context of Classical Greece where women were closely linked to war, but their relationship to warfare has been studied by the scholarly community mainly in relation to tragic works.[1] Studies in ancient Greek warfare always revolve around men because they are the ones who fight and who are directly involved in conflict. Soldiers, weapons, armour and tactical military formations have all been the subject of much discussion to the extent that we know more about adult men's experiences of war than we do of any other member of the population of ancient Greece.[2] This is especially true for the Classical period, from which we have plenty of evidence to testify for soldiers' experiences.

But throughout Classical literature we also have accounts of women involved in conflict. These accounts appear sporadically in our fifth- and fourth-century written sources, and they are often short and distributed over a range of genres. Women's relationship to war can be seen in almost every genre there is, from historiography to philosophy and from comedy to tragedy. But their appearances are not always straightforward. Our sources constantly use the general term γυναῖκας without any further clarification when writing about women in the context of war. Sometimes the word is not used at all, but by looking at the context of the conflict, it is occasionally possible to discern whether or not there were women involved. For example, when the word ἀνδράποδα is used to refer to captives and these were taken after a city fell to enemy hands, we know that women were likely to be included amongst these captives, especially if the

1 For attempts at a somewhat historical understanding, see Schaps (1982); Graf (1984); Barry (1996); Loman (2004); Wintjes (2012); Fabre-Serris and Keith (2015). For women as victims, see Gaca (2010, 2011a, 2011b, 2011c, 2014, 2015); Rabinowitz (2014). Martinez Morales (2015, 2019) addresses this historical gap in scholarship. For women in tragedy, see n. 6 below.

2 For example, *WWW*; *HCGBE*; *GWMR*; Rawlings (2007).

 | DOI:10.1163/9789004501751_006

city was not evacuated beforehand.[3] Our sources also write of both prominent women and ordinary women alike who were involved or affected in one way or another by warfare.

Yet, in spite of the wealth of evidence modern studies often address only one part of women in war. Such studies tend to focus on three aspects: (1) prominent individual women, (2) women as 'fighters', and (3) women as victims.[4] Generalisations are then made from particular episodes which may (or may not) be representative of women's experiences of war. Artemisia, for example, has received much consideration for her role in the battle of Salamis.[5] But hers is a unique case; from a Greek perspective, she was an extraordinary individual whose experiences in war were not representative of other Greek women, and from a Persian perspective, she was another local ruler under the Persian empire who contributed (as any other ruler would) to a military naval expedition. The Amazons and the women from tragedy likewise take centre stage in modern studies, but this leaves us with a partial insight into the ordinary woman's world during conflict.[6]

This chapter, therefore, pays particular attention to the different historical wartime experiences of ordinary women. At the moment, women's participation in war is seen as unusual, as happening only in the most extreme of emergencies and as a breach of gender boundaries.[7] This traditional view

3 See, for example, the case of the people of the city of Methymna in Lesbos (Xen. *Hell.* 1.6.13–15). For recent work on the etymology of the word *andrapoda*, see Gaca (2010).

4 For individual women, see Munson (1988). For women as 'fighters', see Loman (2004) 45, who claims that 'Although Xenophon says that (Greek) women did not fight in wars, being biologically unfit for this, there *were* a few individual women who actually fought or led armies in the battles of Classical Antiquity' [his italics]; Wintjes (2012). For women as victims, see n. 1 above.

5 Munson (1988).

6 Seaford (1990). For women and war in tragedy, see Rabinowitz (2014). The role of women in warfare has been studied at length in the context of tragedy. Euripidean tragedy, for instance, has received much attention because of the playwright's emphasis on prominent female roles, especially in the context of war. This chapter, by contrast, focuses on historical wars and women. But the reader can be directed to the following works which offer wonderful insights in Euripidean tragedy: Due (2006) provides good analyses on the lamentations of captive women in Euripides' *Troades* and *Hecuba*. Recent reception studies are exploring the impact Euripides' tragedies have on contemporary women experiencing the trauma of war and conflict – see Eberwine (2019), for example, on the reinterpretation of *Troades* through a modern lens on Syrian women. There are some studies on Homeric epic as well that deal with women in war. Foley (2002, 92) for instance, argues that 'concubines take center stage in Homeric epic only in the absence of wives due to war'. See also Millington, this volume.

7 Schaps (1982); Cartledge (1993); Barry (1996); Loman (2004); Hornblower (2007); Wintjes (2012).

is challenged here by arguing that women's involvement in war was not extraordinary, that we can see women involved in both emergencies and non-emergencies, and that gender boundaries existed in war but were not enforced in the same way as in peacetime. Furthermore, this chapter also calls for a reassessment of the modern terminology frequently used by scholars to refer to women in war by arguing that women should not be labelled as 'fighters' in modern studies because they were not considered 'fighters' by their contemporaries. Finally, this chapter argues for a wider meaning of the word 'impact' when it refers to the impact of war on ancient Greek women. Wartime rape and sexual violence are regularly seen as the quintessential experiences for women on the losing side of Greek warfare. As we shall see, even though wartime rape and sexual violence occurred, and happened more than our sources care to let us know, the impact of war on women, especially the social impact, took place in much more insidious ways.

The first thing we have to acknowledge when studying the relationship between women and war in the ancient world is the concept of diversity. There is no single experience of war for women and there is no single role of women in war. The complexity of war means that many of its facets are unpredictable, not least its effects on the population. Warfare brings outside variables which can never be predicted: time (when), the type of conflict (what), the men fighting (who) and the reason for the conflict (why). These variables need to be taken into consideration when assessing women in war. Women from different poleis and from different social backgrounds were involved in conflict in different ways and their experiences of war were going to differ greatly.

The most common image we have of women in the context of war is that of throwing roof tiles and stones.[8] Thucydides reports that during the Thebans' surprise attack on Plataia in 431 the women and slaves of the city pelted the enemy with stones and tiles from their houses, while at the same time raising the *ololuge* (Thuc. 2.4.2). During the civil war in Korkyra in 427, the women also pelted the enemy with tiles, leading Thucydides to reflect on their behaviour by saying that they withstood the noise in a manner 'against their nature' (παρὰ φύσιν) (Thuc. 3.74.1). When the Carthaginians were besieging Selinous in 409, Diodoros similarly reports that the women of the city threw tiles and stones from their houses (Diod. Sic. 13.56.7). But this is only one side to women's involvement in war. A brief survey shows a rich diversity in the wartime contributions of women: they could be found disguised as men on city walls (Aen. Tact. 40.4), bringing water to their victorious men (Xen. *Hell.* 7.2.9), cooking in military garrisons (Thuc. 2.78.3), carrying secret messages (Aen. Tact. 31.7),

8 On women in Classical Greek sieges, see Martinez Morales (2019).

building and repairing city walls before and during war, during the day and at night (Thuc. 1.90.3, 5.82.5; Diod. Sic. 13.108.8), entertaining soldiers in army camps (Xen. *An.* 6.1.11–13), spying for others (Plut. *Alex.* 48; Arist. *Pol.* 1313b11–16), and distributing food and missiles to their men (Diod. Sic. 13.55.4–5).[9]

Scholars have rationalised these female actions as being extraordinary and only acceptable because war brought unusual circumstances which allowed women to behave in such a way. Cartledge, for instance, argues that women's behaviour in war is 'transgressive' and '(at best) tolerable in the topsy-turvy world of στάσις'.[10] Barry, similarly, claims that 'the evidence of tile-throwing women would seem, however, to constitute an almost routine breakdown of this gender boundary'.[11] These interpretations usually stem from the misunderstanding of Thucydides' phrase παρὰ φύσιν as a negative evaluation of women's actions in wartime. But as Harvey has quite rightly shown, this phrase 'probably indicates guarded admiration rather than disapproval'.[12] In fact, there is only one instance in the whole Greek corpus that suggests women's actions in war were somehow exceptional and breaching gender boundaries. Diodoros, when addressing the actions of the women of Selinous in 409 who provided their men with food and missiles, states that they counted 'as naught the modesty and the sense of shame which they cherished in time of peace' (Diod. Sic. 13.55.4–5). This, however, is a reflective comment of the author (or his source) that evaluates the women's actions from a later perspective (and date) and is not representative of fifth- and fourth-century standards. Taken out of context, phrases like these may at first glance give the impression that the men of the Classical period thought that their women should stay at home when fighting was taking place.[13] But when the enemy invaded the city and the livelihood of communities was at stake, everyone, irrespective of gender, will tackle the threat collectively. As Aineias Taktikos so eloquently observed:

---

9 This survey is by no means comprehensive, but it shows the range of women's contributions to war and how they were not limited to merely throwing roof tiles and stones. Cf. Lee (2020) 43, 47, which was unavailable to the author at the time of composition.

10 Cartledge (1993) 129. He specifically refers to those women who in 427 threw tiles at the democrats in Korkyra (Thuc. 3.74.2).

11 Barry (1996) 68. The gender boundary addressed here is that of war and violence for men and the household for women. See also Wiedemann (1983) 169.

12 Harvey (1985) 83. He further argues that the women 'were acting better than might be expected; they endured the noise and danger of battle surprisingly well for people who, by definition, could not be ἀνδρεῖοι'.

13 The later account of the siege of Sparta in 273–2 contained in Plutarch, *Life of Pyrrhus* 26.8–29.6, shows how women beyond the time period examined in this chapter engage in similar wartime tasks and they are still not considered as breaching wartime norms. The women there are said to return to their *oikoi* once the city did not need them any longer.

> Ὅσοις τῶν ἀνθρώπων ἐκ τῆς αὑτῶν ὁρμωμένοις χώρας ὑπερόριοί τε ἀγῶνες καὶ κίνδυνοι συμβαίνουσιν, ἄν τι σφάλμα γένηται κατὰ γῆν ἢ κατὰ θάλασσαν, ὑπολείπεται τοῖς περιγιγνομένοις αὐτῶν οἰκεία τε χώρα καὶ πόλις καὶ πατρίς, ὥστε οὐκ ἂν ἄρδην πάντες ἀναιρεθείησαν· τοῖς δὲ ὑπὲρ τῶν μεγίστων μέλλουσι κινδυνεύειν, ἱερῶν καὶ πατρίδος καὶ γονέων καὶ τέκνων καὶ τῶν ἄλλων, οὐκ ἴσος οὐδὲ ὅμοιος ἀγών ἐστιν, ἀλλὰ σωθεῖσι μὲν καὶ καλῶς ἀμυναμένοις τοὺς πολεμίους φοβεροὺς τοῖς ἐναντίοις καὶ δυσεπιθέτους εἰς τὸν λοιπὸν χρόνον εἶναι, κακῶς δὲ προσενεχθεῖσι πρὸς τοὺς κινδύνους οὐδεμία ἐλπὶς σωτηρίας ὑπάρξει.
>
> When men leave their own territory to meet combat and danger beyond its borders, the survivors of any disaster which strikes them, on land or at sea, still have their native soil and state and fatherland between them and utter destruction. But when it is in defence of the fundamentals – shrines and fatherland and parents and children and so on – that the risks are to be run, the struggle is not the same, or even similar. A successful repulse of the enemy means safety, intimidated opponents, and the unlikelihood of attack in the future, whereas a poor showing in the face of danger leaves no hope for salvation.
>
> Preface.1–2[14]

More often than not, old-fashioned paradigms of male *versus* female and public *versus* private are inserted into contexts where they have little (if any) relevance.[15] One of these contexts is war. This is because war is not considered a crucial phase for ancient women and women, likewise, are not considered a vital agent in warfare. By raising the *ololuge*, pelting the enemy with objects and aiding in wartime related tasks, these women were already carrying out and indeed enforcing fifth- and fourth-century standards, not going against them. When women did not conform to wartime standards, they were strongly criticised. The women of Sparta are a case in point.

During the Theban invasion of Lakonia in 370/369, the women of Sparta are said to have behaved erratically causing 'more confusion than the enemy':

> χρησίμου δ' οὔσης τῆς θρασύτητος πρὸς οὐδὲν τῶν ἐγκυκλίων, ἀλλ' εἴπερ, πρὸς τὸν πόλεμον, βλαβερώταται καὶ πρὸς ταῦθ' αἱ τῶν Λακώνων ἦσαν· ἐδήλωσαν δ' ἐπὶ τῆς τῶν Θηβαίων ἐμβολῆς, χρήσιμοι μὲν γὰρ οὐδὲν ἦσαν, ὥσπερ ἐν ἑτέραις πόλεσιν, θόρυβον δὲ παρεῖχον πλείω τῶν πολεμίων.

14 Translation by Whitehead (2001) 45.

15 For instance, Strauss (2007) 246, who claims that 'one of the few good things about life under siege was a slight relaxation of the usual restrictions on women's freedom'.

> And although bravery is of service for none of the regular duties of life, but if at all, in war, even in this respect the Spartans' women were most harmful; and they showed this at the time of the Theban invasion, for they rendered no useful service, like the women in other states, while they caused more confusion than the enemy.
>
> ARIST. *Pol.* 1269b[16]

Aristotle is not alone in his critique. Different sources attest to the women's behaviour: Plutarch says that they were καὶ τῶν γυναικῶν οὐ δυναμένων ἡσυχάζειν, ἀλλὰ παντάπασιν ἐκφρόνων οὐσῶν πρός τε τὴν κραυγὴν καὶ τὸ πῦρ τῶν πολεμίων ('unable to keep quiet but utterly beside themselves when they heard the shouts and saw the fires of the enemy,' Plut. *Ages.* 31.4). He describes a visual state of the city at the time: there was noise, shouting, and people running around (θόρυβοι καὶ κραυγαὶ καὶ διαδρομαὶ) to the point where the occasion troubled the elders. Xenophon is the only one who comes to the women's defence by saying that until that moment they had never seen an enemy (Xen. *Hell.* 6.5.28).[17] His defence of the women strongly suggests that their behaviour became known quickly throughout the Greek world.[18] All tell a similar story: that Sparta's women behaved *differently* from the women of other cities at war and that for some reason they were not useful in war.[19] The women of Sparta are a clear example of women not behaving as they should in war. Implied in these sources is the belief that whole communities were supposed to be proactive in wartime, not just men. It also shows how unexpected the women's actions were in the context of this particular conflict (suggesting that the opposite was perhaps the norm because other women knew what war involved). It appears,

16 On this event, see Ducat (1999) 165–67. See also Powell (2004) who identified another passage in Plato that may (indirectly) refer to this episode (Pl. *Leg.* 806a–b). I follow here Powell's interpretation of the meaning of the clause (ὥσπερ ἐν ἑτέραις πόλεσιν) to be that of 'unlike the women of other *poleis*'. See Powell (2004) 139–40. Aristotle is, of course, writing at a time when the Spartan mirage was already in place. The comment on the women of Sparta forms part of his overall critique of Sparta's constitution. On Aristotle and Sparta, see Shütrumpf (1994).

17 On Xenophon and Spartan women, see Pomeroy (2004). On Xenophon and women, see Oost (1977); Humble (2004).

18 Mainly for two reasons (not necessarily distinct from one another): because these women were from Sparta – one of the most notable poleis in the Greek world – and because they behaved in a distinctively harmful manner.

19 This difference is also emphasised by Powell (2004). This is the only derogatory comment about women during wartime in Classical Greek literature. Today we only hear of these women not because they were the only ones who behaved in such a way, but because they were from Sparta – a much criticised polis at the time. There must have been other women who behaved in a similar fashion, but they never made it into our sources.

then, that women doing nothing proactive, or rather, engaging in too much (harmful) erratic activity, was what was considered as breaching boundaries. All women were expected to act erratically in wartime, especially when the enemy attacked their city, but erratic activity became a problem when it was harmful to the city's defence.

One has to remember that women did not go out of their way to engage the enemy in open combat. As Hornblower has recently stated 'nobody in their right mind would want to suggest that unknown to modern historians women might after all have fought in the hoplite phalanx, in one of the set-piece battles which naturally attracted the historians of antiquity'.[20] In fact, women should not even be addressed as 'fighters' in modern studies since for the men of the Classical period, the actions of the women above did not constitute fighting.[21] In order for an individual to be considered a 'fighter', he first had to be male, be of direct military service to the polis, and engage in combat. Women simply did not meet any of these criteria.

First, women's gender prevented them from being considered as 'fighters'. Women, by nature, were thought to be different from men: they were weaker and less able to endure the outside elements (Xen. *Oec.* 7.23). Men also thought women had a greater propensity for fear (Xen. *Oec.* 7.25) and this emotion is useless in warfare.[22] Aineias best exemplifies men's attitudes towards women in the context of war when he stated that the women of Sinope were told not to throw anything from the city walls because 'a woman is recognizable a long way off by the way she throws' (40.5). This remark is unique in Greek sources because it sheds light into men's attitudes towards their women in an area that is not usually associated with them, namely, war. This remark, although not direct evidence, nevertheless suggests that (1) women perhaps threw objects from city walls, (2) they were known to be weaker than men in doing so, and (3) they could be useful individuals for deceitful military strategies.[23] In light of this, perhaps the woman on the Nereid Monument depicted in figure 5.1 is

20 Hornblower (2007) 44.

21 See Loraux (1995) 232, who says that the women 'simply fight, with the help of slaves, alongside the men for the common salvation of the city or *dēmos*'; Loman (2004); Wintjes (2012, 32) claims that 'Beyond non-combatant support there is enough evidence for women taking an active role in combat to assume that for the Greeks it was nothing terribly unusual.'

22 On fear in Greek warfare, see Echeverria (2014) 8–9.

23 On deception in Classical Greek warfare, see Wheeler (1988); Krentz (2000).

FIGURE 5.1 Slab of the Nereid Monument, showing a woman on the walls of a besieged city *ca.* 390–380 BCE (British Museum 1848,1020.65)

not showing signs of distress as is commonly thought but engaging in another wartime related task.[24]

Secondly, the wartime actions of women also prevented their men from considering them as 'fighters'. Women's actions at war, as seen by the brief survey above, were as far away from pitched battle as they could be. Throwing objects from afar – especially from houses – was not considered 'fighting' by the ancient Greeks. Some have stated that 'the roof-tile is characteristically the woman's weapon', but this can hardly be the case.[25] One can recognise women's importance to the city at war while not aggrandising their actions

24 Smith (1900) 22; Powell (2004) 146; *GWMR* plate 1; Jenkins (2006) 192–93; Chaniotis (2013) 439. It is frequently presumed she is showing signs of distress only because she is a woman depicted in a siege (this perhaps stems from the original publication where she is described as such). However, when compared to other battle scenes in the rest of the monument, it seems odd that she will be showing signs of distress when the rest of individuals are engaged in wartime tasks; some are handing over stones to soldiers on ladders while others are engaging in combat.

25 Harvey (1985) 73.

at the same time. One much overlooked woman made her contributions to war in her own space: Hellas the wife of Gongylos of Eretria. Mentioned by Xenophon in his *Anabasis*, Hellas told Xenophon whom to attack, when he should attack, how many men he would need, what he could hope to capture, and what guides he would need in order to be successful – all from the comfort of her own household (Xen. *An.* 7.8.9). In order for women to be useful during wartime they need not engage in direct actions as the ones above (in the city). The modern view is that in order for women to contribute they need to do so in a direct manner, but women's contributions to the war effort were most likely to be what we would call today 'passive' roles such as pre-war ritual libations in the household and religious dedications.[26] Furthermore, it is hard to see the particular extraordinary circumstances in the wartime scenarios where women are involved. When the walls of Athens were being reinforced at Themistokles' request in 478, it was a pre-emptive measure (Thuc. 1.90.3–4); when the women of Argos were building their walls in 417, there was no particular emergency either (Thuc. 5.82.5). The only obvious emergency is that of the *stasis* of Korkyra, but apart from the context of that conflict, the rest of events where women are involved are merely a normal part of war.

The above are all positive contributions to the war effort. However, women could contribute in a negative way to war as well and these types of contributions have been so far ignored by scholarship. There is at least one woman whose actions can be classified as those of a wartime traitor. During the surprise attack on Plataia in 431, Thucydides reports that a woman gave the invading Thebans an axe and allowed them to escape (2.4.4). Harvey argues that 'we cannot be certain that she handed the axe over voluntarily', but her actions

26 See, for instance, Loman (2004) who argues for the view of 'active' women in war. This modern view is inherited from the current understanding about what warfare is and what it entails. It is frequently assumed that in order for an individual to contribute to war, they need to fight. But, as modern conflicts continue to show, war is much more than fighting; it involves entire communities and various groups of individuals, including women. For religious participation of women which is closely related to war, see the epigram on the prayers carried out by the women of Corinth in the Persian Wars (Theopomp. BNJ 115 F 285b), Page (1981) 206–13. *Contra* Graf (1984) 245, who claims that women 'did not play a role in the rituals surrounding warfare [and that] they did not participate in the prayers and sacrifices before and during the departure of the army', but this is only because he has in mind those rituals specifically connected to the battlefield like sacrifices before battle which were always performed by men. However, before the expedition to Sicily in 415 Thucydides reports how the people on the shore joined in the prayers with the soldiers before the latter set sail; the people on the shore included women even though they are not mentioned explicitly (Thuc. 6.32.2).

and the vocabulary employed by Thucydides suggests otherwise.[27] She was inside the city when the attack took place, and she provided a group of men with the means for escape. These men were clearly identifiable as the enemy and as outsiders since they were running around the city precisely because they did not know the layout of the place (Aen. Tact. 2.6; Diod. Sic. 2.4.2). The word δούσης suggests that she willingly gave the axe to the Thebans and that she was at no point compelled to do so by a threat of violence. If there is one woman whose actions can be adequately seen as extraordinary in our sources, it is this one; not because her actions are unique (surely there existed other female wartime traitors in Greek history), but because she is the only one explicitly mentioned as such.

If one maps the geography of women's contributions to war, it quickly becomes clear that women did not have a specific area from which they made their contributions to the war effort; they could be found on houses, on city walls, in military garrisons, by city gates,[28] and in army camps. The same diversity is to be found in the impact of war on women. Women were evacuated from their homes, mostly from the countryside into the walled city (e.g., Hdt. 8.41; Thuc. 2.6.4); they sought out wartime news of their men and experienced grief and loss (Plut. *Lyc.* 1.40); they had to find paid employment because of war (Dem. 57.35, 45); and when abroad as followers of armies, they were direct witnesses to the effects and vicissitudes of conflict (Xen. *An.* 4.3.20, 5.4.33–4).

The unnamed suppliant daughter of Hegetorides, often neglected in studies about women and war in the ancient world, allows for a brief case study where one can see different experiences for women in war. After the battle of Plataia, this unnamed woman approached the Spartan king Pausanias as a suppliant and asked to be saved from captive slavery (αἰχμαλώτου δουλοσύνης) (Hdt. 9.76). She had been forcibly taken from her native city of Kos by the Persian Pharandates and was now his unwilling (ἄκουσαν) concubine (παλλακή, *pallakai*) (Hdt. 9.76; Paus. 3.4.9–10). Her story, as portrayed by Herodotos, represents her as being passed along from man to man. Her expectations are that once the Greeks are victorious, she will again be passed around to another man. Apart from the obvious Greek *versus* barbarian paradigm at work here, it seems that her experiences may not have been exceptional. Briant argues that 'Most *pallakai* were brought to ... Persian nobles' households as prisoners of war ...'[29] It seems that continuous movement, becoming the property of

27 Harvey (1985) 86 n. 23.
28 The woman with the axe at Plataia was by the city gates when the majority of the women were on the rooftops.
29 Briant (2002) 279.

different soldiers, and sometimes engaging in forced work as concubines were regular experiences for some women taken in war during the Classical period. It is often assumed that the women affected by conflicts endured a homogenised experience of this type of wartime violence.[30] Gaca, for example, sees rape as intrinsic to Greek warfare by claiming that the 'deliberate, degrading sexual violence perpetrated by armed males against women and girl captives was a habitual objective and practice of ancient warfare'.[31] War provides the circumstances for both rape and sexual violence to take place, and women indeed suffered these terrible acts after war (e.g., Hdt. 8.33), but both sexual violence and rape should not be seen as a universal consequence of Classical warfare in the same way it may be today.[32] There is no evidence which suggests that soldiers specifically targeted women as a method of warfare. Gaca's recent analysis on the word *andrapoda* and its related verb forms is incredibly valuable in that it demonstrates the process by which women and girls were directly and physically affected after war and at last we can finally put a word to the cryptic actions described in our sources.[33] But the fact that rape and sexual violence happened when 'andrapodizing' was carried out does not mean that it was 'habitual' in conflicts. Warfare was waged against men, and it was honourable to die fighting in battle against men.[34] The ultimate test of *andreia* was warfare itself. The closest one can find is the consistent method of siege warfare which largely depended on the stoppage of supplies to the city under

30 The literature on rape in Classical Greece is diverse, but mostly concerns itself with legal procedures and myth; it is rarely about wartime rape. See, for instance, Harris (1990); Lefkowitz (1993); Carey (1995); Deacy and Pierce (1997); Omitowoju (2002); Rabinowitz (2011, 2014); Robson (2013). Cf. Lee (2020) 47.

31 Gaca (2011a) 80.

32 On the relative silence on wartime rape in our sources, see Schaps (1982) 203. Humble (2004) 169 n. 19 claims that there is only one instance of rape in our sources, but see the following passages. Non-Greeks who rape women in war: Persians: Hdt. 8.33; Carthaginians: Diod. Sic. 13.58.1; Thracians: Plut. *Alex.* 12. Greeks who rape women in war: Dem. 23.141; Thuc. 8.74.3, 86.3 (imagined rape); Macedonians: Din. 1.24; Plut. *Alex.* 22; Diod. Sic. 17.36.1–4, 70.6; Curt. 3.11.21, 5.6.8, 10.1.4.

33 See Gaca's (2010) study on what she denominates the 'andrapodizing of women and girls' in ancient warfare. Gaca (2010) 117 defines the process of 'andrapodizing' as: '… a type of premeditated and semi-lethal warfare aimed expressively against inhabitants not trained, too old or too young, to fight back.' However, this view is problematic because Gaca claims that 'andrapodizing' applies only to females and not to males when there is in fact evidence that it applied to both (e.g., Hdt. 1.66).

34 Hence the passing references and jokes in our written sources about enemy soldiers who are like women (e.g., Persians) or 'womanly' men (e.g., Persians again). These stem from the common idea that fighting was done against men only but also from the ideology of a worthy adversary. On manliness, especially *andreia*, see Hobbs (2000); Rosen and Sluiter (2003).

attack in order to force it into submission, which in turn affected women as part of the population of the city in question, but there is no evidence that implies or even suggests that a method of fifth- and fourth-century warfare systematically and habitually included an attack upon women. Rape and sexual violence did happen, and it certainly happened more than our sources care to let us know, but in the context of Classical Greek warfare it was not the systematic weapon of war that it can be today.

Attitudes towards wartime rape and sexual violence against women in Classical Greek warfare are rare in our written sources.[35] Attitudes to rape, however, are implicit in the broader status of women in antiquity. For example, there existed a *topos* of the breaching of walls in the metaphorical language of female virginity, which seems to fit into a broader acceptance of what happened to women when cities fell: inevitable sexual vulnerability. Its general absence in historical sources may well be because it was taken for granted and because it incurred shame and dishonour to households. That the mistreatment of women is often found in narratives where non-Greeks appear as the perpetrators does not suggest that Greeks themselves did not engage in these acts. In fact, if anything, this representational strategy on the part of our sources suggests that the Greeks (ideally and in theory) knew that this was not appropriate behaviour and by portraying non-Greeks as carrying out these acts, they are shifting the blame to others. It is commonly assumed that women after war should expect to be sold for sexual exploitation, yet, similar to those women who throw roof tiles and stones from houses, this is only one side of the picture.[36]

After a city fell there were women who ended up working in prostitution. The women and girls of the fallen city of Olynthos are a clear example: they were given as gifts (Dem. 19.305–306), they were brought into Athens for *hybris* (Dem. 19.309) and forced prostitution (Din. 1.23), and they suffered physical abuse and shameful maltreatment at the hands of men because being now spoils of war, they had lost any rights they once had as free-born women.[37]

35 There are a few occasions where wartime rape is punished by death, especially in the narratives about Alexander (Plut. *Alex.* 12, 22; Curt. 10.1.1–4; Diod. Sic. 17.108.4). Notice the particular lack of mention of Greek soldiers committing these acts against women, whether Greek or foreign. It is not a coincidence that the majority of the instances of the wartime rape of women are only said to be committed by non-Greeks (i.e., Persians, Thracians, Carthaginians).

36 Trundle (2010) 243–44; Robson (2013) 72.

37 The women's dishonour (or *hybris*) should be understood here as work or treatment demeaning for a free-born girl or woman, most likely prostitution in this context. The loss of civic rights is clearly illustrated by Diodoros' imagining how the women of Selinous viewed their situation after the fall of the city in 409 (Diod. Sic. 13.58.2).

At a Macedonian *symposion* attended by Aeschines an Olynthian captive woman was brought in and was told to sit down amongst men and sing for them. When she refused, she was undressed (Dem. 19.197), grabbed by the hair (Aeschin. 2.157), and flogged with a whip almost to death (Dem. 19.197; Aeschin. 2.157). Apart from these experiences, we might consider that other women ended up in more banal occupations such as forced agriculture and household work (Xen. *Mem.* 2.7.1–14). The story of Satyros at the *symposion* with Philip suggests this may have been the case. It concerns young unmarried girls from Pydna who were moved to Olynthos after their father's assassination in the hopes of a safer future. However, after Philip destroyed Olynthos, the girls became captives (αἰχμάλωτοι) 'digging in Philip's vineyard' and were being asked for by Satyros as a gift from Philip (Dem. 19.193–195; Aeschin. 2.156; Diod. Sic. 16.55.3). This episode shows how one of the outcomes for women after war was to work for someone else, not in prostitution as is commonly assumed, but in forced labour in agriculture. In fact, the women of tragedies often imagine themselves performing menial household tasks after being sent far away from their homes because of war. In Euripides' *Troades*, for instance, the imagined jobs for an old female captive like Hekuba are keeping watch by the door of an *oikos* and taking care of children (*Tro.* 190–196). The imagined jobs for the chorus of young women, by contrast, are that of sharing the bed of a Greek and drawing water from springs (*Tro.* 202–206). Rape and sexual violence are, therefore, only one negative experience of war and our sources tell us that different women experienced war in different ways.

On one end of the spectrum are the economic impacts brought about by war and male wartime absence. The family of Euxitheos represents one example.[38] He says that his mother (Nikarete) worked as a wet nurse during the Peloponnesian War and that many other women engaged in similar paid employment because war compelled them (Dem. 57.30, 34, 35, 41–42, 45).[39] Two things are crucial here: first, that a woman was forced to work because of wartime difficulties, and secondly, that many other women from the same polis

38 Represented by Demosthenes, Euxitheos was wrongfully expelled from the citizen roll of his deme of Halimous in Attika and was appealing his case in Athens. Those who accused him used his father's accent (18) and the fact that his mother Nikarete worked as a wet-nurse and sold ribbons in the agora (30, 34, 35, 41–42) to say that he was not of citizen parents and therefore could not be a citizen himself.

39 Lacey (1980). See also Brock (1994) 344, who claims that we should not take at face value Euxitheos' claims because they may have been working in the ribbon business for a while. However, the accused was under scrutiny not for the ribbon business but for the wet-nursing job which is the one that got the family into trouble in the first place (although both types of employment were indeed used in the accusation).

were forced to work for exactly the same reason. As the speech develops, we learn more personal details about Nikarete's specific wartime circumstances. Her husband Thoukritos was absent on military service with the general Thrasyboulos[40] and she found herself 'in hard straits' (ἀπορίαις) with two children of her own when she took a boy named Kleinias to nurse (Dem. 42). The absence of husbands due to military expeditions was a regular feature of the wartime lives of all women, not just Athenian ones. Aristophanes makes this a consistent female complaint in his *Lysistrata*: when Lysistrata asks the women if they do not miss the fathers of their children, they all quickly respond – each taking her turn – the specific length of time their husbands have been away. Kalonike, an Athenian woman says that her husband has been away for five months on the Thracian front; Myrrhine from an Attic deme says that hers has been away at Pylos for seven months; and Spartan Lampito that as soon as her husband comes home, he promptly leaves again (Ar. *Lys*. 99–110).

At the other end of the spectrum are the more insidious ways in which women were affected by war. The women of those men denominated 'tremblers' (*tresantes*) in Sparta are one valuable case study where we can see the effects of war in women's social and economic lives.[41] Various sources address the social and legal sanctions on those men who showed cowardice in battle, but two sources in particular address a different kind of sanction which concerns the *tresas*' household. Xenophon says that it is the *tresas*' responsibility to support his female relatives at home (who are in an unmarried state) and that he must also pay the fine for being unmarried (*Lac*. 9.5). Plutarch adds that no one will want to give the *tresas* a wife or receive one from him (Plut. *Ages*. 30.3). The female relatives here are being affected by war without doing anything; they are totally dependent on their male's actions at war. Their lack of control over their own destinies perhaps suggests why Xenophon says that the *tresas* must 'bear the responsibility' (αἰτίαν ὑφεκτέον) for the women's ἀνανδρεία. As Ducat has argued, the women might 'belabour him with reproaches'.[42] This suggests that the shame incurred in war was passed on to his *oikos* as well.

The social ramifications for these women might have been as varied as that of the trembler himself. The nature of the trembler's exclusion may give an

40 An Athenian general who led different campaigns during the Peloponnesian War. He died in 388, thus giving us a *terminus ante quem* for Thoukritos' military service. For Thrasyboulos' campaigns, see Buck (1998).

41 On the *tresantes*, see Ducat (2006) who offers a comprehensive list of instances where they are mentioned in our sources and a close analysis. Aristodemos (Hdt. 7.231) is the most well-known (named) trembler, however, there is also the messenger Pantites (Hdt. 7.232) and the many unnamed men after the battle of Leuktra (Plut. *Ages*. 30.2).

42 Ducat (2006) 21.

indication to the women's exclusion as well. No one addressed him in public (Hdt. 7.231), he was picked last for community ball games, in choruses he gets the 'demeaning positions' (ἐπονειδίστους χώρας ἀπελαύνεται), and he has to make allowances for his inferiors in the streets (Xen. *Lac.* 9.5). Given that these are all public penalties, one may propose that the women's exclusion took form in a similar manner, but within their own female areas. It is possible, then, that given how crucial religion was at Sparta, they may have faced similar public shaming in women-only religious cults (for example).[43] Spartan girls with the *tresas*' reputation shadowing them were essentially unmarriageable. One of the most important categories which characterised women in the ancient world was their marital status. This is reflected in the Greek terms for women *gynē* (woman/married) and *parthenos* (young girl/unmarried). This is also linked with the ability to have children, which in many ways was the defining social and cultural feature of women in antiquity (and beyond). The economic ramifications for an unmarried girl in Sparta were significant. Hodkinson argues that 'a girl on her marriage received a significant transfer of land from her parents'.[44] In the case of the female relatives of tremblers, then, it affected them greatly if they could not be given away in marriage. Without the possibility of marrying, the girls could not become mothers, an all-important aspect for Spartan women.[45] In the case of the wife of an already married trembler, on the other hand, perhaps one should not imagine her abandoning his *oikos*, but instead suffering from social exclusion and bearing the shame of her husband. Ducat claims that the 'marriage would have been dissolved'.[46] But there is no evidence for the dissolution of marriages in Sparta in our sources, and if the context of the sanction was in any way a legal one, this would have been very unlikely given that – if one goes by the extant evidence – Classical Greek laws were not retrospective. In Athens, for instance, the citizenship law introduced by Perikles was not effective retrospectively and only affected those from its institution date onwards.[47] The complications of having such a marriage annulled are far greater than if the marriage was, for instance, kept as it was but with a fine imposed on it. And given Sparta's propensity to fine its citizens,

---

43 On Spartan women and religion, see Pomeroy (2002) 105–23; Richer (2007) 250.

44 Hodkinson (2009) 99. I follow here Hodkinson's theory of female universal inheritance at Sparta.

45 See Pomeroy (2002) 51.

46 Ducat (2006) 22. There are few known cases of divorce at Sparta involving kings and are from exceptional circumstances where property and the hereditary royal line was at stake (Hdt. 5.40). Ultimately, as Powell (2009) 411–12 argues, 'we do not know whether a Spartan woman could – on her own initiative or that of others – divorce her husband'.

47 See, for example, Osborne (2010) 246–47.

this is not hard to believe.[48] If we accept MacDowell's argument that even though we do not know what penalties were imposed on κακογάμιον at Sparta 'they did not include dissolution of the marriage', then it is far more likely that the marriage of a trembler was not dissolved but affected by a number of other measures which Xenophon left unsaid.[49] Surely, if the wives of *tresantes* had to leave their husbands, this would have been just as worthy of reporting as the *tresantes*' fines for not marrying in the first place. Thus, we are seeing here the insidious effects of war on the female population of a particular city which took law and warfare seriously.

The post-war impact on women was not limited to Sparta only. As we saw with Nikarete, the absence of husbands had a profound effect on ancient Athenian women. The absence and resulting death of a man (or husband) in a household would always have negative consequences for a woman (or wife), on both social and economic levels. The garland-seller in Aristophanes' *Thesmophoriazusae* had to engage in trade in the agora because her husband died in Cyprus (443–458). Even though it is not immediately clear whether she is a war widow, the death of her husband still forces her to work in paid employment.[50] No doubt a common reality for many women whose husbands had either died in war or been taken captive. Although presented through the lens of comedy, this episode shows common (economic) concerns at the time when the conflict was still on-going. The concern is often with one type of woman: the war widow. The topic of war widows in ancient Greece makes one question whether Greek cities offered any material support to these women. Raaflaub has rightly stated how 'it is astonishing that we know nothing about

48 Fines were imposed on the overall population and even on Spartan kings: Figueira (2002) 156–57. See also Thuc. 5.63.2–4; Ar. *Nubes*. 859; Xen. *Hell.* 5.2.32; Plut. *Pel.* 6.1; Plut. *Ages.* 34.7.

49 MacDowell (1986) 74. On bad marriages, see Plut. *Lys.* 30.7. MacDowell concludes that 'κακογάμιον' 'presumably meant marrying the daughter of a man who either committed an offence or was not a Spartiate'. Being a trembler in Sparta was definitely an offence (*atimia*); therefore, there is nothing to stop one from assuming that if and when the daughter of a 'trembler' was married, then this probably would have constituted a 'κακογάμιον', as well as the marriage of the trembler himself.

50 Aristophanes does not say that her husband died in war, yet her circumstances are strikingly similar to those of Nikarete. Austin and Olson (2004, 191) claim that 'had Kritylla's husband died in the fighting then, [Cyprus' conflict of 449] her children would be in their late 30's and 40's by now and she would have no need to support them'. But I think, following Sommerstein (1994) 185, that 'her words do not necessarily imply that the children are still young or that she is still responsible for them'. Therefore, it is likely that she is indeed a war widow.

material support for war widows'.[51] There is evidence of Greek cities offering material support to war orphans and wartime refugees, but the equivalent for war widows is nowhere to be found.[52] Women are among both of these groups either as unmarried orphans or refugees in another host city. But no official stipulation has ever surfaced that addresses only war widows. Perhaps we are better off understanding this absence of evidence not as a concern for the widow (or woman) as such, but as an anxiety for the stability of the *oikos*.

One should bear in mind that 'woman', like 'man' is not a ubiquitous category in antiquity, and there were differences of wealth, class, family, and age. Women are often referred to as belonging to one single social category, but this is not the case. War was a social activity, and it affected the entire society, of which half happened to be women. The impact of war took on many forms; some affected entire communities (deportations and land devastation), but others affected women in a much more direct way (wartime evacuations). Unfortunately, as is often the case with ancient evidence, when one attempts to look for the social status of women and identify how the post-war impact affected young, old, widow, wife, slave, or metic in their own ways, one comes to the realisation that very little (explicit) evidence exists. One rare occasion where we can see the different social status of women is during the wartime evacuation of complete households in 480 (Hdt. 8.41, 142, 144; Thuc. 1.89.3; Plut. *Them.* 1–6; Paus. 2.31.7). Herodotos says that the Athenians 'issued a proclamation' for everyone to remove their children and households to safety; individuals preferred Troizen, Aigina, and Salamis (8.41). The so-called 'Decree

51 Raaflaub (2014) 34. During the Archaic period, however, there seems to have been some local concern (if one trusts the source) in Samos for the mothers of men who died in war (Douris of Samos BNJ 76 F 63). The reference describes the origin of a proverb as explained by Douris. See Van Wees (2013) 81, and n. 54, 100. Perhaps it is not too farfetched to consider that this sentiment extended to other parts of the Greek world later.

52 State support for war orphans: Thuc. 2.46.1; *Ath. Pol.* 24.3, and the Decree of Theozotides where war orphans receive an obol per day; for this see Stroud (1971) and in a recent online edition, Lambert and Osborne (2016). The decree lists only male orphans, but (according to the editors, n. 4) one cannot exclude the possibility for unmarried female orphans to have been supported as well. Besides Cudjoe (2000) 223, who argues that state support was only for war orphans, not for war widows, there is not much scholarship that explores the subject. State support for refugees: for the Decree of Nikagoras, to the Athenian women and children evacuated to Troizen during the Persian Wars (Plut. *Them.* 10.2–4), see Frost (1978). Support consisted of two obols per day per family, teachers for male children, and the freedom to pick fruits. The people of Akragas supported Selinuntian refugees after the Carthaginian destruction; food at public expense, division of the refugees amongst the residents' homes, and the state asking its citizens to provide anything the refugees might need were all provided (Diod. Sic. 13.58.3).

of Themistokles' – with all the controversy it has generated over the years[53] – sheds some light on the impact of war on two distinct groups of women: the general female population (line 8) and the priestesses of the Akropolis (lines 11–12).

> ·'Αθηναίου[ς δ' ἅπ]α̣[ντας καὶ τοὺς ξένο]υς τοὺς οἰκοῦντας Ἀθήνησι [τὰ τέκ]ν[α καὶ τὰς γυναῖκ]α̣ς ε[ἰς] Τροιζῆνα καταθέσθαι...τοὺς δὲ ταμίας καὶ τ]ὰς ἱερέας ἐν τῆι ἀκροπόλε[ι μένειν φυλάττοντας τὰ τῶ]ν θεῶν·
>
> The Athenians themselves and the foreigners who live in Athens are to remove their women and children to Troizen ... The treasurers and the priestesses are to remain in the acropolis protecting the possessions of the gods.
>
> Lines 8, 11–12[54]

These much-overlooked lines illustrate how dangerous it is to oversimplify our use of the word 'women' in war contexts. One is simply left to imagine the aftermath.[55] The status and social position of these women are closely intertwined with the polis at large. The priestesses have a sacred duty to the polis and should remain, while the rest of the female population is allowed to seek safety – different wartime treatment for women of different social statuses. But we should not imagine that the female population evacuated to other cities are a homogeneous group either. Amongst these women Pausanias tells us that in his day there still existed in Troizen some statues of the 'most important' or 'most high-ranking women' (ὁπόσαι δὲ ἀξιώματι προεῖχον, τούτων εἰκόνας ἀναθεῖναι μόνων) (2.31.7). If we briefly return to the women of the tremblers, we can see that wealthy Spartan women were not exempt from their men's cowardly behaviour. After the battle of Leuktra, Plutarch tells us that at Sparta 'they

53 See Jameson (1960, 1962); Dow (1962); Meiggs and Lewis (1969) 48–52; Henderson (1977); Frost (1980) 101–105, 117–19; Hammond (1988) 558–61; Demand (1990) 183 n. 6. On this evacuation see also Thuc. 1.18.2; Isoc. 6.43, 83; 15.233; Lys. 2.33, 40; Dem. 6.11, 18.204, 19.303; Plut. *Cim.* 5.2.

54 I use here Jameson's (1960) translation of the inscription.

55 Are we to imagine that the priestesses met a violent death? Seen in this light, the conflicting accounts of what happened to the people who remained in the Akropolis makes more sense: Herodotos says that some committed suicide and that the rest (who took refuge as suppliants) were all killed by the Persians (8.53), while Ktesias says that they escaped during the night (*FGrH* 688 F13 (30)). *Contra* Jordan who believes that the priestesses did not stay in the Akropolis because he thinks Jameson's restoration contradicts Herodotos 8.51; see Jordan (1979) 77–80. I, however, following Henderson (1977) 98–103, see no contradiction.

hesitated to inflict the disabilities required by the laws, since the men were numerous and powerful (πολλοῖς οὖσι καὶ δυνατοῖς), for fear that they might stir up a revolution' (Plut. *Ages.* 30.2). With so many 'powerful' men liable to Sparta's wartime sanctions, one can see why the laws were 'allowed to sleep for that day' (Plut. *Ages.* 30.4). These examples show how priestesses, ordinary, and wealthy women were all affected by war in different ways (often even in the same conflict like the evacuation of the Persian Wars).

The evidence analysed in this chapter shows how ordinary women never fought and never took to the battlefield, but this did not make them detached from the world of warfare. The evidence we have tells us that Greek women contributed to the war effort in diverse ways and that these contributions were by no means considered as 'breaching' gender boundaries by their contemporaries. Women were not fighters and should not be addressed as such in modern studies because they lacked the main elements in order to be considered as such, starting with their gender, but they engaged in a plethora of activities. Women's engagement with war (during and after conflict) was multifaceted and has not been fully appreciated by modern scholars. When it comes to the impact of war on women, it soon becomes clear that it is not always straightforward. Not only did women endure wartime rape and sexual violence, but they were also affected socially and economically by their men's cowardly actions at war and through their husbands' absence due to war. Sometimes, as with the mother of Euxitheos and the women of Athens, they were forced to find paid employment to support their families. The impact of Classical Greek warfare, therefore, went beyond physical mistreatment of women and happened in much more insidious and subtle ways.[56]

Bibliography

Austin, C.F.L. and Olson, S.D. (2004) *Thesmophoriazusae*. Oxford.

Barry, W.D. (1996) 'Roof Tiles and Urban Violence in the Ancient World'. *GRBS* 37: 55–74.

Briant, P. (2002) *From Cyrus to Alexander: A History of the Persian Empire*. Winona Lake, IN.

Brock, R. (1994) 'The Labour of Women in Classical Athens'. *CQ* 44: 336–46.

56 I would like to express my gratitude to the editors of this volume for inviting me to contribute and for their helpful comments. Any errors that remain are my own. Translations are from the Loeb Classical library unless otherwise stated.

Buck, R.J. (1998) *Thrasybulus and the Athenian Democracy: The Life of an Athenian Statesman*. Stuttgart.

Carey, C. (1995) 'Rape and Adultery in Athenian Law'. *CQ* 45: 407–17.

Cartledge, P. (1993) 'The Silent Women of Thucydides: 2.45.2 Re-Viewed', in Rosen and Farrell (1993) 125–32.

Chaniotis, A. (2013) 'Greeks Under Siege: Challenges, Experiences, and Emotions', in *WCW* 438–56.

Cudjoe, R.V. (2000) 'The Social and Legal Position of Widows and Orphans in Classical Athens'. PhD Thesis. University of Glasgow.

Deacy, S. and Pierce, K.F. (eds) (1997) *Rape in Antiquity: Sexual Violence in the Greek and Roman Worlds*. London.

Demand, N.H. (1990) *Urban Relocation in Archaic and Classical Greece: Flight and Consolidation*. Bristol.

Dow, S. (1962) 'The Purported Decree of Themistokles. Stele and Inscription'. *AJA* 66: 353–68.

Ducat, J. (1999) 'La femme de Sparte et la guerre'. *Pallas* 51: 159–71.

Ducat, J. (2006) 'The Spartan "Tremblers"', in Hodkinson and Powell (2006) 1–55.

Due, C. (2006) *The Captive Woman's Lament in Greek Tragedy*. Austin.

Eberwine, P. (2019) '"Music for the Wretched": Euripides' *Trojan Women* as Refugee Theatre'. *CRJ* 11: 194–210.

Echeverria, F.R. (2014) 'El miedo en la guerra Griega antigua y su conceptualizacion en las fuentes. Una introduccion'. *De Rebus Antiquis* 4: 1–24.

Fabre-Serris, J. and Keith, A. (eds) (2015) *Women and War in Antiquity*. Baltimore.

Fagan, G.G. and Trundle, M. (eds) (2010) *New Perspectives on Ancient Warfare*. Leiden.

Figueira, T.J. (2002) 'Iron Money and the Ideology of Consumption in Laconia', in Powell and Hodkinson (2002) 137–70.

Figueira, T.J. (ed) (2004) *Spartan Society*. Swansea.

Foley, H.P. (2002) *Female Acts in Greek Tragedy*. Princeton.

Frost, F.J. (1978) 'Troizen and the Persian War: Some New Data'. *AJA* 82: 105–107.

Frost, F.J. (1980) *Plutarch's Themistocles: A Historical Commentary*. Princeton.

Gaca, K.L. (2010) 'The Andrapodizing of War Captives in Greek Historical Memory'. *TAPhA* 140: 117–61.

Gaca, K.L. (2011a) 'Girls, Women, and the Significance of Sexual Violence in Ancient Warfare', in Heineman (2011) 73–88.

Gaca, K.L. (2011b) 'Manhandled and Kicked around: Reinterpreting the Etymology and Significance of Andrapoda'. *IF* 116: 110–46.

Gaca, K.L. (2011c) 'Telling the Girls from the Boys and Children: Interpreting Παῖδες in the Sexual Violence of Populace-Ravaging Ancient Warfare'. *ICS* 35–36: 85–109.

Gaca, K.L. (2014) 'Martial Rape, Pulsating Fear, and the Sexual Maltreatment of Girls (παῖδες), Virgins (παρθένοι), and Women (γυναῖκες) in Antiquity'. *AJPh* 135: 303–57.

Gaca, K.L. (2015) 'Ancient Warfare and the Ravaging Martial Rape of Girls and Women: Evidence from Homeric Epic and Greek Drama', in Masterson, Rabinowitz and Robson (2015) 278–97.

Graf, F. (1984) 'Women, War, and Warlike Divinities'. *ZPE* 55: 245–54.

Hacker, B.C. and Vining, M. (eds) (2012) *A Companion to Women's Military History*. Leiden.

Hammond, N.G.L. (1988) 'The Expedition of Xerxes', in *CAH*^{2} 4: 518–91.

Harris, E.M. (1990) 'Did the Athenians Regard Seduction as a Worse Crime than Rape?'. *CQ* 40: 370–77.

Harvey, D. (1985) 'Women in Thucydides'. *Arethusa* 18: 67–90.

Heineman, E.D. (ed) (2011) *Sexual Violence in Conflict Zones: From the Ancient World to the Era of Human Rights*. Philadelphia.

Henderson, M.M. (1977) 'The Decree of Themistocles'. *ACD* 20: 85–103.

Herfst, P. (1922) *Le travail de la femme dans la Grèce ancienne*. Utrecht.

Hobbs, A. (2000) *Plato and the Hero: Courage, Manliness, and the Impersonal Good*. Cambridge.

Hodkinson, S. (2009) *Property and Wealth in Classical Sparta*. Swansea.

Hodkinson, S. and Powell, A. (eds) (2006) *Sparta and War*. Swansea.

Hodkinson, S. and Powell, A. (eds) (2009) *Sparta: New Perspectives*. Swansea.

Hornblower, S. (2007) 'Warfare in Ancient Literature: The Paradox of War', in *CHGRW* 22–53.

Humble, N. (2004) 'Reality and Ideology in the Representation of Women and War in Xenophon'. *AncW* 35: 166–83.

Jameson, M.H. (1960) 'A Decree of Themistokles from Troizen'. *Hesperia* 29: 198–223.

Jameson, M.H. (1962) 'A Revised Text of the Decree of Themistokles from Troizen'. *Hesperia* 31: 310–15.

Jenkins, I. (2006) *Greek Architecture and its Sculpture*. London.

Jordan, B. (1979) *Servants of the Gods: A Study in the Religion, History and Literature of Fifth Century Athens*. Göttingen.

Kallet, L. (1993) 'Thucydides 2.45.2 and the Status of War Widows in Periclean Athens', in Rosen and Farrell (1993) 133–43.

Kern, P.B. (1999) *Ancient Siege Warfare*. Bloomington.

Krentz, P. (2000) 'Deception in Archaic and Classical Greek Warfare', in *W&V* 167–200.

Lacey, W.K. (1980) 'The Family of Euxitheus Demosthenes LVII'. *CQ* 30: 57–61.

Laiou, A. (ed) (1993) *Consent, Coercion to Sex and Marriage in Ancient and Medieval Societies*. Washington, D.C.

Lambert, S. and Osborne, R. (2016) 'Support for Orphans of Men Killed under the Oligarchy, ca. 410 BC?'. Available at: https://www.atticinscriptions.com/inscription/OR/178 (Accessed: 15 May 2016).

Lee, J.W.I. (2020) 'Daily Life in Classical Greek Armies 500–330 BCE', in *NAGRW* 39–51.

Lefkowitz, M.R. (1993) 'Seduction and Rape in Greek Myth', in Laiou (1993) 17–37.
Loman, P. (2004) 'No Woman No War: Women's Participation in Ancient Greek Warfare'. *G&R* 51: 34–54.
Loraux, N. (1995) *The Experiences of Tiresias: The Feminine and the Greek Man*. Princeton.
MacDowell, D.M. (1986) *Spartan Law*. Edinburgh.
Martinez Morales, J. (2015) 'Women and War in Classical Greece'. PhD Thesis. University of Liverpool.
Martinez Morales, J. (2019) 'Women on the Walls? The Role and Impact of Women in Classical Greek Sieges', in *Sieges* 150–68.
Masterson, M., Rabinowitz, N.S., and Robson, J. (eds) (2015) *Sex in Antiquity: Exploring Gender and Sexuality in the Ancient World*. London.
Meiggs, R. and Lewis, D. (1969) *A Selection of Greek Historical Inscriptions*. Oxford.
Meineck, P. and Konstan, D. (eds) (2014) *Combat Trauma and the Ancient Greeks*. New York.
Munson, R.V. (1988) 'Artemisia in Herodotus'. *ClAnt* 7: 91–106.
Ogden, D. (ed) (2007) *A Companion to Greek Religion*. Oxford.
Omitowoju, R. (2002) *Rape and the Politics of Consent in Classical Athens*. Cambridge.
Oost, S.I. (1977) 'Xenophon's Attitude Toward Women'. *CW* 71: 225–36.
Osborne, R. (2010) *Athens and Athenian Democracy*. Cambridge.
Page, E. (ed) (1981) *Further Greek Epigrams*. Cambridge.
Pomeroy, S. (2002) *Spartan Women*. New York.
Pomeroy, S. (2004) 'Xenophon's Spartan Women', in Tuplin (2004) 201–213.
Powell, A. (2004) 'The Women of Sparta – And of Other Cities – At War', in Figueira (2004) 137–50.
Powell, A. (2009) 'Spartan Women Assertive in Politics? Plutarch's *Lives* of Agis and Kleomenes', in Hodkinson and Powell (2009) 393–419.
Powell, A. and Hodkinson, S. (eds) (1994) *The Shadow of Sparta*. London.
Powell, A. and Hodkinson, S. (eds) (2002) *Sparta: Beyond the Mirage*. London.
Raaflaub, K.A. (2014) 'War and the City: The Brutality of War and its Impact on the Community', in Meineck and Konstan (2014) 15–46.
Rabinowitz, N.S. (2011) 'Greek Tragedy: A Rape Culture?'. *EuGeStA* 1:1–21.
Rabinowitz, N.S. (2014) 'Women and War in Tragedy', in Meineck and Konstan (2014) 185–206.
Rawlings, L. (2007) *The Ancient Greeks at War*. Manchester.
Richer, N. (2007) 'The Religious System at Sparta', in Ogden (2007) 236–52.
Robson, J. (2013) *Sex and Sexuality in Classical Athens*. Edinburgh.
Rosen, R.M. and Sluiter, I. (eds) (2003) *Andreia: Studies in Manliness and Courage in Classical Antiquity*. Leiden.

Rosen, R.M. and Farrell, J. (eds) (1993) *Nomodeiktes: Greek Studies in Honor of Martin Ostwald*. Ann Arbor, MI.

Schaps, D. (1982) 'The Women of Greece in Wartime'. *CPh* 77: 193–213.

Seaford, R. (1990) 'The Imprisonment of Women in Greek Tragedy'. *JHS* 110: 76–90.

Shütrumpf, E. (1994) 'Aristotle on Sparta', in Powell and Hodkinson (1994) 323–45.

Smith, A.H. (1900) *The Nereid Monument and Later Lycian Sculptures in the British Museum*. London.

Sommerstein, A. (1994) *Thesmophoriazusae*. Warminster.

Strauss, B. (2007) 'Naval Battles and Sieges', in *CHGRW* 223–47.

Stroud, R.S. (1971) 'Greek Inscriptions Theozotides and the Athenian Orphans'. *Hesperia* 40: 280–301.

Trundle, M. (2010) 'Coinage and the Transformation of Greek Warfare', in Fagan and Trundle (2010) 227–52.

Tuplin, C. (ed) (2004) *Xenophon and His World. Papers from a Conference Held in Liverpool in July 1999*. Stuttgart.

Van Wees, H. (2013) *Ships and Silver, Taxes and Tribute: A Fiscal History of Archaic Athens*. London.

Villafane, C. (2015) 'The *Perioikoi*: A Social, Economic and Military Study of the Other Lacedaemonians'. PhD Thesis. University of Liverpool.

Wheeler, E. (1988) *Stratagem and the Vocabulary of Military Trickery*. Leiden.

Whitehead, D. (2001) *Aineias the Tactician: How to Survive under Siege. A Historical Commentary with Translation and Introduction*. London.

Wiedemann, T.E.J. (1983) 'ἐλάχιστον…ἐν τοῖς ἄρσερι κλέος: Thucydides, Women, and the Limits of Rational Analysis'. *G&R* 30: 163–70.

Wintjes, J. (2012) "Keep the Women Out of the Camp!': Women and Military Institutions in the Classical World,' in Hacker and Vining (2012) 17–59.

CHAPTER 6

Worshipping Violence

Alexander T. Millington

As with any other aspect of history, whenever we think or talk or write about Greek warfare, we explore two co-dependent worlds: the physical and the imaginary. On the one hand, we seek to understand Greek warfare in terms of mass and material and motion: the length of spears; the weight of shields; the mechanics of the mass shove (ὠθισμός); who marched where, and which musical instruments guided their rhythm. Simultaneously, we examine the decision-making processes that guided Greek generals, and the ways in which they understood and represented war. Our access to the physical is mediated by the imaginary: Greek storytelling, in verse, prose, and picture, comes together with archaeology to mould our visions of how Greek soldiers fought. While much of this volume explores the significance of divergences between the physical and the imaginary, this chapter focuses on variation within the imaginary: on how the Greeks developed a sprawling complex of logically discordant, but practically and psychologically complementary conceptions of warfare and the nature of the warrior, which included as its lynchpin the unique concept of a divine figure who personified both war and the warrior-archetype. In the ways in which they expressed, explored, and interacted with the nature of this divine figure, through narrative and epithet and cult, can be discerned much of what makes Greek thought about war so unique, and so valuable. This divine figure is of course the Olympian god Ares.

1 Causality in War

It is uncontroversial to assert that throughout Greek poetry, both acts and events may be considered overdetermined. The responsibility of any act or event may be attributed to one or more of a cluster of mundane and divine causes which many modern minds have found fundamentally incompatible.[1] The classic example of this is the description of the death of Patroklos in the *Iliad*, where responsibility for the hero's fall is attributed in turn to Zeus, Apollo,

1 See e.g., Harrison (2000) 228; Lateiner (1989) 196.

 | DOI:10.1163/9789004501751_007

Euphorbos, Hektor, and fate.[2] All events within the *Iliad* occur according to Zeus' plan, which provides a purpose beyond the understanding of the mortal characters. The first mortal wound is inflicted by the opportunistic Euphorbos and the final death-blow by mighty Hektor, but their successes are made possible by the intervention of Apollo, their divine protector. All of this occurs according to fate, which is beyond both men and gods.

Thus the myriad mundane causal factors which the modern historian is accustomed to exploring may be supplemented by a variety of divine factors. On the mundane side, successes and failures in battle may be attributed to the decisions, characteristics, and virtues (mental, moral, or physical) of mortals, in addition to the nature of the terrain and other situational factors. Divine causes range from the named Olympian gods through references to 'the divine' as a whole to the irresistible will of fate.[3]

This overdetermined causality is not simply a generic convention within Greek poetry; rather, it is central to the way in which the Greeks thought and talked about war and structured their practical responses to its dangers. The last few decades have seen the publication of a series of excellent studies exploring the ways in which this mode of thinking manifests within Greek historiography. Fate, necessity, divine envy, and divine vengeance are all key elements of the causal frameworks which underpin the narratives of Herodotos, Thucydides, and Xenophon.[4] The world imagined by the historians is therefore not alien to that imagined by the tragedians.[5] These prose authors do not, however, tend to directly describe the influence of individual named gods, as the poets do. Instead, they limit themselves to those instances which were attested

2 The idea of overdetermination was first applied in a Greek context by Dodds (1951) 7. Cf. Versnel (2011) 174.

3 For further discussion of layers of causality, and in particular of divine causality in the Homeric poems, see Lesky (2001) 176–84; Dietrich (1965); Willcock (1970) 6–7; Braswell (1971) 24–25; Clarke (1999) 277–82; Fowler (1987) 5–6; Lovatt (2013) 87. On overdetermination in Attic tragedy, see Mastronarde (2005) 321–22; Lesky (1983) 22; the discussion between Griffiths and Parker in Jouanna and Montanari (2009) 156–57. On Athena and Apollo specifically as helpers and harmers of mortals in the *Iliad*, see Tsagarakis (1977) 34–56.

4 For full discussions of divine causation in Herodotos, see Mikalson (2003); Harrison (2000); Lachenaud (1978). A concise outline is laid out by Lateiner (1989) 196–203. See also the observations of Fornara (1971) 77–78. On divine causation in Thucydides, see Marinatos (1981). For discussion of divine causation in Xenophon, see Dillery (1995) 179–94; Anderson (1974) 35. Many of these studies react against a tradition of scholarship which wants the founders of Greek historiography to share its own atheistic rationality, e.g., Cartledge (1987) 64–65; Soulis (1972) 53.

5 On Thucydides' debts to tragedy, see Cornford (1907) 129–52. On Herodotos and epic, see e.g., Gould (1989) 48.

to by eyewitnesses (in the case of epiphanies), oracles, and other independent authorities.

This does not suggest that any of the historians did not believe in the existence or power of these named, anthropomorphic gods. Rather, this reflects the fact that, as François Hartog has pointed out, the knowledge-claims made by prose writers differ significantly from those made by poets: the poet gains his knowledge through the divine Muse, while the historian's knowledge is the product of his own enquiries.[6] Although the Greeks possessed a substantial store of knowledge regarding what the gods had done in the past, and therefore about the sort of things that they might do, their day-to-day relationship with the gods, as Hugh Bowden has noted, was fundamentally based on ignorance.[7] Lacking divine inspiration, the historian can only remind his audience of the effects of the powers that effect all equally, unless he can cite a suitable authority. The poet, as Aristotle observed, has greater freedom: not to lie, but to express fundamental truths (Arist. *Poet.* 9.1–3).[8]

The overdetermination of causality is something Greek authors carry out within their storytelling, but not something they explicitly discuss on a theoretical level. We can only speculate as to whether their underlying conception of causality rested on an implicit assumption of synergy,[9] or whether they were able, as Paul Veyne has suggested, to think only of the most psychologically convenient explanation for any given event via some process of code-switching, 'blinking', or 'mental balkanisation'.[10] What is clear is that the Greeks possessed a varied and flexible causal vocabulary which enabled them to adapt their narratives to satisfy a wide range of intertwined social, political, and psychological needs. This causal vocabulary is intertwined with the varied ways in which the Greeks gained knowledge about the nature and causes of events.[11] Epiphanies, where victory went hand in hand with manifest divine intervention, could explain unlikely reverses and boost the confidence of

6 Hartog (1988) 276–77, see also Fowler (2010) 333–34.

7 Bowden (2004) 230.

8 I do not intend to imply that the poet's authority was, therefore, beyond criticism. Knowledge-claims do not equal knowledge. Cf. Plato's *Ion*.

9 The idea of synergistic causality, within which mortal and divine agencies are inextricably linked, is prominent within Biblical scholarship: e.g., Miller (1973) 156; Lind (1980) 24–26.

10 Veyne (1983) 56–57 and 84; Versnel (2011).

11 For discussion of the diverse ways in which the Greeks commemorated victories, and apportioned credit for those victories, see Lonis (1979) 265–316: human courage and divine aid are celebrated side by side.

soldiers.[12] Divination and the consultation of oracles gave insights into the will of the divine, which could be used to bolster confidence when attacking, or to justify hesitation or retreat while avoiding hints of weakness or indecision.[13]

Moreover, the Greeks did not merely tell stories using an overdetermined causality: they reacted to it directly. Just as they responded to mundane causal factors by, for example, encouraging their soldiers to be physically fit,[14] so they responded to divine factors through prayers, dedications, and festivals.[15] Propitiatory sacrifices were carried out to ward off divine displeasure.[16] On the flipside, regular festivals in honour of military deities and cult taxes imposed on military pay sought to cultivate their goodwill.[17] Dedications of precious things, including of arms and armour captured in battle and of the proceeds from the sale of such items, expressed gratitude for both past and future favours, and further developed the relationships between cities and the gods who they hoped would help them, or at least refrain from hindering them.[18]

These elements of the Greek causal vocabulary and the actions which respond to them reflect the ways in which the Greeks hoped to exert a certain amount of control over war. The rest of this paper, however, will focus on a divine causal factor which strikingly reflects the Greek recognition of and attempts to cope with the effects of the dangerous unpredictability of war: the god Ares. The ways in which Ares was used within Greek causal narratives and interacted with through cult reflect some important aspects of the ways in which the Greeks conceptualised war. I will focus in particular on two especially profound and well-developed Classical Athenian meditations on the nature of both Ares and war: Aischylos' *Seven against Thebes*, and Euripides' *Phoenician Women*. Before delving into the ways in which Aischylos and Euripides explored the nature of the god and his realm in these two plays, I will briefly outline the Archaic context from which the Attic tragedians' depictions of Ares' character and causal role developed.[19]

12 See *GSW*3 11–46.
13 See *GSW*3 78–83 and 296–321; Lonis (1979) 103–104.
14 E.g., *GSW*2 208–31.
15 For full surveys of Greek interaction with the divine in martial contexts, see *GSW*3; Lonis (1979).
16 See *GSW*3 83–90; Lonis (1979) 107–10.
17 See *GSW*3 154–229 and 324–26.
18 See *GSW*3 240–95; Lonis (1979) 147, 157–95.
19 For a more general treatment of Ares' functions in Classical Attic tragedy and in Athenian cult, see Millington (2014) 136–213.

2 Ares as a Causal Force in War

Athena, Apollo, Artemis, Hermes, Herakles, the Dioskouroi, and many other gods were imagined to roam the battlefields of the Greek world. They had the power to protect those they loved and to harm those who had offended them. Every city had its own divine protector, whose power was identified with that of its armies. Individuals, too, could gain confidence from their personal relationships with the gods. Similarly diverse collections of warrior gods can be found scattered all across the Aegean and Near East. In Mesopotamia, the warriors of a city might be protected by Marduk, Ninurta, Nergal, or Ishtar, among others. In Egypt, Reshep, Montu, and Sekhmet could carry out the same role.[20] Most of these gods could act as helpers and protectors in many aspects of life, both personal and civic, beyond the battlefield.[21] Defeat could be explained in terms of a broader divine plan, as with Enlil's destruction of his city Ur at the hands of the Elamites,[22] or as the result of mortal failure to give adequate worship to divine protectors.[23] Likewise, to the Greeks, the inextricably intertwined influence of Zeus and Fate permeated the whole of a mortal's existence.

The Greek god Ares, on the other hand, was concerned almost exclusively with war and warriors. Moreover, while Ares does occasionally feature in the *Iliad* and other Greek literary texts as an anthropomorphised agent on the battlefield, helping and harming just as other gods do,[24] this role, both in the *Iliad* and elsewhere, is peripheral to his core functions.

Ares' name is often used in ways that make it clear that war is his realm, just as the sea is the realm of Poseidon, the underworld the realm of Hades, and the sky the realm of Zeus.[25] In the *Iliad*, several characters refer to war as 'the turmoil of Ares'.[26] Nestor refers to a great clash of armies as a 'deed of Ares'. The works of war are the works of Ares (Hom. *Il.* 11.732–36). The poet refers to war as 'the strife of Ares' (Hom. *Il.* 5.861; 14.149). It is at the hands of Ares

20 For copious examples of gods giving assistance in war in the ancient Near East, see e.g., Trimm (2017) 567–88, 617–25; Green (2003) 258–88; Kang (1989) 15–19, 51–56, 63–69.

21 For a specific example, on Montu's role as a protector in battle and giver of victory, see Werner (1986) 54–59, with 26 on his more general role as a divine patron.

22 Michalowski (1989) 58–59.

23 See e.g., Bodi (1991). On the ubiquity of this pattern see, e.g., Shipley (1993) 27.

24 See Leaf (1902) 35; Pötscher (1959).

25 Hom. *Il.* 15.190–91. See Dowden (2006) 54–57 and Cook (1940) 451–52 on the direct identification of Zeus with rain and Cook (1940) 943 on Zeus' control over the sky and all meteorological phenomena. For direct identifications of the sky with Zeus, see Guthrie (1950) 38. On Poseidon, see Farnell (1907) 4–7. For examples of dying being described as going to Hades or being given to Hades, see, e.g., Hom. *Il.* 16.625, 856.

26 Nestor: Hom. *Il.* 7.147; Achilles: Hom. *Il.* 16.241–45; Thetis: Hom. *Il.* 18.134–35.

that battles are endured (Hom. *Il.* 3.126–28), and it is by Ares that the combatants are slain: Nestor speaks of the many long-haired Achaians whose blood keen Ares has spilled around the river Skamander (Hom. *Il.* 7.328–30), while Priam tells both his household and Achilles that Ares has slain his many fine sons (Hom. *Il.* 24.253–62, 498).[27] When the spears are flying, Ares guides their paths. The poet tells us how Ares causes spears to fall short from Meriones and Automedon (Hom. *Il.* 16.610–13, 17.525–29), just as he allows the spear to embed itself in the chest of Alkathous, robbing it of its momentum only once the man has been mortally wounded (Hom. *Il.* 13.443–44).[28]

In similar fashion, the sixth century poet Anakreon,[29] as quoted in the *Palatine Anthology*, describes the battle-dead as being slain by Ares (*Anth. Pal.* 7.226 = Anac. fr.100D Diehl),[30] as do the seventh-century Spartan poet Tyrtaios (Stob. *Flor.* 4.10. 6 = Tyrtaios fr.12 West)[31] and a pair of Archaic inscribed epitaphs.[32] To Anakreon, those who survive a battle do so because they have been spared by Ares (*Anth. Pal.* 7.160 = Anac. fr.100D Diehl).

Ares was unique amongst all the gods of the ancient Aegean in being a god whose name was used as a metonym for war. While divine warriors, patrons of warriors, protectors in battle, and bringers of destruction could be found throughout the Near East, only Ares was identified directly with war.

The Classical Attic tragedians inherited the idea that Ares was both synonymous with war as its anthropomorphic personification and also its divine ruler, responsible for all deaths in battle and for the survival of all those who walked away. In this, Ares filled a distinct niche, more specific than Zeus and Fate, who ruled over all walks of life, and less partisan than the many warrior-protectors who stalked each battlefield. Stories about and cults for Ares therefore enabled the Greeks to engage with the idea of war as an independent force with a distinctive character.[33]

---

27 For further examples of Ares as a bringer of death in the *Iliad*, see 19.31; 22.71–72; 24.414–15.

28 Cf. Neal (2006) 28–29.

29 *Suda* s.v. Anakreon: he lived and worked in the cities of Teos and Abdera.

30 Hunter (2010) 278–82 discusses whether this is likely to be a genuine sixth century poem. Hunter remains agnostic but does show that the ideas within the verse would have appealed to an audience of the fourth century.

31 *Suda* s.v. Tyrtaios: he flourished *ca.* 640–637.

32 One comes from Attika and is thought to date from *ca.* 540–30 – see Hansen (1983) 19–20. The second inscription comes from Corcyra and is thought to date from *ca.* 600 – see Hansen (1983) 80.

33 For a full discussion of Ares' functions in pre-Classical Greek poetry, see Millington (2014) 21–135.

3 *Seven against Thebes*: Causality and Gendered Perspective[34]

Aischylos foregrounds the question of military causality at the very beginning of *Seven against Thebes*. In the speech that opens the play, Eteokles observes that should the Thebans defeat the Argives, the god would be given credit, while should the city fall, he himself would be held responsible. Here, Eteokles is unspecific as regards which god may save the city (Aesch. *Sept.* 4–6). As a mortal unguided by prophetic or poetic inspiration, he cannot safely claim such knowledge. Taken in isolation, Eteokles' statement could easily be taken as a wry comment on his own relationship with the Theban people. Eteokles immediately proceeds, however, to pray for assistance to Zeus (Aesch. *Sept.* 8–9),[35] and a bit later in the same speech, he gives credit for prior successes to the gods (Aesch. *Sept.* 21–23). He explicitly presents himself as buying into this world-view.[36] This rapid affirmation of the centrality of the divine in all mortal success could be read as, in whole or in part, a counter to claims of impiety or hubris that could arise from an uncharitable interpretation of his opening remark. It seems more straightforward, however, to read Eteokles as asserting that the gods help those who help themselves: that although only the gods may grant victory (it would be hubristic to suggest otherwise), their aid is contingent on mortals earning their support. Should the people fail to fight with courage, or their leaders fail to display wisdom, then they will deserve to be defeated. This chimes both with Eteokles' assertion that he himself must take care to be constantly vigilant and to offer sound advice (Aesch. *Sept.* 1–3), and also with his rallying cry to the people. The god has thus far, Eteokles declares, given support to Thebes, as is evidenced by prior successes (Aesch. *Sept.* 21–23). In order both to earn and to give thanks for continued divine support, the people, young and old alike, must protect the city, and the altars of their native gods which it contains (Aesch. *Sept.* 10–20).[37] Should the people hurry in their armour to the battlements and gates, stand firm, and have confidence in in their own strength, then the god will surely bring victory (Aesch. *Sept.* 30–38).

---

34 There has been extensive scholarly discussion of the debate between Eteokles and the chorus of Theban women over the last few decades, but never with an explicit focus on modes of interaction with the gods or models of causality. The interpretation that I develop here is intended to complement rather than challenge other readings. Among the most stimulating are Bruit Zaidman (2015); Foley (1993) 129–43; Caldwell (1973). See also Bakewell (2016) 124; and Bachvarova (2006) 29, for contextualisation of the choral odes in this play alongside the broader ancient Near Eastern tradition of city-laments.

35 See also ll.69–76.

36 Cf. Hutchinson (1985) 42; Conacher (1996) 42.

37 See also ll.76–77.

The attraction of this synergistic world-view is clear: if we, through our courage, commitment, and cunning, can earn the assistance of the god, then the god's power will overwhelm our enemies. If we are victorious, then we must ensure future divine support through giving appropriate thanks to the god. If we are defeated, then we may apprehend the possibility of future success in the idea that our failure came not on account of our inferior numbers or effort, or on account of our gods being weaker, but on account of our failure to earn divine aid. All we must do is find enough courage, wisdom, and piety to please the god, and victory will come. In victory, we find moral validation, and in defeat, hope.

Eteokles has spoken as the leader of his people, responsible both for the outbreak of war and for the defence of the city. His aim and duty is to give hope, and to motivate the Thebans. The play's second speaker, an unnamed scout, arrives with a rather different agenda: to make clear the magnitude of the threat faced by the city. He therefore introduces a new, terrifying factor to the causal web: Ares. The god of war has been invoked, alongside his companions Enyo and blood-loving Phobos, by the seven Argive chieftains. By these dread names, they have sworn either to sack the city of Thebes or to die in the attempt (Aesch. *Sept.* 42–48).[38] To embrace Ares is to accept war's deadly unpredictability: the Argives have prepared themselves for the worst, attaching mementoes of themselves to the chariot of Adrastos, which are to be sent back to their parents should they fall (Aesch. *Sept.* 49–51). Having sworn this oath and accepted its consequences, the seven become like lions, with the light of Ares in their eyes. The spirit of Ares now possesses them, giving them unnatural courage (Aesch. *Sept.* 52–53). Their army has become akin to an uncontrollable natural disaster: a roaring land-wave, driven by the squalls of Ares (Aesch. *Sept.* 62–64).[39] Later in the play, we are reminded of this oath, and of Ares' possession of the army, when Hippomedon, one of the seven Argive captains, is explicitly described as being possessed by Ares, raging like the Maenads who are possessed by Dionysos, with Phobos in his eyes, boasting at the gate like Phobos himself (Aesch. Sept. 497–500). Eteokles responds to this threat by once again crying out to the gods of the city for aid, by praying to the curse and fury of his father that they might refrain from crushing the city alongside the household of Oedipus, and by appealing to the common interest of man and

38 A striking and significant trio, but not a monstrous one – see Hutchinson (1985) 49.

39 Padel (1992) 90, observes that, 'Given general Greek resonances of possession as an incoming divine breath, this suggests that Ares breathes *into* the warrior.'

god: when a city achieves success, it gives due honour to the gods who bring it victory (Aesch. *Sept.* 69–77).[40]

A third point of view is provided by the chorus of Theban women, who respond to the scout's account with an extended and intensely emotional plea for divine aid. They begin by calling upon the gods in general (Aesch. *Sept.* 87–88), before asking which of the gods might answer their plea (Aesch. *Sept.* 92–94). They prostrate themselves before the images of the gods, and embrace them, speaking of adorning them with robes and garlands, reminding us of the physical nature of Greek religion, where each god was incarnated in its statue, and in that manifestation, closely associated with a particular place (Aesch. *Sept.* 95–102). They invoke the gods who dwell in their city one by one, by their attributes and epithets – Pallas, Kypris, the Wolf-God, the lord of horses and ruler of the sea with his fish-spearing weapon, Leto's bow-wielding maiden daughter, and Onka – and by their names – Hera, Artemis, and Apollo.[41] They pray that each might grant protection to themselves, and woe to the Argives. They acknowledge the power of the gods and remind them of their faithful worship (Aesch. *Sept.* 108–181). One god stands out in this long catalogue: Ares. The Theban women beg the war-god to protect the city (Aesch. *Sept.* 106–107; 136–37), name him among the ancient gods of their land (Aesch. *Sept.* 104–105), and refer to his past favour (Aesch. *Sept.* 106–107), while also asking if he has now turned traitor, as his spears and shields clatter around the walls in Argive hands (Aesch. *Sept.* 100–105). Both the Argive generals and Theban women agree that the onset of war puts them all into the hands of unpredictable Ares.

Eteokles returns to the stage to respond directly to this choral ode and to debate the chorus regarding the proper approach to seeking divine aid, repeating his assertion that the gods help those who help themselves (Aesch. *Sept.* 223–25). He castigates the women for praying in demoralising panic, and he demands a calm and measured approach (Aesch. *Sept.* 182–202; 236–38). While the women, terrified, continue to beg for divine salvation (Aesch. *Sept.* 211–15, 249, 251, 255, 258–59, 287–88), Eteokles suggests a better prayer: instead of weeping and wailing, and hoping for a miracle, they should pray with hope and an expectation of triumph for the gods of the land to fight *alongside* the people – as allies, not saviours. Eteokles then promises the spoils of this anticipated victory to the gods (Aesch. *Sept.* 264–81). Ares, Eteokles has already observed, feeds upon the slaughter of human beings, and so weeping and wailing at the consequences of battle can only offend the god. The chorus' pleas to him are, therefore, particularly futile (Aesch. *Sept.* 242–44). Eteokles' final

40 Cf. Tucker (1908) 40.

41 On the possible ritual significance of this list, Hutchinson (1985) 67.

contribution to this debate adds one final causal nuance: acceptance of the inevitability of fate (Aesch. *Sept.* 279–81). Calm fatalism now leavens the hope that personal courage, skill, and piety will earn divine assistance. This attitude is particularly appropriate for Eteokles, since he is fully aware of the looming presence of his father's curse (Aesch. *Sept.* 70).

This debate between Eteokles and the chorus is a dispute not about the nature of military causality, but about whose emotional needs ought to be prioritised, and therefore about whose perspective on the causal complex should be favoured. To men of military age, who were expected to take up arms in defence of their city and its interests, it was essential to focus on the role played by their own actions. The women of a city, on the other hand, generally lacked direct military agency (the handful of stories told by Plutarch about women successfully defending their cities are exceptions that prove the rule).[42] Since they could not earn divine support through entering into battle and fighting the enemy hand-to-hand, one key option for women who wished to combat their feelings of helplessness was to seek divine aid through prayer and ritual.[43] Eteokles accepts this to an extent but reprimands the women of Thebes for displaying excessive emotion, which might damage the morale of his warriors (Aesch. *Sept.* 262). Women, Eteokles implicitly argues, must suppress their own needs in order to fully support their menfolk. The women understand his point, but their terror forces them to reject it (Aesch. *Sept.* 287–303). They proceed to paint a yet more horrifying word-picture of the fate that will befall the city if it falls (Aesch. *Sept.* 321–68), reminding the gods that these horrors can only come to pass with their permission (Aesch. *Sept.* 324). They support their prayers for divine protection by seeking the gods' pity and by appealing to their self-interest (Aesch. *Sept.* 302–311; 321). At the heart of the chorus' lament stands Ares, the raging tamer of peoples, who breathes disaster on falling cities (Aesch. *Sept.* 344–45). When the Theban women earlier prayed to Ares for protection, they did so in the knowledge that should their city fall, he would be responsible.

Upon his return to the stage, Eteokles reiterates this message: that Ares is central to military causality. He also confirms the idea that Ares threatens both sides equally, observing, as he sends Melanippos out to face Tydeus, that although Melanippos may be driven by justice and has virtue and courage on

42 Graf (1984). See also Barry (1996) 68, for the idea that the throwing of roof-tiles in extremis could be conceptualised as women fighting 'from the domestic space'.

43 See Martinez (this volume) for a thought-provoking discussion of other ways in which women could support their city in war, including logistical support, and of the effects of war upon women's lives. See also Ducrey (2015); Georgoudi (2015); Payen (2015).

his side, Ares will nevertheless decide the issue with his dice (Aesch. *Sept.* 414). Melanippos' virtue *is* a factor but not necessarily a decisive one. Kapaneus, on the other hand, is deemed by Eteokles to have doomed himself by hubristically declaring that he will sacrifice the city whether or not Zeus and the other gods will it. Polyphontes will stop him, with the aid of Artemis and the other gods (Aesch. *Sept.* 422–51). Eteokles later levels this charge of hubris and impiety at all of the Argive chieftains, with the exception of Oikles, predicting that all will be punished by the gods (Aesch. *Sept.* 597–625). All men, Eteokles reminds us, depend on the gods for all good fortune (Aesch. *Sept.* 625). It is to this key sentiment, contrasting with his prior emphasis on the importance of mortal agency, that Eteokles returns before leaving for his final fatal confrontation with his brother. In the intervening lines, Eteokles' self-assurance and rationality has finally crumbled away under the pressure of the situation. Now it is the chorus who beg *him* to have faith in his own agency, while Eteokles despairs that the gods have already abandoned both him and Thebes (Aesch. *Sept.* 702). As he sets out to meet his doom, Eteokles grimly observes that when the gods give evil, it cannot be escaped (Aesch. *Sept.* 719).

As the messenger arrives to announce the end of the battle, and the passing of the danger that threatened Thebes, new perspectives on military causality develop. The messenger differentiates between the victories of the six Theban heroes over the Argives chieftains, for which he credits their own competence (Aesch. *Sept.* 797–98), and the fates of Eteokles and Polyneikes, for which he holds Apollo responsible (Aesch. *Sept.* 800–802; 812). The chorus, meanwhile, hold Zeus and the other gods of the city equally responsible for the salvation of the city, for which they are due praise, and for the tragic fates of the brothers (Aesch. *Sept.* 822–28). As their lamentations build, the chorus return once more to the role and nature of Ares. Although Apollo and Oedipus' curse may have ordained the deaths of Eteokles and Polyneikes, it is harsh Ares, 'that evil divider of property', who has made the curse come true (Aesch. *Sept.* 944–46). It is Ares in particular whom the chorus cannot find pleasing, despite the Theban victory over Argos (Aesch. *Sept.* 908–910). In the end, although Ares did not choose to destroy Thebes, we find that Zeus, the blessed ones, and Eteokles himself, in a striking validation of his own synergistic worldview, are those whom the chorus explicitly hold responsible for the salvation of the city (Aesch. *Sept.* 1072–78).[44]

In *Seven against Thebes*, which contemporaries dubbed 'a play full of Ares' (Ar. *Ran.* 1021; Plut. *Mor.* 715e), Aischylos explored both the nature of war,

44 Even if the ending of the play was a later addition, there is no reason to believe that this interpolation did not take place during the Classical period.

and how people can and must respond to it. It is striking that when Eteokles debates the chorus regarding the proper way for a city's women to respond to the terrors of an impending conflict, both sides yield some ground. Eteokles, despite his scornful misogyny, accepts the women's right and duty to pray to the gods of the city, while the women acknowledge his fear that their laments might undermine the morale of their men. Through this dialogue, in which both parties acknowledge each other's emotional needs, Aischylos created a narrative space within which both men and women could find their own perspective on war, its terrors, and the ways in which they could seek to take back at least a marginal feeling of control over its outcome, if only through strengthening their relationships with the inscrutable gods.[45] This discussion about how to respond to conflict is set within a divine context which, through the figure of Ares, emphasises the harsh, unpredictable, two-edged nature of war, which possesses people, driving them mad, and turning them into the waves of a natural disaster. The protectors of the city, both mortal and divine, deserve praise for their aid, but Ares, war itself, is not to be celebrated, however much the people might plead with him for mercy as battle approaches, and however central his role may be in the apportionment of victory and defeat, life and death.

4 *Phoenician Women*: Can War Be Avoided?

Throughout his play *Phoenician Women*, Euripides directly engages with Aischylos' *Seven against Thebes*, and develops new approaches to many of the themes which Aischylos explored.[46] Euripides' chorus of Phoenician women

45 The dialogue between Eteokles and the Theban women may also be understood as opening a narrative space within which the citizens of Athens could consider the perspective of their mothers and wives and daughters and gain a greater understanding of their concerns and emotional needs. While it can sometimes be helpful to consider how female experience within tragedy may be used to explore 'feminine' aspects of the male psyche – see Suter (2006b) 167–69 – this should not preclude more straightforward readings. On the vexed question of whether women attended plays at Athens, see Roselli (2011) 158–94; Goldhill (1994); Henderson (1991); Podlecki (1990). The evidence is too sparse and circumstantial to support a firm conclusion. Zeitlin (1985 66–67 and 89 n.5) argues persuasively that the ritual and civic context suggest that tragedies were primarily aimed at male citizens, even if women were present in the audience. This does not invalidate the exploration of possible female reception of these texts. On other tragic explorations of female experience, cf., e.g., Rabinowitz (2014).

46 Cf. Torrance (2013) 94–133; Masaracchia (1987) 70; Mastronarde (1994) 9–10, 17–30. See also Grube (1941) on the way in which Euripides opens up spaces for discussion, rather than putting forward a clear position.

provide a voice which is sympathetic to Thebes, since they have come from Kadmos' homeland, but neutral as regards the conflict between the brothers.[47] Unlike the people of Thebes, they are not complicit in Eteokles' coup. Their combination of deep concern for the city and distant impartiality between the brothers serves as a dramatic counterweight to Jocasta, whose characterisation revolves around her deep emotional attachment to both brothers. It is the relationships within the family of Oedipus, and especially between Eteokles and Polyneikes, which provide the central focus of Euripides' play. Aischylos' background becomes Euripides' foreground. While Aischylos, who had stood against the invading Persians, carefully explores the effects of the Argive assault on Thebes on different sections of a city's population, and their responses to it, Euripides, writing at a time when his native Athens' long war with Sparta was looking increasingly futile, broaches the question of whether the Argive-Theban war could have been avoided entirely.

Euripides begins *Phoenician Women* with a long monologue delivered by Jocasta, who dies well before the Argive attack on Thebes in other versions of this story. This monologue explicitly sets out her purpose within the play: to seek to mediate peace between the brothers (Eur. *Phoen.* 1–83). The first third of the play is largely concerned with the parlay between Eteokles and Polyneikes engineered by Jocasta (Eur. *Phoen.* 1–637).[48] Once this last-ditch attempt to resolve the brothers' dispute has failed, the play gradually evolves into a series of laments, both collective and personal, as the calamitous consequences of the decisions made by the sons of Oedipus are worked out.

While the characters lament their personal losses, the chorus of Phoenician maidens embark on a sustained meditation on the nature of Ares and war. As A. Masaracchia has pointed out, Euripides' plays each tend to revolve around a single god: Apollo in *Electra*, Hera in *Herakles*, Athena in *Iphigenia in Tauris*, Aphrodite (in competition with Artemis) in *Hippolytus*, and Dionysus in the *Bacchae*. In *Phoenician Women*, it is Ares' turn to take center stage.[49]

The first choral ode begins with an account of the chorus' journey to Thebes, of their joyful mission to serve the god, and of what they hoped to find: a city distinguished by sacred dancing-places in the mountains, rich vines, and flowing wine: the gifts of Dionysos and Apollo (Eur. *Phoen.* 202–238). Instead, however, they found a land ruled by Ares, who is now introduced into the ode

47 On the relative calmness of the chorus of Phoenician maidens, see Lamari (2010) 42 n. 198.

48 As Michael Lloyd (1992, 5) points out, *Phoenician Women* is unusual among plays featuring formal debates in that the competing speeches of Eteokles and Polyneikes are overshadowed by Jocasta's arbitration speech. The focus is on Jocasta's attempt to resolve the conflict, rather than on the merits of either son's case.

49 Masaracchia (1987) 180. On Ares' centrality to this play, see also Arthur (1977) 168.

with jarring force. Ares, an agent of the Furies who have ordained woe for the sons of Oedipus, has come before the walls of Thebes, and sets the turmoil and bloodshed of battle ablaze for the city, about which flashes a thick cloud of shields – the sign of murderous battle. The chorus lament the coming of the war-god, remarking that as friends and kin to the people of Thebes, they share her troubles and pain (Eur. *Phoen.* 239–255). The presence of Ares – of the personification of destructive, unpredictable war – as a witness or arbitrator of the contest between Eteokles and Polyneikes is not a welcome one. Ares' role becomes particularly clear once the full purpose of the chorus is made known: they have come as a thank-offering to Apollo (Eur. *Phoen.* 280–282). Apollo, not Ares, therefore, is the giver of victory in war. Ares brings only bloodshed, fire, and murder. As in Aischylos, the divine protector is contrasted with the incarnation and guiding spirit of war itself. The chorus, still hoping for a reconciliation between the brothers, and thus for a non-violent arbitration, pray that Ares' onslaught may yet be averted (Eur. *Phoen.* 239–242).

The second choral ode comes in the wake of the failed parlay between the sons of Oedipus. The Phoenician women sing of Kadmos and the founding of Thebes, with its wheat-bearing fields, lovely rivers, and grassy plains, and of the birth of Dionysos. But they also sing of how Ares was there even before the birth of the city; of how Kadmos found Ares' serpent coiling around a spring when he went in search of water for a lustration and brought strife to the land as he sowed the teeth of the slain dragon at the command of Pallas. Ares-born slaughter caused the earth to be wetted with blood before the first stones were laid (Eur. *Phoen.* 638–75).

The war feels increasingly inevitable now and rushes on apace. The city born in part from the teeth of Ares' serpentine child cannot escape his arbitration. Disregarding the possibility of peaceful solutions, Eteokles begins to plan the defence of the city in earnest. So that nothing might be said to have been left undone, he sends Kreon's son Menoikeus to fetch Teiresias, but he states that he has no intention of speaking directly to the seer, or of listening to his words. Instead, he departs, calling for his weapons and armour (Eur. *Phoen.* 690–773).

The choral ode that this act sparks opens with a direct address to Ares, and comprises a sustained critique of the war to which Eteokles has committed not only himself, but also his city. Within both of their earlier odes, the chorus set the peaceful revels of Dionysos against the horrors promised by Ares. Now, with the clash of arms growing ever closer, the comparison becomes strikingly explicit.

The chorus address Ares not simply as the ruler or divine personification of war, but as a representative of warriors. When they describe the god as 'much-toiling' and ask him why he is possessed by blood and death, they could well be addressing either of the sons of Oedipus, or any of the Argive leaders (Eur. *Phoen.* 784–86).

The chorus presents Ares as afflicting both sides equally. He both breathes the Argive army upon those of Theban blood (Eur. *Phoen.* 789–90) and breathes the race of the Sown Men against the Argives (Eur. *Phoen.* 794–95). Both sides, it is implied, will partake of the bloodshed and death that he brings. In a grotesque parody of a normal religious festival, both sides are made complicit in the revelry which is unfit for the pipe, and both are members of this violent antithesis to a peaceful, joyous Dionysiac chorus.

Although the Argives are the invading force, the chorus highlights the long association between Ares and the Sown Men of Thebes. It is the Theban Spartoi, and not the Argives, whose city stands highest among the crowns of Ares (Eur. *Phoen.* 831–32). This phrase concludes the ode. The precise significance of the final three lines of the ode, and in particular the reference to the ἄκροι Ἀρηΐοι στεφάνοι, has been the subject of much scholarly debate, from the scholiasts onward: στεφάνος can refer either to a crown (with its connotations of victory and glory) or to a hostile encirclement, and so these lines could either celebrate Thebes' military success, or warn of imminent danger. Some level of ambiguity was clearly intended.[50] Read within the broader context of a litany of past blessings juxtaposed with current threat, the core meaning is clear: those who have long benefited from Ares' favour now stand at risk of being destroyed by him, while still hoping for his continuing support. The ambivalence of Euripides' phrasing reflects the inherent ambivalence both of the god and of war.

As the chorus ask the audience to hearken back to Thebes' bloodstained past, and to reflect on the city's complex relationship with Ares, Menoikeus, son of Kreon and descendant of the Sown Men, re-enters the stage, assisting the ancient prophet Teiresias. Teiresias vividly describes the horrors to come: corpses falling around corpses, the Argives and Thebans mingling together in death, as the brothers slay one another amidst bitter laments (Eur. *Phoen.* 880–83). The city itself will fall, says the prophet, unless Kreon sacrifices Menoikeus (Eur. *Phoen.* 884–85; 911–14). Kreon's son must die at the chamber where the earthborn serpent came to life. Through his death, Menoikeus will expiate the ancestral crime committed by Kadmos against both Earth and Ares when he killed the dragon, and thus regain them as allies for Thebes (Eur. *Phoen.*

50 Cf. Amiech (2004) 430; Craik (1998) 215; Mastronarde (1994) 390; Bremer (1980); Parry (1967); Powell (1911) 190; Pearson (1909) 141.

930–41).[51] Now that Jocasta's attempt to resolve the conflict through reason and negotiation has failed (although Jocasta herself does not lose hope while her sons live),[52] the city born in blood and strife, high in the crowns of Ares, must pour out more of its blood to restore its relationship with the war-god, and bring hope of victory. Menoikeus himself, as he prepares to seek the battlements, reflects upon the fact that it was Ares, god of slaughter, who established the men sown from the dragon's teeth as rulers in the land of Thebes (Eur. *Phoen.* 1006–1008). Menoikeus, therefore, acknowledges his people's debt to the god who now demands his death, reaffirming Ares' ambivalence.

Eteokles, whose hubristic arrogance, greed, and pride caused the conflict, departs before Teiresias' arrival, declining to wait to hear the seer's words. Eteokles does not stay to hear of the suffering and danger brought upon the city by his actions, but rather he heads straight into the battle that he has brought. Eteokles will die, but it is not his death that is necessary for the city's salvation, but that of Menoikeus, whose ancestors on both sides were born from the soil of Thebes. Menoikeus and Kreon may be seen as representatives of all those sons and fathers who must pay the price when their chosen leaders choose to go to war. Once Ares has been unleashed, the young men of the city must sacrifice their lives to protect their people, and their aged fathers must watch them go, and mourn. Within this reading, the fact that Menoikeus does not simply risk his life in war, but kills himself at the behest of the gods, makes the symbolism particularly stark and uncompromising.

Yet, as Helen Foley has pointed out, the necessity of Menoikeus' sacrifice lies contingent on an earlier choice, made by the people of Thebes, not by the brothers: the Thebans chose to accept Eteokles' leadership. By exiling both the brothers, all of this pain could have been averted.[53] To an audience in democratic Athens, this striking evocation of the consequences of a people's failure to reject the leadership of those advocating conflict, whether within the city or with her neighbours, would have been particularly resonant. Menoikeus' selfless courage stands in dramatic contrast to the selfishness of all those that the city accepts as leaders – not only that of Eteokles and Polyneikes, but also that of Kreon.

And yet, as has often been observed, the play swiftly refocuses on the familial drama of the house of Oedipus, obscuring the significance of Menoikeus' fate.[54] Menoikeus, the young, innocent victim,[55] self-sacrificed to Ares to

51 Cf. Aelion (1983) 201. On the motifs of human sacrifice and self-sacrifice on behalf of a city or army in tragedy, see Wilkins (1990) 177–79.

52 Cf. Eur. *Phoen.* 1278–83.

53 Foley (1985) 109, re: Eur. *Phoen.* 886–90.

54 E.g., Dunn (1996) 188, 192; Foley (1985) 110; Conacher (1967) 241.

55 On Menoeceus' relative moral purity, see Lamari (2010) 89.

forestall the war-god's rage and save his city, is quickly forgotten. Only Kreon mourns. Both Teiresias and the messenger who brings news to Kreon of his son's death make the causal centrality of Menoikeus' self-sacrifice clear: Menoikeus saved the city. As a result of his death, Kadmeian Ares defeated the Mycenaean spear (Eur. *Phoen.* 1090–92; 1081). Despite this, he rapidly becomes peripheral to the drama. Overdetermination quickly muddies the waters: Kapaneus foolishly challenged Zeus himself and was struck down by lightning; it was Zeus' omen that caused Adrastos to retreat, and the Thebans to take heart and press forward (Eur. *Phoen.* 1172–95). The duel between the brothers is relegated to a sideshow.[56] It is to Zeus that the Thebans raise a trophy (Eur. *Phoen.* 1473). As in *Seven against Thebes*, Ares may bring victory, but he is not celebrated. The young innocent dies saving his city from the forces unleashed by its leaders,[57] but those self-same leaders swiftly monopolise the attention of the people. The chorus' cries of grief are reserved for the fate of the sons of Oedipus, despite describing them as beasts with murderous souls (Eur. *Phoen.* 1284–1306).[58]

In contrast to the death of Menoikeus, however, the blood of Eteokles and Polyneikes is described by Antigone as being poured out as a libation not to Ares, but by Ares to Hades (Eur. *Phoen.* 1574–76). Ares has saved the city from continued civil strife by slaying both of the sons of Oedipus, just as the chorus has prophesied (Eur. *Phoen.* 255). Eteokles and Polyneikes chose to enter into Ares' realm, forcing their people to make a terrible sacrifice in order to obtain the god's aid. Ares protected the Thebans, but he did so by killing their beloved princes. Again, complex feelings about war are explored through the ambivalent character of Ares. Ares has made Thebes great, but the Thebans' relationship with the god is forged in Theban blood. Euripides' *Phoenician Women* therefore extends and develops the narrative space opened up by Aischylos' *Seven against Thebes* in order to encourage both further discussion of the ways in which the formation of a relationship with Ares may be both necessary and deadly, and also consideration of the question of when and how war can and should be avoided. It is essential to be prepared for war, and to have a good relationship with the god, but war should not be celebrated, and Ares is not honoured through song and dance.

---

56 Cf. Dunn (1996) 188; Conacher (1967) 242; Grube (1941) 372.

57 Cf. Aelion (1983) 225.

58 The contrast between the outpouring of grief for the sons of Oedipus and the failure of the people to mourn for Menoikeus, for whose death the two princes were responsible, gives Kreon a much stronger and more personal motivation for refusing to allow Antigone to mourn her brothers and bury Polyneikes than that present in Sophocles' *Antigone*.

5 Worshipping Violence

Most ways of conceptualising and discussing causality in war lend themselves to straightforward binary narratives of praise and blame. The mundane historical analyses provided by Thucydides and Xenophon explain victory and defeat in terms of courage and cowardice, strength and weakness, strategic skill and tactical ineptitude. Tales of divine intervention in war generally focus on the superior strength of one set of divine supporters, and the superior virtue and piety of their worshippers. These narratives naturally appeal to those who seek to raise the morale of their warriors before a war or to celebrate a victory. Appeals to the vagaries of Fate and the ineffable plans of Zeus, on the other hand, make causality inexplicable to mortals, removing fault from the defeated. This sort of narrative facilitates hope of future victory.

Ares represents a unique alternative to both these religious narratives (which the Greeks shared with the other cultures of the ancient Aegean and East Mediterranean) and the analytical accounts of the Greek prose writers. As I have discussed, Aischylos' *Seven against Thebes* and Euripides' *Phoenician Women* use the divine figure of Ares, incarnate ruler of war, as a way of articulating a distinctly ambivalent and unusually nuanced vision of his realm. These plays suggest that a people may choose between peace and war: between rejecting and invoking Ares. They suggest that once Ares has been unleashed, it is wise to beg his aid and seek to avoid his wrath, but that his support cannot be relied upon. Thus, these plays admit war's fundamental unpredictability and destructiveness. Ares is unreliable, and so is war, however brave and strong the warriors. Past victories are no guarantees of future success. Whoever wins, both sides will have suffered. The divine saviours of the city may deserve praise and glory in victory, but war itself does not.

Ares was not merely a narrative device or conceptual tool. Ares was a god: a recipient of cult. Within tragedy, the idea of Ares allowed the Greeks to develop and discuss a distinctive and important conception of the nature of warfare. Through Ares' cult, Greeks were able to form a relationship with this dangerous force. The nature of Ares' cult closely reflects the way in which his nature is represented in *Seven against Thebes* and *Phoenician Women*.

In his *Laws*, Plato implies that all Greek cities could be expected to have a sanctuary of Ares somewhere within their territory when he suggests that standard training for heavy infantry in any Greek state should include running in full armour to the nearest shrine of Ares and back (Pl. *Leg.* 833b).[59] Ares was the patron of all hoplites, not just those of particular cities.

---

59 I here adopt a reading justified by England (1921) 335, *contra* that implied by Brisson and Pradeau (2006) 317.

Monumental temples of Ares, on the other hand, appear to have been few and far between. The majority of cult-sites of Ares appear to have consisted of small shrines and altars, sometimes in rural areas, in addition to sacred groves.[60] Sites of this nature are less likely to be found by archaeologists. They were, moreover, less likely to attract the attention of Pausanias and Strabo (our main literary sources for religious topography), with the exception of those instances where the god bore an unusual epithet, or where an interesting aetiological tale had developed. As Anthony Snodgrass has remarked, Pausanias rarely ventured far from cities and roads.[61]

The perishable nature of Ares' cult sites is demonstrated by the fact that none of those described by Pausanias have been reliably identified, although traces of walls have been detected in the appropriate places. Even the sanctuary in the Athenian Agora, which housed a statue by Alkamenes, continues to evade excavation.[62] Nonetheless, it is clear that the cult of Ares was active at Athens in the Classical period: an inscription from Acharnai dating from the third quarter of the fourth century records the deme's decision, following consultation of an oracle, to build altars for Ares and Athena Areia at public expense.[63]

The picture that develops from analysis of the limited evidence available is of a widespread but low-key cult for a god whom most cities chose not to celebrate extravagantly but were wary of neglecting outright. This chimes with the way in which *Seven against Thebes* and *Phoenician Women* suggest that a celebratory festival for Ares might be inappropriate, but that careful attention to his cult is an essential part of keeping a city safe. Ares was a god who must be respected and feared, but not loved.

Why and how did the Greeks develop such a unique deity? It may be that Ares reflects the extent to which the Greek poetry which has survived was composed for consumption not only by elites, but also by mass audiences, whose members were often responsible for the decision to go to war. Most of our preserved poetry from the Ancient Near East has been transmitted through

---

60 For a full survey of the extant evidence, see Gonzales (2004).

61 Snodgrass (1987) 76–77. For further discussion of Pausanias' selectivity, see Alcock (1996) 245–48; Habicht (1985) 22–24; Farnell (1909) 396.

62 Pausanias (1.8.4) describes a monumental temple in the Athenian Agora, on which see Parker (2005) 52 and 398 n. 43; following Korrés (1992–1998) and (1996) 113, n. 70. On the origin of the temple see also Millington (2014) 173–74 esp. n. 906. Cf. more recently, Stewart (2016) and Stewart *et al.* (2019) which were not available to the author at the time of writing.

63 *SEG* 21.519. Dating follows RO 443, who point out that the style of the relief is related to others that are securely dated to the third quarter of the fourth century. The letter forms are generally agreed to be fourth-century. For text, translations, and further discussion, see Robert (1938) 293–94; Gonzales (2004) 221–26.

scribal traditions serving autocrats. Although many individual cities had more communal forms of government, they were brought under the sway of imperial powers at a relatively early stage. In Babylon, Assyria, Egypt, and the Hittite Empire, public religion and literature alike served kings, and so were predisposed towards the justification of impending war, celebration of triumph, and lamentation of defeat. In the Greek world, there was more emphasis on the promotion of public debate. This is not a purely Athenian phenomenon. Ares was worshipped far beyond the long walls, and his character and functions in *Phoenician Women* and *Seven against Thebes* are not radically different from that which he possesses in the *Iliad*, shaped by a world where the kings derived their power from acclamation by the people (λαός). Agamemnon must persuade his people to follow him. Throughout the *Iliad*, key decisions are debated, not dictated by a king speaking on behalf of his divine protector. In this context, a divine figure who represents a more realistic conception of war – as a terrifying force that is often necessary, but always painful – was a natural development. The spread of a cult aimed at placating this power, and forming some kind of relationship with it, while stepping back from overt celebration, both fueled and was fueled by a rich conception of military causality which allowed publicly-performed literature to engage with the fears, concerns, and desires of all sections of the population.

Bibliography

Aelion, R. (1983) *Euripide héritier d'Eschyle*. Paris.

Alcock, S. (1996) 'Landscapes of Memory and the Authority of Pausanias', in Bingen (1996) 241–67.

Amiech, C. (2004) *Les Phéniciennes d'Euripide*. Paris.

Anderson, J.K. (1974) *Xenophon*. London.

Arthur, M.B. (1977) 'The Curse of Civilization: The Choral Odes of the *Phoenissae*'. *HSPh* 81:163–85.

Bachvarova, M.R. (2006) 'Sumerian Gala Priests and East Mediterranean Returning Gods: Tragic Lamentation in Cross-Cultural Perspective', in Suter (2006a) 18–52.

Barry, W.D. (1996) 'Roof Tiles and Urban Violence in the Ancient World'. *GRBS* 37.1: 55–74.

Baumbach, M., Petrovic, A., and Petrovic, I. (eds) (2010) *Archaic and Classical Greek Epigram*. Cambridge.

Bingen, J. (ed) (1996) *Pausanias Historien*. Geneva.

Bodi, D. (1991) *The Book of Ezekiel and the Poem of Erra*. Göttingen.

Boldrini, S. and Della Corte, F. (1987) *Filologia e Forme Letterarie: Studi offerti à Francesco della Corte*, Vol. 1. Urbino.
Bowden, H. (2004) 'Xenophon and the Scientific Study of Religion', in Tuplin (2004) 229–46.
Braswell, B.K. (1971) 'Mythological Innovation in the *Iliad*'. *CQ* N.S. 21: 16–26.
Bremer, J.M. (1980) 'Euripides, *Phoenissae* 830–832'. *Mnemosyne* 33: 278–87.
Bremmer, J.N. and Erskine, A. (eds) (2010) *The Gods of Ancient Greece: Identities and Transformations*. Edinburgh.
Brisson, L. and Pradeau, J.-F. (2006) *Platon: Les Lois: Livres VII à XII*. Paris.
Bruit-Zaidman, L. (2015) 'Women and War: From the Theban Cycle to Greek Tragedy', in Fabre-Serris and Keith (2015) 82–99.
Cairns, D.L. (ed) (2001) *Oxford Readings in Homer's* Iliad. Oxford.
Caldwell, R.S. (1973) 'The Misogyny of Eteokles'. *Arethusa* 6: 197–231.
Cartledge, P. (1987) *Agesilaos and the Crisis of Sparta*. London.
Clarke, M. (1999) *Flesh and Spirit in the Songs of Homer: A Study of Words and Myths*. Oxford.
Conacher, D.J. (1967) *Euripidean Drama: Myth, Theme and Structure*. Toronto.
Conacher, D.J. (1996) *Aeschylus. The Earlier Plays and Related Studies*. Toronto.
Cook, A.B. (1940) *Zeus: A Study in Ancient Religion*. Vol. 3. Cambridge.
Cornford, F.M. (1907) *Thucydides Mythistoricus*. London.
Craik, E. (1998) *Euripides: Phoenician Women*. Warminster.
Dietrich, B.C. (1965) *Death, Fate and the Gods: The Development of a Religious Idea in Greek Popular Belief and in Homer*. London.
Dillery, J. (1995) *Xenophon and the History of His Times*. London.
Dodds, E.R. (1951) *The Greeks and the Irrational*. Berkeley, CA.
Dowden, K. (2006) *Zeus*. London.
Ducrey, P. (2015) 'War in the Feminine in Ancient Greece', in Fabre-Serris and Keith (2015) 181–99.
Dunn, F.M. (1996) *Tragedy's End*. New York.
England, E.B. (1921) *The Laws of Plato*. Vol. 2. Manchester.
Fabre-Serris, J. and Keith, A. (eds) (2015) *Women and War in Antiquity*. Baltimore, MD.
Farnell, L.R. (1907–1909) *The Cults of the Greek States*. Vols. 4–5. Oxford.
Foley, H.P. (1985) *Ritual Irony. Poetry and Sacrifice in Euripides*. Ithaca.
Foley, H.P. (1993) 'The Politics of Tragic Lamentation', in Sommerstein, Halliwell, Henderson and Zimmerman (1993) 101–43.
Fornara, C.W. (1971) *Herodotus: An Interpretative Essay*. Oxford.
Fowler, R.L. (1987) *The Nature of Early Greek Lyric: Three Preliminary Studies*. Toronto.
Fowler, R.L. (2010) 'Gods in Early Greek Historiography', in Bremmer and Erskine (2010) 318–34.

Georgoudi, S. (2015) 'To Act, Not Submit', in Fabre-Serris and Keith (2015) 200–13.

Goldhill, S. (1994) 'Representing Democracy: Women at the Great Dionysia', in Osborne and Hornblower (1994) 347–69.

Gonzales, M.P. (2004) 'Cults and Sanctuaries of Ares and Enyalios: A Survey of the Literary, Epigraphic, and Archaeological Evidence'. PhD thesis. University of California, Berkeley.

Gould, J. (1989) *Herodotus*. London.

Graf, F. (1984) 'Women, War, and Warlike Divinities'. *ZPE* 55: 245–54.

Green, A.R.W. (2003) *The Storm-God in the Ancient Near East*. Winona Lake, IN.

Gregory, J. (ed) (2005) *A Companion to Greek Tragedy*. Oxford.

Grube, G.M.A. (1941) *The Drama of Euripides*. London.

Guthrie, W.K.C. (1950) *The Greeks and their Gods*. London.

Habicht, C. (1985) *Pausanias' Guide to Ancient Greece*. Berkeley, CA.

Hansen, P.A. (1983) *Carmina Epigraphica Graeca*, Vol.1. Berlin.

Harrison, T. (2000) *Divinity and History: The Religion of Herodotus*. Oxford.

Hartog, F. (1988) *The Mirror of Herodotus* (trans. J. Lloyd). Berkeley, CA.

Henderson, J. (1991) 'Women and the Athenian Dramatic Festivals'. *TAPhA* 121: 133–47.

Hunter, R. (2010) 'Language and Interpretation in Greek Epigram', in Baumbach, Petrovic and Petrovic (2010) 265–88.

Hutchinson, G.O. (1985) *Aeschylus: Septem contra Thebas*. Oxford.

Jouanna, J. and Montanari, F. (eds) (2009) *Eschyle à l'aube du théâtre occidental: neuf exposés suivis de discussions: Vandoeuvres – Genève, 25–29 août 2008. Entretiens sur l'Antiquité classique 55*. Geneva.

Kang, S.-M. (1989) *Divine War in the Old Testament and in the Ancient Near East*. Berlin.

Korrés, M. (1992–1998) 'Από τον Σταυρό στην αρχαία Αγορά'. *ΗΟΡΟΣ* 10–12: 83–104.

Korrés, M. (1996) 'Ein Beitrag zur Kenntnis der attisch-ionischen Architektur', in Schwandner (1996) 90–113.

Lachenaud, G. (1978) *Mythologies, religion et philosophie de l'histoire dans Herodote*. Paris.

Lamari, A.A. (2010) *Narrative, Intertext, and Space in Euripides' Phoenissae*. Berlin.

Lateiner, D. (1989) *The Historical Method of Herodotus*. Toronto.

Leaf, W. (1902) *The Iliad: Edited, with Apparatus Criticus, Prolegomena, Notes, and Appendices: Vol.2: Books XIII–XXIV*. 2nd edn. London.

Lesky, A. (1983) 'Decision and Responsibility in the Tragedy of Aeschylus', in Segal (1983) 13–23.

Lesky, A. (2001) 'Divine and Human Causation in Homeric Epic', in Cairns (2001) 170–202.

Lind, M.C. (1980) *Yahweh Is a Warrior: The Theology of Warfare in Ancient Israel*. Scottdale, PA.

Lloyd, M. (1992) *The Agon in Euripides*. Oxford.

Lonis, R. (1979) *Guerre et Religion en Grèce a l'Époque Classique*. Paris.
Lovatt, H. (2013) *The Epic Gaze: Vision, Gender and Narrative in Ancient Epic*. Cambridge.
Marinatos, N. (1981) *Thucydides and Religion*. Königstein.
Masaracchia, A. (1987) 'Ares nelle Fenicie di Euripide', in Boldrini (1987) 169–81.
Mastronarde, D. (1994) *Euripides: Phoenissae*. Cambridge.
Mastronarde, D. (2005) 'The Gods', in Gregory (2005) 321–32.
Michalowski, P. (1989) *The Lamentation over the Destruction of Sumer and Ur*. Winona Lake, IN.
Mikalson, J.D. (2003) *Herodotus and Religion in the Peloponnesian Wars*. Chapel Hill, NC.
Miller, P.D. (1973) *The Divine Warrior in Early Israel*. Cambridge, MA.
Millington, A.T. (2014) 'War and the Warrior: Functions of Ares in Literature and Cult'. PhD thesis. University College London.
Osborne, R. and Hornblower, S. (eds) (1994) *Ritual, Finance, Politics. Athenian Democratic Accounts Presented to David Lewis*. Oxford.
Padel, R. (1992) *In and Out of the Mind: Greek Images of the Tragic Self*. Princeton.
Parker, R. (2005) *Polytheism and Society at Athens*. Oxford.
Parry, H. (1967) 'Lines 830–832 of Euripides' *Phoenissae*'. *Phoenix* 21: 20–26.
Payen, P. (2015) 'Women's Wars, Censored Wars?', in Fabre-Serris and Keith (2015) 214–27.
Pearson, A.C. (1909) *Euripides: The Phoenissae*. Cambridge.
Podlecki, A. (1990) '"Could Women Attend the Theatre in Ancient Athens?" A Collection of Testimonia'. *AncW* 21: 27–43.
Pötscher, W. (1959) 'Ares'. *Gymnasium* 66: 5–14.
Powell, A. (ed) (1990) *Euripides, Women, and Sexuality*. London.
Powell, J.U. (1911) *The Phoenissae of Euripides*. London.
Rabinowitz, N.S. (2014) 'Women and War in Tragedy', in Meineck and Konstan (2014) 185–206.
Robert, L. (1938) *Études épigraphiques et philologiques*. Paris.
Roselli, D. (2011) *Theater of the People: Spectators and Society in Ancient Athens*. Princeton.
Schwandner, E.-L. (ed) (1996) *Säule und Gebälk: zu Struktur und Wandlungsprozess griechisch-römischer Architektur*. Mainz.
Segal, C. (ed) (1983) *Oxford Readings in Greek Tragedy*. Oxford.
Shipley, G. (1993) 'Introduction: The Limits of War', in *W&S* 1–24.
Snodgrass, A.M. (1987) *An Archaeology of Greece: The Present State and Future Scope of a Discipline*. Berkeley, CA.
Sommerstein, A., Halliwell, S., Henderson, J., and Zimmermann, B. (eds) (1993) *Tragedy, Comedy and the Polis*. Bari.

Soulis, E.M. (1972) *Xenophon and Thucydides: A Study on the Historical Methods of Xenophon in the Hellenica with Special Reference to the Influence of Thucydides*. Athens.

Sourvinou-Inwood, C. (2003) *Tragedy and Athenian Religion*. Lanham, MA.

Stewart, A. (2016) 'The Borghese Ares Revisited: New Evidence from the Agora and a Reconstruction of Augustan Cult Group in the Temple of Ares'. *Hesperia* 85: 577–625.

Stewart, A., Driscoll, E., and Estrin, S. (2019) 'Classical Sculpture from the Agora. 2, The Friezes of the Temple of Ares (Temple of Athena Pallenis)'. *Hesperia* 88: 625–705.

Suter, A. (ed) (2006a) *Lament: Studies in the Ancient Mediterranean and Beyond*. Oxford.

Suter, A. (2006b) 'Male Lament in Greek Tragedy', in Suter (2006a) 156–80.

Torrance, I. (2013) *Metapoetry in Euripides*. Oxford.

Trimm, C. (2017) *Fighting for the King and the Gods: A Survey of Warfare in the Ancient Near East*. Atlanta.

Tsagarakis, O. (1976) *Nature and Background of Major Concepts of Divine Power in Homer*. Amsterdam.

Tucker, T.G. (1908) *The Seven Against Thebes of Aeschylus*. Cambridge.

Tuplin, C. (ed) (2004) *Xenophon and his World: Papers from a Conference Held in Liverpool in July 1999*. Stuttgart.

Versnel, H.S. (2011) *Coping with the Gods: Wayward Readings in Greek Theology*. Leiden.

Veyne, P. (1988) *Did the Greeks Believe in their Myths? An Essay on the Constitutive Imagination* (trans. P. Wissing). Chicago.

Werner, E.K. (1985) 'The God Montu: From the Earliest Attestations to the End of the New Kingdom'. PhD Dissertation, Yale University.

Wilkins, J. (1990) 'The State and the Individual: Euripides' Plays of Voluntary Self-Sacrifice', in Powell (1990) 177–94.

Zeitlin, F.I. (1985) 'Playing the Other: Theater, Theatricality, and the Feminine in Greek Drama', *Representations* 11: 63–94.

CHAPTER 7

Cavalry and the Character of Classical Warfare

Roel Konijnendijk

What part did cavalry play in the wars of the Classical Greeks? This question has been asked countless times in the last two centuries of scholarship, but the answer has remained more or less the same. Greek warfare, it is argued, was 'hoplite warfare'; throughout the fifth and fourth centuries, wars continued to be decided by large formations of heavily armoured infantry. The cavalrymen of the Classical period were too few, too technologically impaired, and too primitive in their tactics to defeat hoplites and affect the outcome of major engagements.[1] They are therefore seen as minor support troops, largely irrelevant to discussions on the development of Greek warfare. In most general works, horsemen are treated only briefly; in some, they are ignored altogether. They do not feature in the popular image of Greeks at war.

This is not to say that Greek horsemen have received no attention from scholars. They have been studied since the nineteenth century, and recent decades have seen several new and valuable monographs on the subject.[2] Yet these works have only recently begun to challenge the consensus of the textbooks. Glenn Bugh was mainly interested in cavalry as an economic and political class within Athenian society, as Albert Martin had been a century earlier; neither author claimed a greater military prominence for horsemen than the standard works of their time. Iain Spence and Robert Gaebel, meanwhile, were careful not to claim too much. Spence went to great lengths to hedge his claim that hoplites were vulnerable to cavalry: 'by this I do *not* mean to imply that Greek cavalry was always, or even often, successful against Greek footsoldiers, merely that when well and aggressively led, good quality horse could, in the right circumstances, be very dangerous to infantry'.[3] The guarded tone of these

1 Rüstow and Köchly (1852) 144; Martin (1887) 431–33; Droysen (1889) 94–95; Bauer (1893) 325–27; Grundy (1911) 280; Kromayer and Veith (1928) 84, 87; Tarn (1930) 5; Gomme (1950) 15; Adcock (1956) 11, 14–16, 48–51; Anderson (1961) 129–30, 152–54, (1970) 58, 122; Lazenby (1989) 71–74; Sage (1996) xviii; Hanson (1995) 234–36, 249; (1998) 7–8; Low (2002) 103; *GWMR* 66, 196; Lee (2006) 491–92; Hunt (2008) 119; Crowley (2012) 102; Mann (2013) 20–21; Sears (2019) 44; Wrightson (2019) 104–105.

2 Martin (1887); Helbig (1902); Anderson (1961); Greenhalgh (1973); Bugh (1988); Spence (1993); Worley (1994); Gaebel (2002); Sidnell (2006); Blaineau (2015). Willekes (2016) treats the subject in a much wider ancient context; more directly relevant is Sears and Willekes (2016).

3 Spence (1990) 97; emphasis in the original. See also Spence (2010) 117 and 123–24.

 | DOI:10.1163/9789004501751_008

experts encouraged the common notion of cavalry ineffectiveness to persist. Lesley Worley, Philip Sidnell, and Matthew Sears and Carolyn Willekes have been more assertive; all have claimed a much greater role for ancient cavalry in general than it is traditionally thought to have played. Yet their works have, so far, not achieved a significant shift in the perception of Classical Greek mounted forces.[4]

The common narrative is as follows. The Persians first launched organised units of cavalry onto the stage of Greek warfare when they invaded mainland Greece in the early fifth century. Up to that time, the Greeks had fought their battles exclusively with large hoplite formations. Yet these hoplite masses actually proved superior to Persian cavalry at Plataia, and the Greeks initially saw no reason to change their ways. It was not until the Peloponnesian War that they began to use cavalry to support the hoplite body. These horsemen became increasingly important in the more sophisticated battle plans of the fourth century, but they remained a tentative, skirmishing force; they lacked the ability to fight hoplites head-on and could not play a dominant role in battle. Greek warfare still ultimately revolved around the hoplite. The Macedonians were the first to create genuine shock cavalry, allowing horsemen to become the decisive force on the battlefield.[5]

This chapter will challenge the basic narrative. Its body of evidence is not new; what follows is an attempt to synthesise well-known passages from ancient literature in ways that seem to do them more justice than the common view. I believe even Worley and Sidnell have not gone far enough in asserting the strategic and tactical value of cavalry in Classical Greece. Far from being a supporting force of secondary importance, horsemen dominated the Greek tactical system from the fifth century onwards.

The question how the horse developed from a prestige object of the leisure class to a potent weapon of war at the end of the Archaic period would require a lengthy study of its own, which has been written well by others;[6] I will not pursue this topic in detail here. It will suffice to note that the surge in iconographic evidence for 'true cavalry' in the final quarter of the sixth century also seems to signal the beginning of its rise to military prominence.[7] The period around 500 marked an essential phase in the emergence of the homogenous

4 Hanson (1999, 393) dismissed the conclusions of Bugh, Spence, and Worley by stating that 'there will inevitably be a degree of exaggeration in the claims of modern scholars for cavalry proficiency and dominance'. Blaineau (2015, 21, 64) still takes for granted that horsemen did not play a major role in Classical Greek warfare.

5 For a representative example of this narrative, Lee (2006) 492, 503.

6 See n. 2 above.

7 Greenhalgh (1973) 96–145; Gaebel (2002) 58–60, although this view is rejected by Osborne (2018) 94–95.

hoplite phalanx, and the key to the rise of horsemen may well lie in an increase in numbers and organisation running parallel to, and presumably developing in close interaction with, that of the hoplites.[8] The resulting formations of heavy infantry and cavalry soon acquired the forms and functions that characterised them throughout the Classical period.

During that period, the impact of horsemen on military thought and practice is well-attested – and the evidence suggests that their influence was out of all proportion to their numbers. The presence of cavalry decided battles and wars. Hoplites did not display a confident superiority but a keen awareness of their many vulnerabilities; they depended on cavalry both to obtain victory and to prevent defeat. In their military environment, Agesilaos could be rightly praised for assembling πολλοὺς καὶ πολεμικοὺς ἱππεῖς ἀντὶ δειλῶν ὁπλιτῶν ('many and warlike horsemen instead of worthless hoplites', Plut. *Ages.* 9.4).

1 The Other Warriors

At first glance, these assertions may be difficult to accept. The armies that won the battles of Marathon and Plataia contained no mounted forces, and communities like Sparta appear to have had no organised cavalry until the end of the fifth century. If horsemen were all but absent from Greek warfare, how could they have played a prominent role? Conversely, if they were so important, how could some of the biggest players on the Greek stage afford to do without them for so long?

Yet the picture we get from synoptic histories of the period, with their focus on the Athenian fleet and the Spartan phalanx, is misleading. At the end of the sixth and the beginning of the fifth century, horsemen already formed a major part of the forces fielded by Greek communities in Sicily, Thessaly, Boiotia, and Anatolia. To my knowledge, no one has ever called into question either the prominence or the abilities of cavalry from these parts.[9] As Joshua Hall points out elsewhere in this volume, the Tarentines, too, may have owed several major victories over their local rivals in the early fifth century to their superior

8 Krentz (2002) 34–35, (2010) 59–60, (2013) 42–43; Sheldon (2012) 44–47; Echeverría (2012) 313–15; Van Wees (2013a) 240–44. For more on the military developments of this period, see Kucewicz (this volume).

9 Sicily: Hdt. 7.158.4; Diod. Sic. 11.22.1. Thessaly: Hdt. 8.28; Paus. 10.1.3–4. Boiotia: Hdt. 9.68, 9.69.2. Among the states of Anatolia, particular mention is made of the horsemen of Magnesia (Arist. *Pol.* 1289b.39–40) and Kolophon (Strab. 14.1.28; Polyaenus, *Strat.* 7.2.2). The cavalry of nearby Lydia was renowned for its quality; Herodotos claimed that even the Persians feared it (Hdt. 1.79.3, 1.80.2–4). Even Hanson (1995, 236–37) concedes the presence and value of Greek cavalry in these regions.

cavalry. No modern scholar seems to doubt that the Macedonians already fielded excellent cavalry at this time,[10] and the same goes for the neighbouring Greek communities in Chalkidike and Thrace.[11] Until their defeat at the hands of the Athenians in 507/6, Chalkis and Eretria on Euboia were ruled by cavalry elites whose horses were instruments of military power as well as prestigious commodities.[12] Spence has suggested that Lokris and Phokis organised corps of horsemen in the aftermath of Xerxes' invasion, if they did not already have such units before.[13] Sixth-century Athens already boasted a small cavalry force,[14] and the Peisistratids, having come to power with aid from the horsemen of Eretria, were happy to use allied Thessalian mounted troops as the first line of defence against Spartan invasions.[15] Spence's handy overview of cavalry forces in the Greek world does not demonstrate the rarity of these troops, but their ubiquity. They appear to have been genuinely absent *only* on the Peloponnese and *only* during the late sixth century and the first three quarters of the fifth.[16] By the later Classical period, as Spence himself noted, all major Peloponnesian states had horsemen of their own.[17]

The thunder of all these hooves echoes in the sources. The clearest mark of the presence of horsemen in Greek warfare is the substantial number of references in Classical literary works to the terror they inspired in other warriors. This body of evidence has never been separately considered as a feature of Greek historical or military writing. I will begin, therefore, with a brief survey of these passages, which give an impression of a military environment largely dominated by one particular troop type. This survey will serve as an introduction to a more detailed look at the ways in which cavalry was effectively used at a strategic and tactical level.

Fear of horsemen is already evident in some of the earliest battle descriptions in Greek historiography. In his account of the Greek decision to send an

10 Spence (1993) 22–27; Gaebel (2002) 280; Sidnell (2006) 29; Blaineau (2015) 81–83; note Thuc. 2.100.5.

11 Spence (1993) 27–29; Blaineau (2015) 83–84.

12 Arist. *Pol.* 1289b.35–39; Hdt. 5.77.2; Strab. 10.1.10; Worley (1994) 27; Van Wees (2011) 100.

13 Spence (1993) 17–18, with Paus. 10.1.8; his suggestion is accepted by Blaineau (2015) 88.

14 Bugh (1988) 4–13; Spence (1993) 9–10; Van Wees (2013b) 74.

15 Arist. [*Ath. Pol.*] 15.2–3; Hdt. 5.63.3–4; Spence (2010) 119.

16 Frederiksen (1968) 9; *GWMR* 67; Sidnell (2006) 27–28. Possible reasons for this will be discussed in the final section. Worley (1994, 23–25) and Figueira (2006, 61, 67–68) have argued that the name of the elite Spartan *hippeis* suggests they were once a cavalry unit, but Lazenby considered this 'doubtful' (1985, 12). It may originally have referred to a social or property class, as at Athens and Chalkis.

17 Spence (1993) 1–9. Blaineau (2015, 92–102, 111) notes that the Peloponnese was one of the major centres of equestrian culture throughout the Archaic and Classical periods.

army to Thermopylai, Herodotos stresses the importance of the fact that the Persians would not be able to use their cavalry there (7.177). At Plataia the following year, the constant attacks of Persian mounted troops drove the Greeks out of their position; according to Plutarch, when the hard-pressed Megarians asked for help against the enemy horse, the Greek commander Pausanias realised that ἀμήχανος ἦν πρὸς ἱππότας ἀμύνειν ὁπλιτικῇ φάλαγγι καὶ βαρείᾳ τῇ Σπαρτιατῶν ('he was incapable of defending against horsemen with the heavily equipped and cumbersome Spartiate phalanx', *Arist.* 14.2).[18] The concerns of the gathered allies are revealed by their plan 'to move in the second watch of the night, so that the Persians would not see them and send the horsemen after them to throw them into disorder' (Hdt. 9.51.3). The Spartans even retreated by a circuitous route through broken ground in the foothills of Kithairon out of fear that Persian cavalry might attack them (Hdt. 9.56.2). These were not just Greek responses to a foreign elite force; after the main battle, it was the Boiotian cavalry that covered the flight of the Persians' Greek allies, and their mere presence was enough to keep their fellow citizens safe from harm (Hdt. 9.68).

These experiences from the Persian Wars set the tone for the rest of the Classical period. When the Athenians marched out in force to punish the Thessalians for their betrayal at Tanagra in 457, their invasion ended in complete failure 'because the horsemen of the Thessalians shut them in' (Thuc. 1.111.1). According to Polyainos (*Strat.* 1.35.2) and Frontinus (*Str.* 4.7.21), the Athenians' greater success against the Boiotians at Oinophyta was due to their general Myronides, who persuaded the hoplites to stand their ground despite their fear of the enemy cavalry, pointing out that they would be utterly wiped out if they fled.[19] Hippokrates exhorted the Athenian army at Delion in 424 with the claim that the Spartans would not dare to invade Attika and face the Athenian cavalry any longer if they were deprived of the support of their mounted Boiotian allies (Thuc. 4.95.2).[20] The Athenians sailed for Sicily with great confidence, but they feared for their lives in battle against Syracuse: as Nikias put it, 'we must conquer, or we will not easily get away, for their many horsemen will be upon us' (Thuc. 6.68.3). This is only one of six separate warnings about the power of the Syracusan cavalry from this general alone.[21] Thucydides claims that Nikias' men 'thought it impossible to carry on the war'

18 The domination of the battlefield at Plataia by the Persian cavalry is stressed by Wardman (1959) 56–58; Worley (1994) 56–58; Gaebel (2002) 75–78.

19 Although an inscription (*IG* I² 400) set up by three hipparchs around this time suggests that the Athenians had their own cavalry at the battle; see Bugh (1988) 45–50.

20 The value of the Boiotian cavalry to the Spartans is underlined by Christesen (2006) 57.

21 Thuc. 6.20.4, 6.21.1, 6.22, 6.68.3, 7.11.4, 7.13.2; Stahl (1973) 66.

until horses could be brought in from Athens to give the besieging force a worthy counterpart to the horsemen of Syracuse (6.71.2). During their final retreat, the Athenians' lack of cavalry confined them to their marching formation and prevented them from gathering supplies: οὐ γὰρ ἔτι ἀποχωρεῖν οἷόν τ' ἦν ὑπὸ τῶν ἱππέων ('they could not leave because of the horsemen', Thuc. 7.78.6).[22]

Xenophon learned about the power of mounted troops from personal experience. Scholars often cite the speech in which he tried to lift the spirits of the Ten Thousand by dismissing the Persian cavalry as fragile and weak (*An.* 3.2.18–19), but that speech follows a bleak overview of the challenges faced by the mercenaries, who knew they were left behind οὐδὲ ἱππέα οὐδένα σύμμαχον ἔχοντες ('without a single horseman or ally', *An.* 3.1.2). It is clear from the rest of the *Anabasis* that Xenophon was not giving his honest opinion about the Persian horse in his speech; instead, he was making a desperate attempt to address the troops' greatest fear.[23] In fact, as Thomas put it, 'if there is a single lesson Xenophon would like us to take away from his work, it is the value of good cavalry'.[24] In the *Anabasis* and elsewhere, he actually underlines the terrible warning he puts into the mouth of Klearchos: 'if we need to fight, we have no horsemen to help us, while the enemy's horsemen are very many and very skilled. So if we win, whom could we kill? And if we lose, not one of us can be saved' (*An.* 2.4.6).[25]

This grim assessment is implicit throughout Xenophon's description of Sparta's campaigns in Anatolia in 399–95. Thibron, the first commander of the Spartan expeditionary force, was completely intimidated by the Persian cavalry: 'seeing the horsemen, he did not go down into the plain' (*Hell.* 3.1.5). His successor Derkylidas assumed that the Persians would look down on his forces because of their own strength in cavalry (*Hell.* 3.2.1), and indeed, Xenophon later claims that the satrap Tissaphernes νομίζων ἱκανὸς εἶναι καταπατῆσαι τῇ ἵππῳ τοὺς Ἕλληνας ('thought he was strong enough to trample the Greeks underfoot with his horsemen', *Hell.* 3.4.12). The third Spartan commander, Agesilaos, was forced to conclude that 'unless he acquired a sufficient cavalry force, he would not be able to campaign in the plain' (*Hell.* 3.4.15); he needed horsemen of his

22 For the central role of cavalry in the Athenian defeat, see Frederiksen (1968) 11–12; Stahl (1973) 66–69; Worley (1994) 104–19; Brice (2013) 625, 630–33, 639. It is hard to sustain the view of Steiner (2005, 407–408 n.5) that Thucydides consistently downplays the importance of this troop type.

23 Worley (1994) 124; Lendle (1995) 160; Sidnell (2006) 23; Crowley (2012) 101. Compare Brasidas' rhetoric at Thuc. 4.126.4–6, with Crowley (2012) 101.

24 Thomas (2009) xlvi.

25 See Xen. *An.* 3.1.2; *Cyr.* 4.3.5. See further Xen. *An.* 3.4.24, 6.5.29.

own ὡς μὴ δραπετεύοντα πολεμεῖν δέοι ('so that he would not have to fight a war of running away', Xen. *Ag.* 1.23).[26] His efforts to remedy the situation prompted Plutarch to coin the phrase cited in the introduction. When faced with strong cavalry in open ground, hoplites were indeed worthless.

It was not just the Persian cavalry that could upset the balance of forces so dramatically.[27] In his description of the desperate plight of the Argives preparing for battle against the Spartan alliance in 418, Thucydides highlights that they had no cavalry, while the Boiotian horsemen had them surrounded and were about to crash into their rear (5.58.3). Once the pro-democratic rebels under Thrasyboulos had acquired a group of seventy horsemen to support their hoplites and light-armed troops, only the cavalry of the Thirty remained willing to sally from Athens to attack them (Xen. *Hell.* 2.4.26). One of the reasons why the Spartans decided not to fight the Boiotians for the bodies of Lysander and his men at Haliartos in 395 was that 'on the matter of horsemen, the enemy had many, while they had few' (Xen. *Hell.* 3.5.23). The aggressive action of Theban horsemen in the plain near Thespiai in 378 terrified their Spartan-led enemies and saved the lives of Gorgidas and the Sacred Band (Xen. *Hell.* 5.4.42–45; Polyaenus, *Strat.* 2.5.2). In 370, when Agesilaos was campaigning against Mantineia, the sudden appearance of some peltasts and Phleiasian horsemen 'caused the men to run to their positions' until it became clear that these forces were friendly (Xen. *Hell.* 6.5.17). When the Thebans invaded Thessaly in 364, the Thessalian cavalry fell upon them, and the Thebans, οὔτε μένειν οὔτε προάγειν ἐώμενοι εἰς πολλὴν ἀμηχανίαν ἐνέπιπτον ('being allowed neither to stay nor to advance, were reduced to utter helplessness', Diod. Sic. 15.71.5–6) until Epameinondas took over and organised a counterattack.

In short, the Classical literary sources present a consistent picture, in which other warriors confronted by enemy cavalry are, or perceive themselves as, helpless. This should be the foundation of any assessment of the role horsemen played in Greek warfare. While many of the passages cited above could perhaps be explained away as rhetorical devices in narratives about the military asymmetry between mainland Greeks and their distant enemies, examples of the fear of cavalry extend beyond those contexts, reflecting a military reality in which *all* horsemen were a force to be reckoned with. Persians may have been barbarians, but Syracusans, Boiotians, and Thessalians were Greeks,

---

26 The point is repeated in the account of Nepos (17.3): 'the Spartan king, seeing that the enemy were superior to him in cavalry, never gave them an opportunity of attacking him in the plains, but engaged them in those parts in which infantry would be of greater service'.

27 On Achaemenid cavalry, see Nefedkin (2006); Tuplin (2010); Charles (2015).

and their horsemen were no less dangerous.[28] Scholars who doubt the abilities of Greek cavalry must account for the fact that the Greeks themselves considered them dangerous indeed.

2 The Roles of Horsemen

The orator Lysias took it for granted that any man who 'had never served as a hoplite but had always served as a horseman' would have 'inflicted many evils on the enemy' (πολλὰ κακὰ τοὺς πολεμίους πεποιηκότες, 14.10). What particular evils were these? In what ways did the effectiveness of cavalry manifest itself?

This question deserves to be treated at some length. Contrary to the consensus of modern scholarship, horsemen were neither marginal to Classical Greek warfare nor functionally limited to a small number of specialist tasks. The available historiographical works and military manuals tell us more than just the fact that they were frightening; they also make it easy to understand *why* they were such a daunting element of the tactical system. As I will show, many of their roles were interrelated: the use of cavalry shaped new strategic and tactical situations in which mounted forces were able to play additional roles in turn. Not all the attested roles of horsemen have been given equal recognition in secondary literature. I will explore their most important functions here, moving from the widely accepted to the underrated or ignored.

2.1 *Protecting Territory*

The least controversial role in which horsemen could be effective was what Spence has termed 'mobile defence'.[29] Ravaging was one of the main elements of Greek offensive warfare, but plundering and razing required invading troops to spread out; this made them vulnerable to hit-and-run attacks by mounted forces sallying out from fortified centres to defend their territory. Such attacks were among the main duties for which Xenophon hoped to prepare the Athenian cavalry.[30] The horsemen would avoid the main body of the invading army, but target those of the enemy hoplites, light troops, and camp followers who were scattered about the countryside looking for things to steal or

28 Note Agesilaos' pride in defeating Thessalians (Xen. *Hell.* 4.3.9 = *Ages.* 2.5). The Thessalians intimidated even the Macedonians, who retreated at Lamia in 322 'for fear of the cavalry' (Diod. Sic. 18.15.4–6).

29 Spence (1990) 93, 96–97; see also Martin (1887) 434; Anderson (1970) 57–58; Spence (1993) 122–23, 127–33; Worley (1994) 83–87, 121; Hanson (1998) 122–28; *GWMR* 66; Hunt (2008) 119.

30 Xen. *Eq. mag.* 7.4–10; Garlan (1989) 100.

destroy. Faced by oncoming cavalry, ravagers and foragers would be compelled to rush back to their camp or marching column for safety. If they failed to reach a friendly formation or fortification, they would be massacred. The Athenian light troops ravaging Megarian territory near Nisaia in 424 were easy prey for the Boiotian horsemen, who fell upon them unexpectedly (Thuc. 4.72.2). The Ten Thousand lost as many as 500 men to a Persian cavalry attack on their raiding party in Bithynia (Xen. *An.* 6.4.24); Agesilaos' plundering troops suffered a similar fate near Sardis in 395 (Xen. *Hell.* 3.4.22) and again in Phrygia later that year (Xen. *Hell.* 4.1.17–19). The invaders' need for constant protection against cavalry prevented them from doing serious damage to the defender's territory.

A region defended by well-managed cavalry was all but immune to ravaging. We have seen how the Athenians in Thessaly and on Sicily were practically unable to leave their camp; in the former case, Thucydides stresses that they could neither ravage the land nor seize a city, and were eventually forced to return home, ἔχοντες ἄπρακτοι ('having accomplished nothing', Thuc. 1.111.1). During the Peloponnesian War, the Athenian cavalry played a well-known part in minimising the damage done to Attika. Their actions left the countryside substantially intact, despite yearly invasions, until the establishment of a permanent Peloponnesian garrison at Dekeleia.[31] A similarly active defence was mounted by the Macedonian cavalry when the Thracians invaded in 430/29 (Thuc. 2.100.5). The main reason for the Spartans to raise their own cavalry force in 425/4 was the growing threat of Athenian invasion from Pylos and Kythera, showing Spartan awareness of the effectiveness of cavalry as a countermeasure (Thuc. 4.55.1–2). When the remnants of the Ten Thousand reached Thrace, they found that they could not provide for themselves by plundering, because they no longer possessed horsemen of their own, and the enemy's cavalry were always at their heels (Xen. *An.* 6.6.29).

The importance of this use of cavalry can hardly be overstated. Mobile defence allowed a community to avoid the risks of pitched battle without too great a loss to their homes and crops. The invader's only choice was to try to engage the sallying horsemen, which required a suitable force of friendly cavalry, ideally supported by other arms. Thucydides did not think much of the skirmish between the Boiotian cavalry and that of the Athenians and Thessalians during the first invasion of Attika in 431 (2.22.2), but such clashes must have been common throughout the war. The need for cavalry in this role made it

31 Thuc. 2.22.2, 3.1.2, 7.27.5; *Hell. Oxy.*17.4–5; Ober (1985) 177–81; Spence (1990) 106–107; Hanson (1998) 131–73; Lazenby (2004) 35, 38, 49; Hunter (2005); Schubert and Laspe (2009) 380–81; Spence (2010).

reasonable for the Athenians to believe that the Spartans would cease their incursions if they lost the support of the Boiotian horsemen (Thuc. 4.95.2). Indeed, the same applied to Athenians ravaging the Megarid in 424; when the plundering light troops were suddenly attacked by Boiotian cavalry, it was the timely intervention of their own horsemen that saved them (Thuc. 4.72.3–4). When the Boiotians and Thessalians who invaded Mantineian territory in 362 were suddenly attacked by Athenian horsemen, they accepted the challenge and fought it out at great cost to both sides (Xen. *Hell.* 7.5.15–17). Only after the defeat of the Persian cavalry at Sardis in 395 could Agesilaos march 'without even having his troops drawn up in a hollow square', his men finally free to plunder with impunity (*Hell. Oxy.* 12.2).

2.2 *Pursuing Fleeing Enemies*

The same abilities that made horsemen so effective at containing enemy ravagers and foragers made them uniquely dangerous in the aftermath of pitched battle. When an army collapsed and the defeated hoplites and light troops fled the battlefield, they could be slaughtered with relative ease by the winning side – an opportunity the Greeks habitually exploited with savage enthusiasm.[32] But the hoplites of a victorious battle line, exhausted by combat and weighed down by their equipment, could not pursue fleeing men very well or for very long. This was a role in which cavalry could shine. Swift horsemen were ideally suited to the task of riding down scattered opponents who were no longer able to mount organised resistance.[33]

A determined pursuit could claim a devastating toll, and our sources often point out specifically that it was the cavalry doing the killing. When the Athenians were routed at Delion in 424, the Boiotians set about butchering them, καὶ μάλιστα οἱ ἱππῆς ('the cavalry most of all', Thuc. 4.96.8). During their battle against the Bithynians, the hoplites of the Ten Thousand immediately put their enemies to flight, after which their tiny force of horsemen gave chase and 'killed as many as they could' (Xen. *An.* 6.5.28). During Agesilaos' second invasion of Boiotia in 377, the cavalry of his Olynthian allies caused a massacre among the Theban peltasts as they fled into the hills, 'for going uphill where the riding is good, infantry is quickly overtaken by horsemen' (Xen. *Hell.* 5.4.54). At the Tearless Battle in 368 the Arkadians 'were cut down as they fled, many by the horsemen and many by the Celts' (Xen. *Hell.* 7.1.31). According to

32 For example, Hdt. 3.54.2; Thuc. 3.113.7; Xen. *Hell.* 4.4.12, 5.3.6; Diod. Sic. 12.10.1, 15.72.3. The point has been made at length in recent scholarship: Krentz (2002) 30–31; Dayton (2005) 73–76; Echeverría (2011) 72–73; Van Wees (2011) 71–76; Konijnendijk (2018) 188–205.

33 Martin (1887) 432; Spence (1993) 157–59; *GWMR* 66, 196; Hunt (2008) 119.

Plutarch, Pelopidas' victorious cavalry at the battle of Kynoskephalai ἐνέπλησαν νεκρῶν τὴν χώραν ('filled the country with corpses', Plut. *Pel.* 32.7), killing as many as 3,000 fleeing men.[34]

Since such slaughter could change the outcome of entire wars, the mounted pursuit of a fleeing enemy was a crucial extension to the efforts of the hoplite phalanx. Without a cavalry chase, victory could even be seen as incomplete. As noted above, Klearchos is made to express the consequence of his lack of cavalry in a simple rhetorical question (Xen. *An.* 2.4.6): 'if we win, whom could we kill?' The implication is that there was no point in winning a battle unless the enemy could be chased and wiped out – which apparently only horsemen could do. Xenophon repeats this sentiment in his own words later on in the *Anabasis* (3.1.2). Then, after he shows how Klearchos' prediction came true in battle against Mithradates (*An.* 3.3.8–9), he himself proposes to raise a cavalry force from among the Ten Thousand specifically τοὺς φεύγοντας ἀνιάσουσιν ('to distress those who are fleeing', Xen. *An.* 3.3.19). Indeed, Xenophon thought it worth repeating the initial warning again in the *Kyroupaideia* (4.3.5): 'when we have routed them, how could we, without horses, catch or kill fleeing horsemen or archers or peltasts?'. Light infantry perhaps sufficed to overtake hoplites in pursuit, but there was only one troop type that could catch all others. Hoplites might put the enemy to flight, but only cavalry could destroy them.

2.3 *Covering Flight*

There was another side to this destructive potential. While cavalry was the most effective troop type for chasing a routed enemy, it was also a defeated army's best hope to prevent such chase and slaughter. When the battle line crumbled, intact units of cavalry could cover its retreat. Containing the pursuing enemy in the same way that they contained a ravaging invasion force, horsemen could give fleeing hoplites and light infantry the opportunity to escape unharmed.[35] To a beaten infantry force, this kind of cavalry support was vital. Even a few hundred horsemen could save entire armies.

At Plataia, when the Persian infantry was routed, their cavalry did not flee along with them, but took up a position between them and the advancing Greeks to protect them from harm. When the Greek centre ran up in disorder to join the pursuit of the fleeing enemy, the horsemen of the Boiotians, fighting on the Persian side, saw their chance and attacked. They killed 600 men and

---

34 Liston's analysis (2020, 83–89) of bones from Chaironeia may demonstrate in graphic detail what cavalry could inflict on fleeing soldiers.

35 Martin (1887) 432; Spence (1993) 159–62; *GWMR* 196; Rawlings (2007) 97–98; Bugh (2020) 67.

pursued the rest back to Mount Kithairon (Hdt. 9.68, 9.69.2). Their charge is a striking example of the importance of cavalry in the aftermath of battle. With the day already lost and the Persian army on the run, these horsemen inflicted more losses on the allied Greeks than they had suffered in the entire engagement up to that point, if Herodotos' numbers (9.70.5) are to be believed.

The counterattack of the Boiotian cavalry may have been unusual for the sheer toll it exacted, but its circumstances were far from unique. At Mantineia in 418, the Athenians on the allied left were nearly surrounded by the Spartans, 'and they would have suffered worse than any part of the army, if not for the cavalry they had with them, which served them well' (Thuc. 5.73.1). During the first battle outside Syracuse in 415, the Athenians routed their opponents, but they could not exploit their victory, 'for the Syracusan cavalry were many and undefeated, and they hemmed them in, attacking and pushing back any of the hoplites they saw pursuing ahead of the rest' (Thuc. 6.70.3). In the aftermath of another Syracusan rout, a swift cavalry charge broke a pursuing unit of Athenian picked troops, allowing the defeated Syracusan hoplites to retreat in safety (Thuc. 6.101.5). In Bithynia, while the cavalry of the Ten Thousand was pursuing and killing at will on the right wing of the army, the hoplites and peltasts on the left achieved little against their fleeing opponents, τὸ γὰρ ἱππικὸν φόβον παρεῖχε τὸ τῶν πολεμίων πολὺ ὄν ('for the horsemen of the enemy, many as they were, inspired fear', Xen. *An.* 6.4.29). At Orchomenos in 370, when the mercenary commander Polytropos fell and his troops scattered, 'very many of the fleeing men would have been killed', if not for the timely arrival of supporting Phleiasian cavalry (Xen. *Hell.* 6.5.14).

The importance of horsemen in covering the retreat allows us to fully appreciate Klearchos' warning to the Ten Thousand. Without cavalry, this mercenary force could no more pursue a broken opponent than they could protect themselves in defeat. If they were to fight a battle against the Persians, victory would be pointless, but defeat would be catastrophic. The examples cited here show that this warning cannot simply be isolated to its Persian context. For all Greek armies, relative cavalry strength determined the degree to which a beaten force could be harmed or preserved.

2.4 *Hounding the Enemy*

The value of cavalry in protecting the backs of fleeing hoplites lay in the effect their presence had on the enemy. Because Greek horsemen could strike at range with javelins, as well as engage in close combat, their opponents were compelled not just to huddle together, but also to move away from the threat. This effect – not just containment, but repulsion – allowed them to do more than merely confine the enemy to their camp or formation. When aggressively led, cavalry could 'herd' entire armies, determining through their mobility and harrying tactics where and when the enemy could move.

Examples of this ability to control enemy movement come in various forms. A spirited sally by the Athenian cavalry drove the entire Peloponnesian army away from the city walls in 411 (Thuc. 8.71). The Olynthians in 381 (Xen. *Hell.* 5.3.3) and the Thebans in 378 (Xen. *Hell.* 5.4.39, 5.4.44–45) used horsemen to shadow their Spartan enemies as they advanced, forcing them to respond to the cavalry's presence, and attacking at the first sign of disorder. The cavalry of the Spartan allies at Leuktra drove the departing Thespiaians and their baggage train back into the main hoplite body (Xen. *Hell.* 6.4.9). In 369, a tiny force of Phleiasian horsemen charged and routed the rearguard of a large Argive army as it was leaving their territory, inflicting many casualties (Xen. *Hell.* 7.2.4). The Thessalians did the same to the Thebans in 364 with a far larger number of mounted troops (Diod. 15.71.5). Yet by far the most impressive example of horsemen 'herding' an army is that of the Syracusan force sent to support the Spartan alliance against the Thebans at Corinth in 369:

> οἱ δὲ παρὰ τοῦ Διονυσίου ἱππεῖς, ὅσοιπερ ἦσαν, οὗτοι διεσκεδασμένοι ἄλλος ἄλλῃ παραθέοντες ἠκόντιζόν τε προσελαύνοντες, καὶ ἐπεὶ ὥρμων ἐπ᾽ αὐτούς, ἀνεχώρουν, καὶ πάλιν ἀναστρέφοντες ἠκόντιζον. […] εἰ δ᾽ αὖ τινες διώξειαν αὐτοὺς πολὺ ἀπὸ τοῦ στρατεύματος, τούτους, ὁπότε ἀποχωροῖεν, ἐπικείμενοι καὶ ἀκοντίζοντες δεινὰ εἰργάζοντο, καὶ πᾶν τὸ στράτευμα ἠνάγκαζον ἑαυτῶν ἕνεκα καὶ προϊέναι καὶ ἀναχωρεῖν.
>
> But the horsemen sent by Dionysios, though they were few, scattering here and there and riding along the enemy line, charged them and threw javelins at them, and when the enemy moved against them, they retreated, and then turned around and threw javelins again. […] If anyone pursued them far from the army, they would charge them when they were retreating, and by throwing javelins wreak havoc among them, and so they forced the entire army, as they wished, either to advance or to fall back.
>
> XEN. *Hell.* 7.1.21

After enduring this for a few days, the Thebans, assessing their current situation and their earlier failure to take Corinth, disbanded the coalition army and marched back home.

There could be no clearer evidence of the disproportionate impact that even a modest force of cavalry could have. The Syracusans involved in this skirmish numbered just fifty, against a Theban allied army that consisted of tens of thousands of hoplites. Dionysios' horsemen could not have hoped to defeat the enemy in detail; yet the very tactics that made them so effective at minimising the damage done by ravaging also allowed them to channel and constrain the movements of forces much larger than their own, and, in the right circumstances, nullify the momentum of whole campaigns.

Again, the only real defence against the menace that cavalry presented was to protect an army with horsemen of its own.[36] While missile troops might be used to keep mounted troops at bay,[37] they were vulnerable if they strayed too far from the shelter of a hoplite formation; the preferred response to enemy cavalry was, rather, an immediate and aggressive counterattack in which cavalry led the way (Xen. *Hell.* 3.4.13–14, 3.4.23, 5.3.5, 5.4.40, 6.5.30–31, 7.5.15–17).

The point is illustrated by the cavalry force Agesilaos created in response to the strategic limitations he faced in Anatolia. After their successful battle against the Persians, this unit continued to serve as the vanguard and rearguard of his army during his march home through Thessaly in 394, where his men were soon shadowed by a large number of hostile Thessalian cavalry. His hoplites were once again forced to march in a hollow square. Xenophon's account of the engagement that followed is instructive of the roles that horsemen were expected to play, both in managing enemy movement and in protecting armies against attempts to do so. The Thessalians, νομίσαντες οὐκ ἐν καλῷ εἶναι πρὸς τοὺς ὁπλίτας ἱππομαχεῖν ('thinking that it would not be a good thing to fight on horseback against hoplites'), did not press the attack, but slowly retired. Agesilaos' rearguard pursued, but cautiously. At this point, 'Agesilaos, recognising what both sides were doing wrong' (γνοὺς δὲ ὁ Ἀγησίλαος ἃ ἑκάτεροι ἡμάρτανον), ordered his personal bodyguard to lead by example and charge straight at the enemy (Xen. *Hell.* 4.3.5–6 = *Ag.* 2.3). What both sides were doing wrong, clearly, was that they were not being aggressive enough. The Thessalians *should* have fought on horseback against hoplites, harassing them and driving them out of their territory; Agesilaos' rearguard *should* have responded with an immediate and sustained attack. If the enemy had been better led, they would have presented an existential threat to the hoplite army – and one that only well-led cavalry could have kept at bay. As we have seen, when the Thebans tried to campaign in Thessaly three decades later, the Thessalian cavalry under Alexander of Pherai hounded them and picked them off until they 'gave up hope of salvation' (τὴν σωτηρίαν ἀπογινωσκόντων), when

36 Spence (1993) 141–45; Sidnell (2006) 50–52. The principles of their placement in a marching column are discussed by Xenophon (*An.* 7.3.37–38).

37 Hdt. 9.22.1; Thuc. 6.22; Arist. *Pol.* 1321a.19–20; Arr. *Anab.* 4.4.5. On light infantry see Kucewicz, Lloyd and Konijnendijk (this volume).

suddenly Epameinondas, elected to the generalship on the spot, gathered the light infantry and the cavalry for a rearguard action that saved the army (Diod. Sic. 15.71.5–6).

2.5 *Controlling Open Ground*

The scene of cavalry 'herding' at Corinth demonstrates how the mobility of horsemen gave them a distinct advantage over all other troop types in open terrain. Modern scholars often assert that hoplites were 'soldiers of the open plain', who habitually, or even exclusively, fought on level ground.[38] But our ancient sources readily acknowledge that it was the horseman – not the hoplite – who ruled the plain. Open ground allowed cavalry forces to roam where they pleased, to survey and control the enemy's movements, to attack from any direction, and to escape from any undesired engagement. By consequence, far from seeking to fight on level terrain, hoplites typically tried to avoid such terrain if they knew enemy horsemen were present.

The correlation between plains and cavalry dominance is clear in Herodotos, who has the Persians choose the flat ground both at Marathon and at Plataia to suit their mounted troops.[39] It is also clear in Thucydides, who describes how the Athenians were paralysed by the presence of Thessalian and Syracusan horsemen operating in their own territory. But it is most abundantly clear in the works of Xenophon. When the Ten Thousand, on their retreat from the battlefield near Babylon, reached the first rugged terrain north of the old Assyrian heartland, εἶδον μὲν τοὺς λόφους ἄσμενοι οἱ Ἕλληνες, ὡς εἰκὸς τῶν πολεμίων ὄντων ἱππέων ('the Greeks were glad to see hills, which made sense, since the enemy was cavalry', Xen. *An.* 3.4.24). During their campaigns in Anatolia, successive Spartan commanders recognised the futility of even marching into the plain while the Persians had more and better horsemen (Xen. *Hell.* 3.1.5, 3.4.15). In his account of Leuktra, Xenophon states simply that the Spartans decided to place their cavalry in front of the phalanx 'since the space between the lines was a plain' (*Hell.* 6.4.10). The implied reasoning, too obvious for Xenophon to spell out, was that horsemen led the way on ground that favoured them. Even Diodoros took this reasoning for granted; he states that at Krannon in 322 the Greeks decided to rely on their cavalry to win 'because the men were brave and the ground was level' (Diod. Sic. 18.17.2). To the Greeks, the superiority of horsemen on open ground was proverbial. In one of Plato's dialogues, a

---

38 The phrase is from Ober (1991) 173; for this view in general, see Grundy (1911) 244; Gomme (1950) 10; Snodgrass (1980) 130; Osborne (1987) 13, 144; Connor (1988) 12, 25; Hanson (2000) 206–208; Lee (2006) 483; Wheeler (2008) 202.

39 Hdt. 6.102, 9.13.3; Krentz (2010) 14–15, 103, 143.

character jokes that 'it is challenging "cavalry in the plain" to challenge Sokrates to an argument' ("Ἱππέας εἰς πεδίον' προκαλῇ Σωκράτη εἰς λόγους προκαλούμενος, Pl. *Tht.* 183d).

The open plain, then, was not the hoplite's comfort zone; on the contrary, on plains where cavalry was present, hoplites were totally outmatched.[40] Compelled to form into a hollow square or retreat to a fortified camp, they could neither ravage nor fight effectively. If they were caught in the open, they risked death. When the Spartans landed at Phaleron in 512, they unwittingly marched into a prepared killing zone, cleared of all trees and obstacles; Athens' allies, a thousand Thessalian horsemen, swooped down on them and destroyed them (Hdt. 5.63.3–4; Arist. [*Ath. Pol.*] 19.5). In 480, when Mardonios pretended to send his cavalry against the recently subjected Phokians to test their resolve, the Phokians abandoned hope and prepared to die where they stood (Hdt. 9.17–19). When the Athenian hoplites sent to Spartolos in 429 were deprived of their cavalry and light infantry support, the Chalkidian horsemen and light troops easily routed them, killing hundreds (Thuc. 2.79.3–7).

The easiest way to avert such disasters was to avoid level ground. We have seen how the Greeks at Plataia planned their retreat through the most rugged terrain to steer clear of the Persian cavalry, and how Thibron's solution to Persian cavalry supremacy in Anatolia was to stay out of the plain altogether. Greeks habitually encamped on hills or behind field fortifications,[41] and often formed up for battle there.[42] Diodoros tells us that even the vaunted Macedonian pikemen, terrified by the Thessalian cavalry at Lamia, 'immediately withdrew from the plain to the rough high ground, to gain safety for themselves' (Diod. Sic. 18.15.4). When Antipater arrived with reinforcements, he still decided to keep to the hills, 'since the enemy were superior in cavalry' (18.15.6).

If entering open ground was inevitable, at least the location could still be chosen with care. The finest example of this is the first battle outside Syracuse: the Athenians 'quietly placed their army in suitable ground, where they could start a battle when they pleased, and where the Syracusan cavalry would have least opportunity to bother them either before or during the action' (Thuc. 6.66.1). The ground in question was a narrow strip of land with rocky cliffs on one side and a swamp and some houses on the other. Only in such restrictive conditions could a hoplite army risk battle against an enemy superior in cavalry.

---

40 As noted already in Kromayer and Veith (1928) 90–92.

41 Xen. *Lac.* 12.1–3; Isoc. 6.74; Polyb. 6.42.2; for historical examples of army camps on hills, see Krentz (2002) 27–28; Echeverría (2011) 49 n.10. Compare Lee (2020) 43–46.

42 Thuc. 5.65.1; Xen. *Hell.* 2.4.10–11, 4.4.9, 5.4.50, 7.5.8, 7.5.11; Diod. 15.32.3–6.

2.6 *Winning Pitched Battles*

The one thing that Greek cavalry could not do, according to most modern scholars, was decide a pitched battle between hoplite phalanxes.[43] In the orthodox interpretation of the character of Greek warfare, this was a critical shortcoming. The assumed inability of horsemen to break heavy infantry formations through direct attack rendered them irrelevant in a way of war that supposedly revolved entirely around open battle. Cavalrymen, reduced to indirect methods by the untouchable wall of spears that a homogenous phalanx presented, became bystanders in the hoplite battles that defined the age.[44]

However, even if this were an accurate characterisation of Greek warfare,[45] the related assessment of the capabilities of cavalry in open battle is incorrect. While only a few authors have been prepared to argue that the head-on shock charge actually *was* part of the arsenal of Classical horsemen,[46] many have pointed out that it did not matter if they could not demolish heavy infantry by frontal assault, since their mobility allowed them to approach the vulnerable flanks or rear of a phalanx instead.[47] It is often unclear from the surviving accounts exactly how they did it, but from the late Archaic period onward, horsemen regularly played a decisive tactical role in 'hoplite' battles.

43 See above, n.1. The more recent experts on mounted warfare are notable exceptions to the rule: see Worley (1994) 171–72; Gaebel (2002) 63–65; Sidnell (2006) 73–74; Sears and Willekes (2016).

44 Martin (1887) 432–33; Greenhalgh (1973) 146–47; Worley (1994) 49–51, 58, 170; Hanson (1995) 234; Rawlings (2007) 86–88; Bugh (2020) 67.

45 Recent studies have shown that it is not: see especially Krentz (1997, 2000); Rawlings (2000), (2007) 66–73; Sheldon (2012); Echeverría (this volume).

46 Kromayer and Veith (1928) 92–93; Worley (1994) 49–50, 123. Spence seems to have wavered on this point: compare (1993) 103–109 and 156–57. On the one hand, there is no doubt that the cavalry of this period was capable of shock action. The Persians nearly broke the Greeks at Plataia by charging into them (Hdt. 9.23); Thucydides praises the Macedonian cavalry for their skill in the charge (2.100.5); Xenophon's complaint about the ease with which Greek horsemen's lances snapped in battle (*Hell.* 3.4.14; *Eq.* 12.12) reveals that many of them were armed with lances rather than javelins, and thus specialised in close combat. Xenophon elsewhere even recommends the use of lances as a display weapon by the Athenian cavalry (*Eq. mag.* 3.2–3). On the other hand, it is important not to overstate the case – even if mounted shock tactics *could* be used against a formed phalanx (as argued in Sears and Willekes 2016, 1029–31, 1033–34) it is difficult to show conclusively that they *were*.

47 Frederiksen (1968) 9; Rahe (1980) 82–83; Spence (1993) 155–56; Gaebel (2002) 302–303; Sidnell (2006) 41–50. Anderson (1961, 152–54) argued that even the famous Macedonian Companion cavalry did not charge enemy heavy infantry head-on, leaving little distinction between Greek and Macedonian uses of horsemen. By contrast, Kromayer saw no evidence for mounted flank attacks in the sources for the Classical period, and therefore assumed Greek horsemen always attacked head-on: Kromayer and Veith (1928) 93.

At Malene in 493 – the final stand of the last remnants of the Ionian Revolt – the outcome hung in the balance 'until the Persian cavalry charged and fell upon the Greeks'. Herodotos forestalls any doubts by repeating that τό τε δὴ ἔργον τῆς ἵππου τοῦτο ἐγένετο ('this was the achievement of the cavalry', 6.29.1). Similarly, the bitter fight between the Athenians and the Corinthians at Solygeia in 425 was finally decided when 'the Athenians, who were helped by their cavalry fighting with them, while the enemy had none, routed the Corinthians' (Thuc. 4.44.1).[48] At Delion the following year, the Theban commander Pagondas sent a few hundred horsemen round behind a hill to support his crumbling left wing; the Athenians, thinking a whole fresh army was bearing down on them, broke and fled (Thuc. 4.96.5). Thanks to their strong position on a hilltop, the hoplites of the forsaken Athenian right wing at Amphipolis were able to repulse several infantry charges, but they collapsed when subjected to a storm of missiles from the enemy's horsemen and peltasts (Thuc. 5.10.9). During Gylippos' second battle as the commander of the Syracusans, he posted his cavalry at the end of his line, where they charged and shattered the Athenian left wing, 'and the rest of the army was, therefore, also defeated by the Syracusans' (Thuc. 7.6.2–3). At Kynoskephalai in 364, after Pelopidas had been killed, his hoplites pinned the enemy down until οἵ τε ἱππεῖς προσελάσαντες ὅλην ἐτρέψαντο τὴν φάλαγγα ('the cavalry rode up and routed the entire phalanx', Plut. *Pel.* 32.7), resulting in the bloody pursuit cited above. When the Boiotian cavalry had chased off the horsemen of their opponents at Second Mantineia in 362, they turned their attention to the Athenians on the left wing of the enemy phalanx; the hard-pressed hoplites were about to break, but in the nick of time a unit of allied Eleian cavalry charged in to save them (Diod. Sic. 15.85.7).

It is true that there were some major battles, like First Mantineia, Nemea, and Koroneia, in which cavalry apparently had no part to play until after the matter was decided. However, in each of these cases, our sources dutifully report the presence of significant numbers of horsemen on both sides. Modern authors have sometimes struggled to explain the failure of these horsemen to act.[49] Indeed, the list of battles cited above raises the question: if cavalry attacks against hoplites were so effective, why did cavalry not *always* decide battles in which they were involved?

One common explanation is that restrictive terrain at times prevented cavalry manoeuvre. If such terrain shaped the field of battle, it was probably not by accident. The Athenians' careful choice of ground for the first battle at

48 Lee (2006) 493.

49 Delbrück (1908) 59, 62, 99–100, 145, 156, 158; Anderson (1970) 213–16; Sidnell (2006) 54.

Syracuse shows that the selection of a 'cavalry-proof' battlefield could be very deliberate. We have seen that this was also a factor in the Greek decision to fight the Persians at Thermopylai (Hdt. 7.177). The people of Kyme intentionally fought the Etruscans on a strip of land between mountains and marshes, where the numerous Etruscan cavalry ἀδύνατοι δ' ὄντες κυκλώσασθαι τοὺς πολεμίους ('were unable to surround their enemies', Dion. Hal. *Ant. Rom.* 7.4.1–2). After the first clash at Kounaxa, the Greeks tried to redeploy with their backs to the Euphrates so they could not be outflanked by the Persian horsemen (Xen. *An.* 1.10.9). When their cavalry support was routed at the battle of Lamia in 322, the Macedonian pikemen withdrew uphill, where their mounted Thessalian enemies could not harm them (Diod. Sic. 18.15.4). Such methods were acknowledged in later military theory. Faced with an enemy superior in cavalry, the Roman-era tactical author Onasander recommends χωρία τραχέα καὶ στενὰ καὶ παρ' ὄρη, ἃ ἥκιστα ἱππάσιμα ('rough and narrow ground near mountains, which is least suitable for horses', *Strat.* 31). A battlefield delineated by cliffs, rivers, walls, or broken ground allowed hoplites to operate with less concern for their usual vulnerabilities.[50]

In cases where cavalry was more at liberty to move, but apparently did not, other factors must have been at play. One explanation for cavalry inactivity is that they were instructed not to engage. The crucial role they played in the immediate aftermath of battle – regardless of whether the phalanx won or lost – means they may have been worth more to an army when held in reserve up to that moment than they would have been if they tried to influence the clash itself. The cavalry of the Boiotians at Plataia and of the Athenians at First Mantineia suggests this practice, with neither making an appearance during the encounter of the battle lines, but both playing a crucial role in covering the retreat (Hdt. 9.69.2; Thuc. 5.73.1). In Xenophon's account of the battle of Koroneia in 394, Agesilaos' vaunted cavalry is conspicuously absent, until they ride up to their commander after the victory to report that some of the fleeing enemy have taken refuge in a nearby sanctuary (Xen. *Hell.* 4.3.20 = *Ag.* 2.13). It is obvious that these men, who played no known part in the battle itself, were actively engaged afterwards in the chase. The most explicit example of cavalry being held back for the final phase of battle is at Gela in 405, where Dionysios gave orders to his horsemen that 'if they should see their comrades winning, they should join in the fighting, but in case they were losing, they should receive any who were in distress' (Diod. Sic. 13.109.5).

Yet the clearest explanation we get from the sources for the failure of cavalry to play a decisive role is that they were prevented by deliberate tactical

50 Rahe (1980) 87; Hanson (2000) 209–10; Gaebel (2002) 70.

countermeasures. When Potidaia revolted from Athens in 432, the Potidaians planned to defeat the Athenians in battle by having the cavalry of their Chalkidian and Macedonian allies charge into the rear of the Athenian phalanx once it was engaged. However, the Athenians anticipated the move, and sent 600 Macedonian cavalry of their own to block the attack (Thuc. 1.62.3–4).[51] At Delion, the Athenians themselves held back 300 horsemen 'to watch for a chance to attack the Boiotians during the battle', but the Boiotians 'placed men to deal with them' (πρὸς τούτους ἀντικατέστησαν τοὺς ἀμυνουμένους), and it was the remainder of *their* cavalry that ended up deciding the battle (Thuc. 4.93.2–3). At Second Mantineia, the Thebans stationed 'horsemen and hoplites' in the hills on which their right flank rested, providing a deterrent against any encircling moves; these troops fought and defeated the Athenian cavalry stationed over against them (Xen. *Hell.* 7.5.24–25). Yet, as noted above, when these Theban flank guards themselves tried to attack their opponents' left wing, the enemy's Eleian cavalry reserve responded quickly and decisively (Diod. Sic. 15.85.4–7).

In pitched battle, then, the best way for a hoplite force to protect itself against enemy cavalry was to field cavalry of its own – just as it was on the march, while ravaging, and while withdrawing from battle. Concerns over relative cavalry strength at Syracuse and Haliartos show that the function of horsemen in battle was to engage their counterparts in the enemy army in order to prevent them from interfering with the clash of the hoplites. If only one side had cavalry, like the Athenians at Solygeia, their advantage was clear. If one side's horsemen were not up to the task of neutralising their opponents, like the weak Spartan cavalry at Leuktra (Xen. *Hell.* 6.4.10–11) or that of Alexander of Pherai at Kynoskephalai (Plut. *Pel.* 32.2), the result could be disastrous. These realities of Greek battle confirm Onasander's belief (*Strat.* 16) that cavalry actually offered far fewer tactical options than a commander would like: 'the general will not deploy his cavalry as he wishes, but as he is compelled; for he will oppose them to the horsemen of the enemy'.

In fact, the success of cavalry in this role – cancelling out its opposite number – may be another explanation for their occasional absence from the battle descriptions we find in the sources. It is often claimed that such descriptions were shaped by their authors' ideological fixation on hoplites,[52] but the fact that the numbers of horsemen and light troops are usually listed at the

51 The Spartans made a similar arrangement with the Boiotian cavalry at Argos in 418: Thuc. 5.58.4, 5.59.5.

52 Spence (1993) 165–69; Van Wees (1995) 162; Pritchard (1998) 44–52; Hanson (2000) 202, 219–20; Cawkwell (2005) 250.

outset makes this difficult to believe. The many accounts cited here of battles decided by cavalry action suggest that the 'hoplitic' paradigm had little influence on the way battle accounts were written. If we assume instead that the Greek historians focused primarily on the forces that *decided the outcome*, it becomes easier to understand why cavalry action was sometimes left out. Where the horsemen of both sides held each other in check, the decision was left to the hoplites, whose actions were consequently described in more detail. The same goes for engagements in which mounted troops were held back on purpose. However, where armies were able to use cavalry to tip the scales in their favour, their actions were duly recorded, and credit was given to the decisive force.

In sum, Classical Greek cavalry played a number of crucial, interlocking roles that largely defined Greek military practice at both the tactical and the strategic level. Their presence was one of the most important factors that determined where an army could move, where it could fight, how much it could achieve in victory, and what its chances of survival were in defeat. They consequently played a significant part in most major battles and wars of the Classical period. They may have occasionally failed to influence particular engagements or campaigns, but this tended to happen only if they were evenly matched, poorly led, sidelined by the terrain, or simply absent. Only under such favourable conditions – often deliberately and carefully secured – could hoplites be said to dominate the field.

3 Greek Warfare as Cavalry Warfare

If this characterisation is right, why have scholars argued for centuries that cavalry played only a secondary, supporting role in the Greek way of war? As I noted in the introduction, none of the evidence I present here is new. Much of it is among the most widely cited literary evidence for Greek military methods. Yet works on this subject have typically given a dismissive assessment of the importance of horsemen.

Part of the answer no doubt lies in an excessive focus on precisely those major engagements – Marathon, First Mantineia, the Nemea, Koroneia, Leuktra – in which cavalry did not have a decisive role. For the reasons given above, surviving ancient accounts do not always allow us to say with certainty why cavalry failed to play a bigger part, leaving modern authors free to argue that horsemen must have been typically or even *deliberately* marginal to Greek pitched battles. But several scholars have recently stressed that such battles actually took a great variety of forms, and that the notion of a 'typical' hoplite battle is 'illusory'.[53] The many engagements in which the presence of cavalry *was* decisive need to be included in any analysis – and all pitched battles need to be seen in a wider context of military operations in which horsemen were often of paramount importance.

A number of other points have been raised. It has been claimed that the geography of Greece restricted the use of cavalry,[54] and, controversially, that their lack of saddles and stirrups would have made Greek horsemen ineffective in combat.[55] Neither argument seems to hold up in light of the passages cited above; there was clearly a wide range of situations in which horsemen could be used to devastating effect, even against heavy infantry. They may not have been able to harm hoplites and light troops in broken ground, but the reverse was also true: infantry could not hope to defeat cavalry in the plain.[56] It could just as easily be argued that the geography of Greece, with its many small but strategically vital plains, restricted the use of *hoplites*. Meanwhile, the ubiquitous presence of native Greek cavalry in surviving accounts of campaigns and battles suggests that moral objections to mounted combat and cavalry service were not particularly influential, as some authors have suggested – except perhaps where there was thought to be a connection between horsemen and oligarchic tendencies.[57]

Perhaps the most compelling argument for a modest appraisal of the role of cavalry is that Greek mounted forces were typically too small to make a difference.[58] In relative terms, horsemen rarely seem to have made up more than a small minority of the men in Greek armies; the common ratio of heavy infantry to cavalry was around 10:1.[59] In absolute terms, only Thessaly and Macedon were capable of fielding mounted forces numbering in the thousands. The horsemen of other regions simply did not have the numerical strength to take on the many thousands of hoplites that coalitions of Greek states regularly put to the field. The result was a way of war in which cavalry may have been present, but infantry ruled supreme.

---

53 Wheeler (1991) 125–26; see also Lendon (2005) 81–83; Rawlings (2007) 81–91; Hanson (2013) 267.

54 Adcock (1957) 48; *GWMR* 66; Lee (2006) 491; Krentz (2010) 61.

55 Martin (1887) 432–33; Adcock (1957) 49–50; Hanson (1995) 234; Hutchinson (2000) 102; *GWMR* 66; Hunt (2008) 119; Crowley (2012) 101 n.266. The argument is addressed and qualified by Spence (1993, 40–49) and dismissed altogether by Sidnell (2006, 312–18).

56 Krentz (2010 139, 143) makes this point specifically with regard to the Persian cavalry at Marathon. The only definite exception to the rule is the battle of Eleusis in 510 (Hdt. 5.64.2).

57 Spence (1993) 164–229; Low (2002); *GWMR* 67; Crowley (2012) 102–103.

58 Adcock (1957) 48; Bugh (1988) 27; Spence (1993) 97–102; *GWMR* 66.

59 The ratios are tabulated in Spence (1993) 97–102; Blaineau (2015) 213.

There are two points to be made against this argument. The first is that small units of cavalry could apparently have a disproportionate impact. Most of the examples cited above involved forces of just a few hundred horsemen. Indeed, it is difficult to overstate the relative importance of units as small as the seventy horsemen of Thrasyboulos' pro-democratic insurgents, or the sixty of Phleious, or the fifty Syracusan cavalry at Corinth, or the forty horsemen of the Ten Thousand. The impression we get from the sources is that judicious use and aggressive tactics counted for more than raw numbers in mounted warfare.

But the second point is that we may wonder whether Greek cavalry forces were indeed small. It is often noted that raising horses for military use was prohibitively expensive; throughout the Classical period, it remained the privilege of the leisure class.[60] It goes without saying that the comparatively tiny city-states and federations of Classical Greece could not have raised mounted forces on the same scale as the large territorial states that dominated the centuries that followed.[61] For many Greek communities, fielding even a few dozen horsemen likely stretched their cavalry potential to its absolute maximum. The large subsidies paid to the Athenian cavalry reveal that their corps of 1,000 horsemen, supplemented by a further 200 horse archers, went far beyond what the socio-economic hierarchy of the city-state was willing to muster without support. Bugh suggested that their earlier force of 300 horsemen was probably closer to what Athens' horse-owning class would have fielded by itself.[62] In that sense, the cavalry forces of Greek states could only be modest, both in absolute terms and relative to the number of available hoplites.

However, the Periklean state funding of the Athenian cavalry is just one prominent example of the apparently general concern of Greek states to raise the largest and best possible mounted units from their relatively small pools of wealthy men. Agesilaos bent the rules of military service to raise as

60 Grundy (1911) 278–79; Anderson (1961) 132–38; Bugh (1988) 37–38; Lazenby (1989) 77; Spence (1993) xix; Low (2002) 102–103; *GWMR* 66; Steiner (2005) 408; Lee (2006) 491; Sidnell (2006) 319–20; Krentz (2010) 61; Toalster (2011) 9, 71; Blaineau (2015) 220–22, 289–90. Hanson (1995, 235, 249) went further than this, arguing that the rise of intensive agriculture on small farms totally supplanted horse rearing on large estates, which polis citizens came to see as a waste of fertile land.

61 A few extreme examples illustrate the point. At Ipsos in 301, both Antigonos and Seleukos fielded in excess of 10,000 horsemen (Diod. Sic. 20.113.4; Plut. *Demetr.* 28.3). Hannibal used 10,000 cavalry during his campaign in Italy (Polyb. 3.72.9, 3.114.5). Antiochos III gathered more than 12,000 mounted troops for the battle of Magnesia (Livy 37.37.9). No cavalry from the Classical period ever approached a five-figure total strength.

62 Bugh (1988) 63, 70. On the social and financial context of the Athenian cavalry, see also Worley (1994) 70–74; Spence (2010); Blaineau (2015) 203–58; Pritchard (2018) 412–15.

large a force of cavalry as he could from the Greeks of Anatolia (Xen. *Ag.* 1.23–24). Iason of Pherai reassessed the obligations of the cities under his rule to increase the already unrivalled number of horsemen at his disposal (Xen. *Hell.* 6.1.19). Xenophon complained that the Spartan method of recruiting horsemen yielded worthless cavalry (*Hell.* 6.4.10–11) and Aristotle argued that the Spartans did themselves a disservice by letting their constitution restrict their potential levy of hoplites and horsemen (*Pol.* 1270a.29–30). Even after the disasters of the Peloponnesian War halved the number of adult male Athenian citizens, the city strived to maintain the cavalry corps at its pre-war strength of 1,000 (Xen. *Eq. Mag.* 9.3). This amounted to a notable enlargement of the cavalry in relative terms; it would have required a sustained effort to persuade young citizens of hoplite status to serve as horsemen instead of taking their place in the phalanx. These young men were apparently thought to be of greater value to their city as horsemen than as hoplites. Indeed, when it became clear that the full strength of the corps could not be maintained from the citizen body, Xenophon urged the Athenians to recruit metics to cover the shortfall – while simultaneously arguing that metics should be excluded from hoplite service (*Eq. mag.* 9.6; *Vect.* 2.2–5). He also advised the acquisition of cavalry mercenaries to increase both the size and the quality of the cavalry, noting that this measure had already proved its effectiveness in Sparta and 'in the other poleis', all of which held their hired horsemen in high regard (*Eq. mag.* 9.3–4).[63] The fighting ability of the Athenian cavalry was another matter of serious concern; both Xenophon (*Eq. mag.* 3.2–14) and the author of the *Athenaion Politeia* (49.1–2) describe the procedures put in place to ensure the quality of riders and their mounts. Plato insisted that the citizens of his ideal state – both male and female – would learn to ride and fight on horseback from age six (*Leg.* 794c, 813d–e). All this reveals a widespread interest in realising each community's full cavalry potential.[64]

The Athenian example shows how this interest was reflected in the numbers of horsemen actually put to the field. Even with its artificially inflated cavalry corps, the ratio of heavy infantry to cavalry in Athens' full citizen levy

---

63 Cavalry mercenaries seem to have been available in substantial numbers; Demosthenes (18.237) boasted that he had gathered a mercenary army of 15,000 infantry and 2,000 cavalry (a ratio of 7.5:1).

64 There may be further evidence of an Athenian desire to mobilise every available horseman in the existence of a unit of cavalry raised on the island of Lemnos, which was commanded by one of the *hipparchoi* elected at Athens (Dem. 4.27; [Arist.] *Ath. Pol.* 61.6). On the effort and expense required to maintain the Athenian cavalry, see now Pritchard (2018) 408–12.

at the beginning of the Peloponnesian War was nearly 24:1.[65] Since horsemen were recruited exclusively from citizens in their twenties or early thirties,[66] every single one of them qualified for service in the field army, but the ratio of citizen hoplites to cavalry in that subcategory remained 13:1 (Thuc. 2.13.6–8). The common 10:1 ratio found in actual Athenian expeditionary forces therefore constitutes a significant overrepresentation of the cavalry component of the levy. Disregarding light troops, the mounted 4.2% of the militia (or 7.7% of the field army) – extracted at great expense from a population that would have yielded significantly fewer horsemen without state support – was frequently called upon to provide 9.1% of a given army.[67] If the discrepancy between the cavalry share in the levy and the typical expeditionary force was less pronounced in the fourth century, it was only because the Athenians did their best to retain the full corps of 1,000 horsemen alongside a much diminished pool of citizen hoplites.

Yet their investments did not raise the Athenians to cavalry supremacy in mainland Greece. The cavalry subsidies merely allowed their mounted corps to compete on a more equal footing with the Boiotians, whose federal levy had a paper strength of 11,000 hoplites and 1,100 cavalry (*Hell. Oxy.* 16.4). On campaign, the Boiotians – like the Athenians – could exceed the infantry-to-cavalry ratio of this full-strength army: they brought 5,000 hoplites and 800 horsemen to the Nemea, which would have required the mounted 9.1% of their militia to supply 13.8% of the expeditionary force (Xen. *Hell.* 4.2.17).[68]

Horsemen were similarly well-represented elsewhere. As early as the 520s, Kyme in Southern Italy was allegedly able to raise 1,800 cavalry against their regional enemies, at an infantry-to-cavalry ratio of 7.5:1 (Dion. Hal. *Ant. Rom.* 7.3.4). Gelon, tyrant of Gela and Syracuse, promised to send 20,000 hoplites

65 Following Van Wees (2004, 241), there were about 23,700 citizen hoplites and 1,000 cavalry. If we add metics and horse archers, the totals are 29,000 hoplites and 1,200 cavalry, for a ratio of 24.2:1.

66 Bugh (1988) 52–53, 64–65.

67 For example, at Spartolos in 429 (2,000 hoplites and 200 cavalry: Thuc. 2.79.1), at Nemea in 394 (6,000 hoplites and 600 cavalry: Xen. *Hell.* 4.2.17), and in Boiotia in 379/8 (5,000 hoplites and 500 cavalry: Diod. Sic. 15.26.2). A few armies from the first years of the Peloponnesian War more closely reflect the 13:1 ratio of the field army (such as the 4,000 hoplites and 300 cavalry sent to Potidaia: Thuc. 2.58.1) but after this, the cavalry contingent consistently forms a larger part of every Athenian army, with a record ratio of 2.5:1 at Megara in 409 (1,000 infantry and 400 horsemen: Diod. Sic. 13.65.1–2). See Blaineau (2015) 213.

68 Around 353, they sent a force of 4,000 hoplites and 500 cavalry to support Megalopolis against Sparta (Diod. Sic. 16.39.2). Again, this would have drawn on 9.1% of the militia to furnish 11.1% of the army.

and 2,000 horsemen against the Persians in 480 (Hdt. 7.158.4).[69] Diodoros (11.21.1) puts the full strength of his field army at 50,000 infantry and 5,000 cavalry, maintaining the 10:1 ratio. The city of Syracuse mustered more than 1,200 mounted troops by itself in the late fifth century (Thuc. 6.67.2), and while the size of the Syracusan hoplite levy is unknown, it is unlikely to have reached a total of ten hoplites for every one of these horsemen. For a joint campaign against Dionysios of Syracuse about 398, Rhegion and Messene gathered 10,000 infantry and 1,000 horsemen between them (Diod. Sic. 14.40.3–4). When they were attacked by the Leukanians in 390, the Thourians were able to field 14,000 infantry and 1,000 cavalry for their defence (Diod. Sic. 14.101.2) – a ratio comparable to that of the Athenian field army in 431. While our sources claim widely disparate numbers for the hoplite levy of the federation that formed around Olynthos in the 380s, they agree that it could raise 1,000 cavalry (Xen. *Hell.* 5.2.14; Dem. 19.266). At the time of Sparta's supremacy in the early fourth century, the Spartan cavalry numbered perhaps 720 at full strength;[70] Spartan *morai* appear to have consisted of hoplites and cavalry at a ratio of 5:1. However, as noted above, Aristotle regarded the size of Sparta's levy as evidence of the failure of its constitution. He believed that its territory could have supported as many as 1,500 horsemen and 30,000 hoplites if it had not fallen into the hands of a few (*Pol.* 1270a.29–30). At a ratio of 20:1, the resulting total levy would notably have featured more cavalry than the full Athenian militia described by Thucydides. Some of the most impressive numbers, however, are found with the Thessalians, who sent 1,000 mounted troops to prop up the tyranny of Hippias in 512 (Hdt. 5.63.3). They could muster 20,000 hoplites and 8,000 horsemen under Iason in the late 370s–a ratio of 2.5:1 (Xen. *Hell.* 6.1.19).[71]

69 If the *hippodromoi psiloi* mentioned by Herodotos are also cavalry, the proposed expeditionary source would include as many as 4,000 mounted troops. For discussion of the passage, see Hall (this volume).

70 Lazenby (1985) 12; Spence (1993) 2–3.

71 The number of horsemen may be exaggerated, but if we choose to favour Iason's own earlier claim of a total of 6,000 (Xen. *Hell.* 6.1.8), or even the much lower figure of 3,000 given by Isokrates (8.118), the Thessalian cavalry remains the largest in the Greek world. If 1,800 Thessalian horsemen went to Asia with Alexander (Diod. Sic. 17.17.4) and a further 2,000 were raised for the Lamian War while the others were still in Asia (Diod. Sic. 18.15.2), the total mounted levy in the final third of the fourth century must have been at least 3,800. The only evidence for a comparable mounted force raised from a single Greek region is the 5,000 cavalry of Gelon's Sicilian army reported by Diodoros (11.21.1) and Strabo's claim (6.3.4) that there had once been a time when Taras was regularly able to muster '3,000 horsemen and 1,000 *hipparchoi*'; for the latter, see Bugh (2020) 70, which was unavailable at the time of writing.

To get a sense of how much emphasis the Greeks placed on their cavalry, we may compare their situation to the practices of Republican Rome around the late third century. According to Polybios, the full levy of Romans and Campanians at the time of the Gallic invasion of 225 consisted of about 250,000 infantry and 23,000 horsemen, or a ratio of about 11:1. However, the number of Romans actually in service was 49,200 infantry and 3,100 cavalry – a ratio of nearly 16:1 (Polyb. 2.24.3–17).[72] In the army of 225, then, the mounted 8.4% of the levy supplied only 5.9% of the troops in the field. Contrary to the Greeks, the Romans' enormous supply of manpower would have allowed them to field as many citizen horsemen as they liked – yet the ratios at which they raised infantry and mounted troops did not match those found in Classical Greece. Far from working to increase the number of horsemen at their disposal, they left their available cavalry reserves largely untapped. Indeed, the Romans frequently seem to have recruited legions with more than the standard number of *infantry*, without expanding the *cavalry* detachment at all.[73] The 'Spartan' 5:1 ratio of heavy infantry to cavalry in certain Roman consular armies of the period was achieved only by requiring the allies to provide three times as much cavalry as the Romans themselves (Polyb. 6.20.8–9, 21.9, 26.7). In short, if the absolute numbers of horsemen found in Classical Greek levies and field armies may strike us as relatively modest, they nevertheless represent a significantly greater focus on acquiring and deploying cavalry than that displayed by Republican Rome. Similarly, even if the Hellenistic kingdoms accorded a greater importance to their mounted troops, the sheer scale of their potential levies makes it unlikely that they would have surpassed the intensity of Classical Greek cavalry mobilisation.[74]

The result of the deliberately strong representation of horsemen in Greek recruitment pools was that, especially from the fourth century onward,

---

72 Brunt (1971) 44–49. This reading of Polybios' figures is controversial, but other interpretations only increase the number of infantry per horseman. Pearson (2016, 103–21) examines the passage and its problems in detail.

73 According to Polybios (3.107.10–11) and Livy (22.36.3), both the infantry and cavalry components of each legion could be enlarged in emergencies – the latter from 200 to 300 per legion. However, Polybios assumes elsewhere (6.20.9) that every legion had a cavalry complement of 300 to begin with, which would imply that only the infantry component was subject to increase. The latter seems generally accepted as a rule of thumb for the strength of Republican legions: see Sumner (1970) 67; Pearson (2016) 69.

74 There seems to have been an eventual return to something like a 10:1 ratio of infantry to cavalry in Hellenistic field armies after the outsized mounted forces of Alexander's immediate successors; see Sekunda (2008) 347; Bugh (2020). Even if they operated on a much larger scale, this would, at best, put the cavalry recruitment of the Hellenistic kings on par with those of Classical Greek states.

coalitions of relatively small Greek states fielded cavalry in significant numbers. As noted above, Gelon of Syracuse may have raised as many as 5,000 horsemen from the Greek cities of Sicily to ward off the Carthaginian threat. At the height of his power in the 390s, Dionysios of Syracuse gathered the forces of the same region into an army that included more than 3,000 cavalry (Diod. Sic. 14.47.7, 58.2, 103.2). Against these, the Greeks in Italy mustered 2,000 cavalry of their own (Diod. Sic. 14.103.6). According to Xenophon, the anti-Spartan alliance at the Nemea boasted 1,550 horsemen in total (*Hell.* 4.2.17).[75] At Second Mantineia, the Spartan coalition's 2,000 cavalry faced the Boiotians' 3,000 (Diod. Sic. 15.84.4). The 2,400 Greek horsemen who crossed to Asia with Alexander were almost as numerous as his Macedonian and Thracian cavalry combined (Diod. Sic. 17.17.4; Plut. *Alex.* 15.1). At the battle of Krannon in 322, even after mass desertion from the army, the Greek alliance could field 3,500 mounted troops against the 5,000 of the Macedonians (Diod. Sic. 18.16.5–17.2). It would seem that, far from regarding horsemen as a troop type of secondary importance, the commanders of the Classical Greek world were broadly alike in striving to bring along as many of them as they could. They may well have agreed with the maxim recorded by Polybios (3.117.5), that it was better to have half the infantry of the enemy, but to be superior in cavalry, than to be evenly matched in both.[76]

What remains of the case against cavalry is the simple observation, highlighted at the beginning of this chapter, that the states of the Peloponnese apparently had no horsemen even as late as the final quarter of the fifth century. The Corinthians had none for the battle of Solygeia in 425 (Thuc. 4.44.1); the Argives had none to counter the Boiotian threat in 418 (Thuc. 5.59.3). The Spartans created theirs παρὰ τὸ εἰωθὸς ('against custom', Thuc. 4.55.2) in 425/4 specifically to fight Athenian raiders. This reality is difficult to explain. The entire Peloponnese – Argos and Corinth in particular – had been famous for equestrian culture since the early Archaic period; cavalry units duly popped up all over the peninsula in the early fourth century.[77] Apparently neither cost nor terrain prevented the establishment of mounted forces. Nothing stood in the way of their creation earlier on. Still, no cavalry was raised until the battleground of the interregional wars of the Classical period shifted to the Peloponnese.

75 Diodoros (14.82.10) claims they had just 500, but this seems unlikely, since the alliance included both Boiotia and Athens. Anderson (1970, 144) accepts the figure given by Xenophon.

76 The Thebans successfully leveraged such ratios against the Spartans at Tegyra (Plut. *Pel.* 17.2–4).

77 Spence (1993) 1–9; Blaineau (2015) 92–102.

This shift itself, however, may be the answer to the riddle. As noted above, many of the roles played by horsemen were interrelated. Cavalry was essential specifically as a response to enemy cavalry, whether on the march, in battle, or in pursuit. It would theoretically have been possible for a regionally shared tactical system to exist entirely without horsemen, so long as none of the participant states decided to introduce them. In such a system, there would have been no cavalry to harass a marching column, and therefore no need for cavalry to protect it; there would have been no mounted troops to pursue a fleeing force, and, therefore, no need for horsemen to cover its flight. Without cavalry to threaten a phalanx in battle, there would have been little need to employ cavalry to protect it. All these tasks could have been left to light infantry – saving horse owners the considerable effort of preparing their horses to face armed crowds.[78]

It is easy to see how the Peloponnese in the first half of the fifth century could have sustained such a system. The final victory at Plataia and the withdrawal of the Peloponnesian states from the continuing war against Persia would have discouraged a hard look at the performance of the region's levies against Mardonios' horsemen. Meanwhile, all the major military events that took place on the peninsula itself were internal affairs, pitting Spartans against Arkadians, Argives, and helots. The region suffered no invasions beyond the naval raids orchestrated by Athens during the First Peloponnesian War, which, of course, involved no horsemen. While the Greeks north of the Isthmus had been forced to adapt to the presence of the Thessalians and other horse lords among their regional rivals, the Peloponnese was sheltered both from these and from the Persian campaigns that encouraged the establishment of mounted forces in parts of Central Greece.

The result was an entire region which, even as late as 431, mustered no cavalry of its own. Warfare on the Peloponnese remained the business of light and heavy infantry for a full century longer than elsewhere in the Greek world.[79] We may get a glimpse of the terror which the unfamiliar sight of organised cavalry inspired among Peloponnesians from an epigram for the Megarian dead of the Persian Wars, which lists those who fell 'in nearby Boiotia, who dared to

---

78 Sears and Willekes (2016) 1033–34. See Kucewicz, Lloyd and Konijnendijk (this volume).

79 The notion of a separate 'Peloponnesian' way of war in the early Classical period may find some support in the fact that conventions regarding the ransom of prisoners seem to have existed in the early fifth-century Peloponnese (Hdt. 6.79.1), and nowhere else in the Greek world: Dayton (2005) 76–79. Despite its apocryphal context, there may also be some significance in Pausanias' remark (4.8.12) that the Messenian Wars saw little cavalry action because 'the Peloponnesians were not good horsemen at that time' – not just the Spartans and Messenians.

come to grips with men fighting on horseback' (*IG* 7.53 = *SEG* 13.312). Indeed, the campaign that culminated in the battle of Tanagra in 457 could have meant a rough awakening for the Peloponnesian coalition army, if the Thessalian cavalry had not decided to abandon their Athenian allies and choose the Spartan side (Thuc. 1.107.7).

It may be objected that Sparta would have had plenty of opportunities to see the usefulness of cavalry. In the late sixth century, the Spartans involved themselves in several campaigns outside the Peloponnese that pitted their forces against mounted troops. We learn from an anecdote in Herodotos (6.108.2) that a Spartan army under Kleomenes was in Boiotia in 519, though their purpose there is unknown. In 512, a Spartan invasion force was destroyed by Thessalian cavalry at Phaleron (Hdt. 5.63.3–4); the Spartans faced the same force again at Eleusis in 510. However, during this second encounter, the Spartan army actually managed to drive off the Thessalians, inflicting more than 40 casualties (Hdt. 5.64.2). It is not clear how they achieved this feat, but it may well have silenced any voices calling for a home-grown Spartan cavalry. Existing methods would have appeared sufficient, and the creation of a potentially costly cavalry force would have been regarded as unnecessary. The victory at Plataia would no doubt have reinforced this view.

A further explanation for Sparta's continued inaction on the cavalry front is a matter of policy rather than tactics; it seems that the rise of organised cavalry in the Greek world coincided with Sparta's geopolitical withdrawal into the Peloponnese. The point of Herodotos' tale about the Spartans in Boiotia in 519 was their unwillingness to make an alliance with Plataia, which they judged too far away to protect. Kleomenes' attempts to interfere with Athenian politics in the years 512–507 finally fell apart when both the members of the Peloponnesian League and the other Spartan king, Demaratos, decided that his personal project did not deserve public support (Hdt. 5.75). After the defeat at Thermopylai, the Spartans proved reluctant to send additional forces north of the Isthmus to fight Xerxes, to the anger and frustration of their Athenian allies (Hdt. 7.139.3–4, 8.40.2, 9.7β.1, 9.9.2); after the battle of Plataia, they marched no further than Thebes, and returned home immediately upon its surrender (Hdt. 9.88). At some point in the early 470s, they ventured north to punish the Thessalians who had taken the Persian side, but their king Leotychidas accepted a bribe and aborted the campaign (Hdt. 6.72.1). Twenty years passed before their next foray beyond Megara – a campaign in support of Doris that ended only inadvertently in battle against the Athenians at Tanagra (Thuc. 1.107.2–5). Their king Pleistoanax was easily persuaded to abandon Sparta's only invasion of Attika during the First Peloponnesian War (Thuc. 1.114.2; Plut. *Per.* 22). In short, whatever the Spartans may have known about the effectiveness of cavalry, their foreign policy in the late Archaic and early Classical periods largely isolated them from the wider Greek tactical system within which they might have been prompted to raise their own. For their invasions of Attika during the Archidamian War, the Spartans were able to rely on the horsemen of the Boiotians, Phokians, and Lokrians (Thuc. 2.9.3), again obviating the need for the Peloponnesian state to establish a native mounted force.

If this interpretation is right, then the probable game-changer was the introduction of horse transports at Athens around 430 (Thuc. 2.56.2). The raiding Athenians were now able to bring their cavalry with them. It helped them in their ravaging expeditions around the Peloponnese and tipped the scales in their favour against Corinth at Solygeia. After Sphakteria, when the Spartans recognised that Athenian raiding had become a permanent problem, they finally embraced the notion of a home-grown cavalry corps, to carry out a mobile defence of the sort that the Athenians had demonstrated to them year after year since the start of the war. This decision, however, necessarily shattered the Peloponnesians' fragile tactical equilibrium. If one state had horsemen, all states had to have horsemen. This imperative went beyond the rivalries of peer-polity interaction; armies on foot were simply too vulnerable in a world where cavalry ruled the plains.

This reconstruction of events is necessarily speculative. The theory of a horse-free tactical system also raises the question what interest was being served by the apparent tacit agreement among Peloponnesian states not to seize a temporary advantage by introducing cavalry at some point before the outbreak of the Peloponnesian War. Perhaps the regional nature of major conflicts discouraged new ideas; perhaps the horse owners of the peninsula were at collective pains to avoid the training needed to make horses into war horses and to prevent their precious possessions being put in harm's way. Perhaps these practical considerations reinforced the normative influence of a 'hoplitic', superficially egalitarian way of thinking. Pritchard has shown that the conceptualisation of warfare in terms of the hoplite and his fighting style was deeply rooted in Athenian public discourse, despite the respect and admiration fifth-century Athenians showered on their cavalry.[80] In the absence of mounted warriors to look up to, a similar paradigm may have been even more powerful on the Peloponnese. Alternatively, we might conclude from the example of Athens and from the steady increase in cavalry numbers throughout the Greek world in the fourth century that state organisation and financial resources were the missing elements that prevented the rise of organised

80 Pritchard (1998) 44–52; on cavalry in particular, 50. Note also Low (2002) 108.

cavalry in the fifth-century Peloponnese.[81] However this may be, the theory presented here may throw some light on the peculiar history of cavalry in this region – its strange absence in the fifth century, followed by its sudden ubiquity in the fourth. In doing so, it helps us understand why modern scholars have tended to downplay the importance of cavalry in Classical Greek warfare.

4 Conclusion

Greek warfare was never purely defined by heavy infantry. There were a considerable number of situations in which even the finest hoplites were categorically outclassed. From the moment they marched into enemy territory, their movements and their options were determined by enemy cavalry. They were forced either to cling to broken ground or to dance to the tune of the horsemen. They had to choose their battlefields with care or run the risk of being trampled. Whenever cavalry acted, they were constrained to respond – with only occasional success. The Macedonians did not revolutionise battle tactics with their focus on large numbers of shock cavalry; they merely built on the tradition of cavalry prominence that had long typified the Greek tactical system.

These observations should have serious consequences for the way we characterise the warfare of the Classical Greeks – as well as the economic, social, moral, and political forces that underpinned it. How does the military supremacy of the most exclusive form of combat relate to distribution of wealth and the shifting constitutions of the Greek city-states? To what extent does the evidence we have for a more broad-based 'hoplite ethic' accommodate a bitter reality in which hoplites did not dominate the field? Fundamental to these and other questions should be the awareness that hoplites were not as central or as significant to the Greek way of war as modern historians have typically assumed. Greek tactical thought centred on combinations of troop types, which covered for the weakness of one by the strengths of another. In this system the hoplite had neither the greatest strengths nor the fewest weaknesses. From the appearance of 'true cavalry' in the late sixth century, it frequently turned hoplites into second-class warriors – 'worthless hoplites' who needed to find safety in numbers, terrain, and the protection offered by horsemen of their own.

81 *GWMR* 235–39.

Bibliography

Adcock, F.E. (1957) *The Greek and Macedonian Art of War*. Berkeley, CA.

Anderson, J.K. (1961) *Ancient Greek Horsemanship*. Berkeley, CA.

Anderson, J.K. (1970) *Military Theory and Practice in the Age of Xenophon*. Berkeley, CA.

Bauer, A. (1893) *Die griechischen Kriegsaltertümer*. Munich.

Blaineau, A. (2015) *Le Cheval de Guerre en Grèce Ancienne*. Rennes.

Brice, L.L. (2013) 'The Athenian Expedition to Sicily', in *WCW* 623–41.

Brunt, P.A. (1971) *Italian Manpower 225 B.C.–A.D. 14*. Oxford.

Bugh, G.R. (1988) *The Horsemen of Athens*. Princeton.

Bugh, G.R. (2020) 'Greek Cavalry in the Hellenistic World: Review and Reappraisal', in *NAGRW* 65–80.

Carey, C. and Edwards, M. (eds) (2013) *Marathon – 2500 Years: Proceedings of the Marathon Conference 2010*. London.

Cawkwell, G. (2005) *The Greek Wars: The Failure of Persia*. New York.

Charles, M.B. (2015) 'Achaemenid Elite Cavalry: From Xerxes to Darius III'. *CQ* 65.1: 14–34.

Connor, W.R. (1988) 'Early Greek Land Warfare as Symbolic Expression'. *P&P* 119: 3–29.

Crowley, J. (2012) *The Psychology of the Athenian Hoplite: The Culture of Combat in Classical Athens*. Cambridge.

Dayton, J.C. (2005) *The Athletes of War: An Evaluation of the Agonistic Elements of Greek Warfare*. Toronto.

Droysen, H. (1889) *Heerwesen und Kriegführung der Griechen*. Freiburg.

Eadie, J.W. and Ober, J. (eds) (1985) *The Craft of the Ancient Historian: Essays in Honor of Chester G. Starr*. Lanham, MD.

Echeverría, F. (2011) '*Taktikè Technè*: The Neglected Element in Classical "Hoplite" Battles'. *AncSoc* 41: 45–82.

Echeverría, F. (2012) 'Hoplite and Phalanx in Archaic and Classical Greece: A Reassessment'. *CPh* 107.4: 291–318.

Fagan, G.G. and Trundle, M. (eds) (2010) *New Perspectives on Ancient Warfare*. Leiden.

Figueira, T.J. (2006) 'The Spartan *Hippeis*', in Hodkinson and Powell (2006) 57–84.

Frederiksen, M.W. (1968) 'Campanian Cavalry: A Question of Origins'. *DArch* 2: 3–31.

Gaebel, R.E. (2002) *Cavalry Operations in the Ancient Greek World*. Norman, OK.

Garlan, Y. (1989) *Guerre et Économie en Grèce Ancienne*. Paris.

Gomme, A.W. (1945) *A Historical Commentary on Thucydides*. Vol. 1. Oxford.

Greenhalgh, P.A.L. (1973) *Early Greek Warfare: Horsemen and Chariots in the Homeric and Archaic Ages*. Cambridge.

Grundy, G.B. (1911) *Thucydides and the History of his Age*. London.

Hackett, J. (ed) (1989) *Warfare in the Ancient World*. London.

Hamilton, C.D. and Krentz, P. (eds) (1997) *Polis and Polemos: Essays on Politics, War, and History in Ancient Greece in Honor of Donald Kagan*. Claremont, CA.

Hanson, V.D. (1995) *The Other Greeks: The Family Farm and the Agrarian Roots of Western Civilization*. New York.

Hanson, V.D. (1998) *Warfare and Agriculture in Classical Greece*. 2nd edn. Berkeley, CA.

Hanson, V.D. (1999b) 'The Status of Ancient Military History: Traditional Work, Recent Research, and On-Going Controversies', *Journal of Military History* 63.2: 379–413.

Hanson, V.D. (2000) 'Hoplite Battle as Ancient Greek Warfare: When, Where, and Why?' in *W&V* 201–32.

Hanson, V.D. (2013) 'The Hoplite Narrative', in *MoB* 256–75.

Helbig, W. (1902) *Les ἱππεῖς Athéniens*. Paris.

Hodkinson, S. and Powell, A. (eds) (2006) *Sparta and War*. Swansea.

Hunt, P. (2008) 'Military Forces', in *CHGRW* 1:108–46.

Hunter, J.H. (2005) 'Pericles' Cavalry Strategy'. *QUCC* n.s. 81.3: 101–108.

Hutchinson, G. (2000) *Xenophon and the Art of Command*. London.

Kinzl, K.H. (2006) *A Companion to the Classical Greek World*. Malden, MA.

Konijnendijk, R. (2018) *Classical Greek Tactics: A Cultural History*. Leiden.

Krentz, P. (1997) 'The Strategic Culture of Periclean Athens', in Hamilton and Krentz (1997) 55–72.

Krentz, P. (2000) 'Deception in Archaic and Classical Greek Warfare', in *W&V* 167–200.

Krentz, P. (2002) 'Fighting by the Rules: The Invention of the Hoplite Agôn'. *Hesperia* 71: 23–39.

Krentz, P. (2010) *The Battle of Marathon*. London.

Krentz, P. (2013) 'Marathon and the Development of the Exclusive Hoplite Phalanx', in Carey and Edwards (2013) 35–44.

Kromayer, J. and Veith, G. (1928) *Heerwesen und Kriegführung der Griechen und Römer*. Munich.

Lazenby, J.F. (1985) *The Spartan Army*. Mechanicsburg.

Lazenby, J.F. (1989) 'Hoplite Warfare', in Hackett (1989) 54–81.

Lazenby, J.F. (2004) *The Peloponnesian War: A Military Study*. London.

Lee, J.W.I. (2006) 'Warfare in the Classical Age', in Kinzl (2006) 480–508.

Lee, J.W.I. (2020) 'Daily Life in Classical Greek Armies 500–330 BCE', in *NAGRW* 39–51.

Lendle, O. (1995) *Kommentar zu Xenophons Anabasis (Bücher 1–7)*. Darmstadt.

Lendon, J.E. (2005) *Soldiers and Ghosts: A History of Battle in Classical Antiquity*. London.

Liston, M.A. (2020) 'Skeletal Evidence for the Impact of Battle on Soldiers and Non-Combatants', in *NAGRW* 81–94.

Low, P. (2002) 'Cavalry Identity and Democratic Ideology in Early Fourth-Century Athens'. *PCPhS* n.s. 48: 102–22.

Mann, C. (2013) *Militär und Kriegführung in der Antike*. Munich.

Martin, A. (1887) *Les Cavaliers Athéniens*. Paris.
Nefedkin, A. (2006) 'The Tactical Development of Achaemenid Cavalry'. *Gladius* 26: 5–18.
Ober, J. (1985) 'Thucydides, Pericles, and the Strategy of Defence', in Eadie and Ober (1985) 171–88.
Ober, J. (1991) 'Hoplites and Obstacles', in *HCGBE* 173–96.
Osborne, R. (1987) *Classical Landscape with Figures: The Ancient Greek City and its Countryside*. London.
Osborne, R. (2018) *The Transformation of Athens: Painted Pottery and the Creation of Classical Greece*. Princeton.
Pearson, E. (2016) 'The Development of Army Administration in the Roman Republic'. PhD Thesis. University of Manchester.
Powell, A. (ed) (1995) *The Greek World*. London.
Pritchard, D.M. (1998) '"The Fractured Imaginary": Popular Thinking on Military Matters in Fifth-Century Athens'. *AH* 28.1: 38–61.
Pritchard, D.M. (ed) (2010) *War, Democracy and Culture in Classical Athens*. Cambridge.
Pritchard, D.M. (2018) 'The Horsemen of Classical Athens: Some Considerations on their Recruitment and Social Background'. *Athenaeum* 106.2: 405–19.
Rahe, P.A. (1980) 'The Military Situation in Western Asia on the Eve of Cunaxa'. *AJPh* 101.1: 79–96.
Rawlings, L. (2000) 'Alternative Agonies: Hoplite Martial and Combat Experiences Beyond the Phalanx', in *W&V* 233–59.
Rawlings, L. (2007) *The Ancient Greeks at War*. Manchester.
Rüstow, W. and Köchly H. (1852) *Geschichte des Griechischen Kriegswesens von der Ältesten Zeit bis auf Pyrrhos*. Aarau.
Sage, M.M. (1996) *Warfare in Ancient Greece: A Sourcebook*. London.
Schubert, C. and Laspe, D. (2009) 'Perikles' defensiver Kriegsplan: eine thukydideische Erfindung?' *Historia* 58.4: 373–94.
Sears, M.A. (2019) *Understanding Greek Warfare*. New York.
Sears, M.A. and Willekes, C. (2016) 'Alexander's Cavalry Charge at Chaeronea, 338 BCE'. *Journal of Military History* 80: 1017–35.
Sekunda, N. (2008) 'Military Forces: Land Forces', in *CHGRW* 1: 325–57.
Sheldon, R.M. (2012) *Ambush: Surprise Attack in Ancient Greek Warfare*. Barnsley, UK.
Sidnell, P. (2006) *Warhorse: Cavalry in Ancient Warfare*. London.
Snodgrass, A.M. (1980) *Archaic Greece: The Age of Experiment*. Berkeley, CA.
Spence, I.G. (1990) 'Perikles and the Defence of Attika During the Peloponnesian War'. *JHS* 110: 91–109.
Spence, I.G. (1993) *The Cavalry of Classical Greece*. Oxford.
Spence, I.G. (2010) 'Cavalry, Democracy and Military Thinking in Classical Athens', in Pritchard (2010) 111–38.

Stadter, P.A. (ed) (1973) *The Speeches in Thucydides*. Charlotte, NC.

Stahl, H.-P. (1973) 'Speeches and Course of Events in Books Six and Seven of Thucydides', in Stadter (1973) 60–77.

Steiner, D. (2005) 'For Want of a Horse: Thucydides 6.30–2 and Reversals in the Athenian Civic Ideal'. *CQ* 55.2: 407–22.

Strassler, R.B. (ed) (2009) *The Landmark Xenophon's* Hellenika. New York.

Sumner, G.V. (1970) 'The Legion and the Centuriate Organization'. *JRS* 60: 67–78.

Tarn, W.W. (1930) *Hellenistic Military and Naval Developments*. Cambridge.

Tausend, S. (ed) (2011) *"Böser Krieg": Exzessive Gewalt in der antiken Kriegsführung und Strategien zu deren Vermeidung*. Bayreuth.

Thomas, D. (2009) 'Introduction', in Strassler (2009) ix–lxvi.

Toalster, D. (2011) *Unzeitgemäße Feldherren: Der Hipparch als Prototyp des erfolgreichen Feldherrn in Xenophons Hellenika*. Gutenberg.

Tuplin, C. (2010) 'All the King's Horse', in Fagan and Trundle (2010) 101–82.

Van Wees, H. (1995) 'Politics and the Battlefield: Ideology in Greek Warfare', in Powell (1995) 153–78.

Van Wees, H. (2011) 'Defeat and Destruction: The Ethics of Ancient Greek Warfare', in Tausend (2011) 69–110.

Van Wees, H. (2013a) 'Farmers and Hoplites: Models of Historical Development', in *MoB* 222–55.

Van Wees, H. (2013b) *Ships and Silver, Taxes and Tribute: A Fiscal History of Archaic Athens*. London.

Wardman, A.E. (1959) 'Tactics and the Tradition of the Persian Wars'. *Historia* 8.1: 49–60.

Wheeler, E.L. (1991) 'The General as Hoplite', in *HCGBE* 121–70.

Wheeler, E.L. (2008) 'Land Battles', in *CHGRW* 1:186–223.

Willekes, C. (2016) *The Horse in the Ancient World: From Bucephalus to the Hippodrome*. London.

Worley, L.J. (1994) *Hippeis: The Cavalry of Ancient Greece*. Boulder, CO.

CHAPTER 8

'Not Many Bows'?

Light-Armed Fighters of the Tenth through Fourth Centuries

Cezary Kucewicz, Matthew Lloyd, and Roel Konijnendijk

Light-armed fighters are a nebulous concept throughout much of the period covered by this volume. The concept, which encompasses those who fight with long-range weapons and little armour, emerges after the Early Iron Age, when armour is rare in the archaeological record. The reappearance of body armour in the late eighth century, alongside the double-grip *aspis* and Corinthian helmet, allowed that 'light' and 'heavy' armed fighters are truly distinguished. Yet this distinction remains fluid for several centuries. Archers are certainly among Tyrtaios' light-armed *gymnetes*; the javelin, however, remains the preserve of heavily-armed spearmen until the rise of the javelin-thrower as a separate type of light infantry in the fifth century. Slingers, meanwhile, remain elusive throughout.

In contrast to Archilochos' statement that 'Not many bows indeed will be stretched tight, nor frequent slings be whirled' between the lords of Euboia (Fr. 3 West),[1] archers, slingers, and javelin-throwers can be found in texts, images, and archaeological evidence throughout the Greek world of the tenth through fourth centuries. However, they always remain overshadowed by the evidence for other fighters, the necessity of interpretation, or the outright bias of literary authors. From the Homeric epics onward, there is evidence of anger towards enemies, though not allies, who fought from a distance rather than standing in the front line, which grows into general contempt for the roles of the light-armed in the fifth and fourth centuries. This contempt is also present in some modern scholarship, resulting in the marginalisation of the light-armed in the study of Greek warfare. Nevertheless, they can always be found by those willing to look.

1 Translation by Perrin (1914).

 | DOI:10.1163/9789004501751_009

1 The 'Aristocratic' Archer? The Light-Armed in the Early Iron Age and Homer

In the EIA, there is no apparent distinction between heavy- and light-armed fighters. Swords appear to be the most common close-combat weapon, while spearheads may be javelins – long-range throwing weapons – rather than lances – medium-range thrusting weapons. Distinguishing between lances and javelins in the archaeological record is difficult without a visual repertoire to indicate how they were used; nevertheless, it is apparent that in the later palatial and post-palatial Bronze Age spearheads get smaller, and all of those known from the EIA are small enough to be javelins.[2] On the depictions of 'sailor-warriors' from the twelfth-century Euboian gulf, warriors prepare to throw javelins between boats.[3] In burials, spearheads are sometimes found in pairs or groups of three and often alongside swords, and warriors depicted in Late Geometric (*ca.* 760–700) vase painting frequently carry multiple spears and a sword. As multiple spears are useless in a close-combat situation, this suggests that they were javelins carried to be thrown prior to engaging in close combat.

The bow presents a much more complicated question. Evidence for archery is far from ubiquitous and yet it occurs in significant contexts that suggests it was an important part of the lifestyle of EIA elites. Without literary evidence, investigations into prehistoric archery are guided by a seemly simple, yet in practice frustrating, question: were bows used for hunting or fighting? In the pre-palatial and palatial contexts of the Late Bronze Age, there are extremely few depictions of the bow used in combat, but many of it used in the hunt;[4] thus, the thousands of arrowheads mentioned in Linear B tablets may or may not relate to warfare.[5] In the post-palatial Bronze Age of the twelfth century, archery is invisible in iconography and burials-with-weapons, and remains so until the early first millennium.

Among the dozens of burials with weapons in Early Iron Age central Greece, only a handful contain arrowheads – but this handful is significant. The most notable burial with arrowheads at Lefkandi on Euboia is the so-called

2 Dickinson (1994) 206, (2006) 158; Lloyd (2020) 503–505.

3 See Lloyd, this volume.

4 The sixteenth-century Shaft Grave IV in Grave Circle A at Mycenae revealed a scene of the bow in combat on the silver 'Siege Rhyton', and in the hunt on a bronze dagger with inlaid silver and gold decoration, illustrated in Brouwers (2013) 17. The decoration of the 'Siege Rhyton' also includes a rare depiction of slingers.

5 Ventris and Chadwick (1973) 360–61.

Warrior-Trader, Toumba Tomb 79.[6] Buried in the mid-ninth century, Tomb 79 is part of a tradition in which the cremated remains of individuals are wrapped in cloth and buried in metal urns, known throughout the Aegean and further afield from the twelfth to the seventh century (and beyond) and comparable to practices in the Homeric epics (*Il.* 23.250–57; *Od.* 24.71–79).[7] In Tomb 79, the cremated remains were placed in a bronze cauldron and buried with an iron sword, a spearhead, two iron knives, and a bronze grater – another reference with Homeric parallels.[8] Twenty-five iron arrowheads were found in the niche in which the urn was placed, and nine more were found in the shaft. The shaft also contained a North Syrian haematite cylinder seal, Phoenician and Cypriot jugs, and a substantial amount of eating and drinking ware;[9] like the arrowheads, numerous balance weights were found throughout the niche and shaft, as well as a bronze object interpreted as a weighing scale, hence the moniker 'warrior-trader'.[10] This grave clearly belonged to an individual granted significant social prestige. The only other metal urn cremation at Lefkandi is the burial of a man in the Toumba building, otherwise lacking in the elaborate offerings which seem to have gone to his female companion.

There is one other burial with arrowheads at Lefkandi, Toumba Tomb 26, which is a late tenth century (Late Protogeometric period, LPG) inhumation burial in which the deceased was laid out in a supine position surrounded by grave goods. A cluster of ten arrowheads were found at the left of the pelvis, and a sword lay along the left leg. Hans van Wees has suggested that burials with weapons represent men buried as they were dressed in the EIA.[11] While there are several problems with this hypothesis in some burials with weapons,[12] in the case of Tomb 26 it seems reasonable. This burial is far less extravagant than Tomb 79, and although the presence of any metal objects still indicates that it is prestigious, it may be that this inhumated individual regularly wore his sword and his quiver at his belt. Hector Catling also suggested there was perhaps originally a bow in this burial.[13]

6 Popham and Lemos (1995).
7 Crielaard (2016) esp. 59–62; Popham *et al.* (1982); for EIA metal urn cremations in Athens, Dalsoglio (2019).
8 Ridgeway (1997); West (1998).
9 On the significance of overseas connections and feasting in elite identification in the EIA, see Lloyd in this volume.
10 Popham and Lemos (1995) 152–54.
11 Van Wees (1998) 338.
12 In particular burials such as the metal urn cremations at the West Gate in Eretria, some of which contained several swords; see Lloyd this volume.
13 Catling (1985) 21.

How can we tell if these arrowheads were used for warfare, hunting, or both? Representational art is limited at this time, but offers some clues. The bronze vessel used as the urn in the male burial in the Toumba building is decorated with scenes of hunting large fauna – bulls and lions – in a characteristically elite mode of hunting.[14] This might indicate that the bow was symbolically connected with this elite form of hunting for sport, rather than the more commonplace hunting for food. However, we must recognize that the vessel in the Toumba building is an antique from Cyprus, some two centuries old at its time of deposition.[15] Whatever influence such art may have had in the context of the vessel's deposition, it was not created in the same context.

While contemporary, locally-produced artwork is extremely limited, what there is suggests that human combat was recognised as the appropriate arena for archery. A tiny hydria from Tomb 51 in the Skoubris cemetery at Lefkandi, dated to the Middle Protogeometric Period and thus roughly contemporary with the Toumba building, depicts a pair of confronted archers and no animals.[16] In Late Geometric pictorial vase painting, several centuries later, archers are dotted across the battlefield, but in scenes of hunting spears and swords are more commonly used.[17] While the latter of these are more clearly mythological, depicting battles with lions and mythical creatures, the absence of bows suggests they were not closely associated with elite hunting.[18]

Other EIA evidence points to the use of the bow in human combat. Two burials from Athens contained arrowheads. The earlier grave, Grave 147 on the south bank of the Eridanos in the Kerameikos, is an inhumation dating to the eleventh century (Early Protogeometric), in which an arrowhead was embedded in the shoulder of the skeleton of a child aged between eight and twelve years old. The only other object in this grave was a lekythos. Anna Maria D'Onofrio suggests that this is the continuation of a Late Bronze Age tradition of a 'simple bowman's' burial,[19] although she only refers to four other burials in this tradition, all of them wildly different from Grave 147.[20] Such simplicity would contrast sharply with the metal urn cremation of Toumba Tomb 79 and

14 Catling (1993) plate 19–20.

15 Catling (1993).

16 Popham *et al.* (1980) 127, fig. 4. On the unlikelihood that these images copy those on the Cypriot bronze vessel, see Lemos (2000) 16.

17 Archers in combat: Ahlberg (1971). Hunting with spear and sword: Brouwers (2013) 57; *Od.* 19.348.

18 One of the limited examples of a bow used in combat with a centaur is a seal in Paris' Cabinet de Médailles (M5837), which is perhaps an early depiction of Herakles and Nessos, Langdon (2008) 213–14, fig. 4.13.

19 D'Onofrio (2011) 648.

20 Lloyd (2014) 86–87.

even the inhumed archer in Toumba Tomb 26. However, Anna Lagia, who conducted the anthropological examination of the remains, suggested that it was possible that the arrowhead could have caused a fatal wound that resulted in the child's death.[21] We do not know the context in which this wound occurred – archery is a skill requiring practice, and this may be the result of a training accident. It does, however, point away from the 'simple' archers' burial and perhaps toward the use of the bow in interhuman conflict.

The later burial dates to the second half of the tenth century (LPG) and is a more traditional burial with weapons in the Athenian trench-and-hole style where the cremated remains are placed in a ceramic amphora. In this burial, a sword and a dagger were 'killed' by being bent into a circle and placed around the neck of the urn.[22] The solitary arrowhead was found in the cremation urn alongside the bones of the deceased, which is unusual for weapons in EIA Athenian burials. It is much less clear what occurred here, but it is possible that the arrowhead was inside the individual when they were cremated, and thus ended up in the amphora when the remains were collected. The level of precision evident in the collection of bones in Athens means that this is likely to have been intentional;[23] however, it is unusual to find a weapon inside the cremation urn. Therefore, it is plausible that this arrowhead was the cause of death.

That the bow was used in interhuman conflict does not mean that it could not have been used in hunting; indeed, it almost certainly was. However, the most frequent point of reference within early Greek artwork is of battle, not hunting, suggesting that we should not reconsider the occupant of Tomb 79 to be a 'Warrior-Trader-Hunter'. Indeed, it seems that archery could be associated with quite high-status warriors in the EIA – at least until the eighth century.

Once pictorial iconography and literature resume in the eighth and seventh centuries the situation becomes clearer, although no less complicated. Archers are extremely common in Late Geometric pictorial vase painting, comprising roughly a third of individuals engaged in combat;[24] spears, meanwhile, appear to be almost exclusively javelins (Figure 2.1).[25] LG pictorial scenes of combat are notoriously disorganised, with archers, spearmen, and sword-fighters scattered across the battlefield in most cases. All of these types of warriors may

21 Lagia (2007) 277, 279.

22 On killed swords, Lloyd (2015).

23 The most telling example of this level of care is the presence of the bones of a neonate in the cremation of the 'Rich Athenian Lady' in the Agora, Liston and Papadopoulos (2004) 16–17.

24 Van Wees (1994) 144.

25 Ahlberg (1971) 45.

wear armour – a helmet or a shield – and archers and spearmen may also carry swords.

A few examples deserve close consideration. Among the handful of depictions of chariots in combat known in LG pictorial, one includes a scene of an archer and a javelin-thrower, each in the back of the chariot from which they presumably fight.[26] While it is tempting to see this as evidence for how chariots and archers were used in LG combat, that this is the only such image we know and the similarity of this concept, if not the execution, to depictions of archers in chariots from the eastern Mediterranean, indicates that we should be cautious about over-emphasising it.

More significant is the depiction of three archers confronting three slingers over the body of a dead warrior on the body of an amphora from the *polyandria* at Paroikia on Paros.[27] While this scene has been interpreted as an early representation of the phalanx, the front-line presence of archers and slingers indicates that this is nothing so formal. It is at present our only LG representation of slingers, who do not seem to have ever been held in high regard as light-armed troops before the Peloponnesian War; nevertheless, they stand at the centre of a battle over a fallen warrior – a regular type-scene in the *Iliad* (e.g., *Il.* 16.755–764).[28] The archer at the front of the opposing line, however, carries a shield comparable with those of the spearmen elsewhere on the vessel; this, and their occasional association with chariots on other vessels, may be an indication that archers remain of relatively high status at this time.[29] While the second archer in this line wears nothing more than a helmet, no armour at all is represented on the slingers.

When we turn to the Homeric epics, we find several examples of heroic archers – but also a more ambiguous attitude.[30] In contrast to LG pictorial representations, only a tenth of the weapons used in the poem are bows, although the most common weapon, spears, are thrown as javelins in half of all instances that they are used.[31] We see early evidence of archers and slingers as a specialised unit in the case of the Lokrians (*Il.* 13.712–22), but for the most part they fight individually, often hiding behind the shields of their more heavily-armoured compatriots (*Il.* 8.266–72). While 'archer' is sometimes used

26 Williams (2015) 261–62, fig. 6. On chariots in eighth-century pictorial, see Lloyd in this volume.

27 Zaphiropoulou (2006) fig. 5–8.

28 Kucewicz (2021) 18–30.

29 Brouwers (2013) 60.

30 The date of the Homeric epics, and the combat they describe, is contentious: see Snodgrass (2013); Lloyd, this volume.

31 Van Wees (1994) 144, *GWMR* 265.

as an insult against enemy fighters (*Il.* 11.385, 14.479), Teukros is lavished with praise by Agamemnon for his skill with the bow (*Il.* 8.281–91). While there is little positive said of Alexandros/Paris throughout the poem, he is also an elite warrior who is primarily known for fighting with his bow (e.g., *Il.* 11.369–83). Finally, Odysseus fights with a bow in the final confrontation in the *Odyssey* until he runs out of arrows (*Od.* 22.1–125).

The more popular spears are not only carried in battle, but whenever a hero goes out in public (e.g., *Il.* 19.47–49; *Od.* 15.551–52)[32] – an approach also seen on LG pottery, where the standard armoured figure usually carries two spears (e.g., Figure 2.1, far left). It is clear that, in the epics, there is no general social distinction between heavy- and light-infantry and that elite heroes could perform both roles.

2 Take It All Off: From 'Aristocrats' to Naked Men

That Teukros needs to hide behind his brother Ajax's shield indicates that this is a warrior who is not heavily armed. Alexandros/Paris is described as wearing only a leopard skin before his duel with Menelaos, although he carries a sword and two spears with his bow (*Il.* 3.17–19). Indeed, we may look at the Paros amphora and question whether the first archer is intended to carry his own shield, or if we are to imagine that he stands behind the shield of another, as his companion archer stands behind him. In the seventh century, while javelin-throwers continue to be represented in full armour – perhaps related to the returning prominence of the thrusting spear – archers in combat are more regularly seen fully in the nude. Throughout this century, with the spread of new equipment, the distinction between the heavy- and light-armed warrior began to be more clearly defined and began to take on more strongly the social aspects that would come to define the distinction in later centuries.

The crouching, naked archer appears first in Early Protocorinthian vase painting, typically dated to the late eighth-early seventh centuries. On an aryballos from Lechaion (Corinth CP-2096),[33] a naked archer crouches behind a fully-armed compatriot carrying a long thrusting spear and a large, round shield with a double grip – the *aspis*; it appears that another naked fighter behind this archer is stabbing him in the back. On a later, Middle Protocorinthian aryballos from the sanctuary of Hera at Perachora (Perachora 27), an archer in an unclear state of dress crouches behind a warrior with a Dipylon/Boiotian

32 Van Wees (1998) 335–38.

33 Viggiano and Van Wees (2013) 63, fig. 2.4.

shield.[34] Elsewhere in the Peloponnese, toward the end of the century and beyond, among the lead figurines dedicated at the sanctuary of Artemis Orthia in Sparta are crouching archers (see below), although unlike the Corinthian figures these wear helmets.

There are two possibilities here. The first is that the changes in artistic style mean that Corinthian vase-painters of the late eighth and early seventh century were able and willing to depict more regularly organised battlefields that more accurately depict how light-armed archers were fighting in the preceding centuries. The second is that we are witnessing a genuine change in how the light-armed functioned in combat and a corresponding change in the way in which the light-armed were understood. In both cases, we need to understand that there are also regional factors at play.

A short fragment of poetry by the Parian poet Archilochos indicates that the latter option is more likely. In this poem, which presumably dates to the poet's flourishing in the second quarter of the seventh century, Archilochos describes a forthcoming battle in which the Euboians will prefer to fight at close-quarters rather than using the bow and the sling:

> οὔ τοι πόλλ' ἐπὶ τόξα τανύσσεται οὐδὲ θαμειαὶ
> σφενδόναι, εὖτ' ἂν δὴ μῶλον Ἄρης συνάγῃ
> ἐν πεδίῳ, ξιφέων δὲ πολύστονον ἔσσεται ἔργον:
> ταύτης γὰρ κεῖνοι δαίμονές εἰσι μάχης
> δεσπόται Εὐβοίας δουρικλυτοί.
>
> Not many bows indeed will be stretched tight, nor frequent slings
> Be whirled, when Ares joins men in the moil of war
> Upon the plain, but swords will do their mournful work;
> For this is the warfare wherein those men are expert
> Who lord it over Euboia and are famous with the spear.
>
> ARCHIL. Fr. 3 West[35]

Here we see the apparently remarkable decision on the part of the Euboians to reject the use of long-distance weapons – although javelins are not mentioned – in favour of the sword.

Some scholars of the last century connected this rejection of the light-armed to the war between the central Euboian cities Chalkis and Eretria known

34 Viggiano and Van Wees (2013) 64, fig. 2.5. On the Dipylon/Boiotian shield see Lloyd, this volume.

35 Translation by Perrin (1914).

as the Lelantine War, which is mentioned by Thucydides as the earliest war to divide all of the Greeks on one side or the other (1.15.3).[36] Other scholars disagree about the connection.[37] A decision to ban missile weapons in this conflict was said to have been recorded in a formal inscription associated with that conflict at the sanctuary of Artemis at Amarynthos near Eretria (Strab. 10.1.12), but it is usually accepted that this must be a later invention as such inscriptions do not survive from the seventh century.[38] Indeed, as Jonathan Hall has explained at length, much of what we appear to know about the Lelantine War is a later invention and as such 'we do not know when – or even whether – the Lelantine War occurred'.[39] It is difficult, therefore, to countenance arguments that the Euboians 'solemnly agreed to ban obsolete weapons and use only the newest weapons';[40] rather, we should look to this Archilochos fragment as Plutarch does: as evidence for a specific Euboian way of fighting in the middle of the seventh century.

The Archilochos fragment points toward changing attitudes toward archers and other light-armed warriors in the seventh century; a perspective also represented in the poetry of Tyrtaios. It is from Tyrtaios that we first hear of a contrast between the heavily-armoured *panoploi* and the light-armed *gymnetes*, literally 'naked men' (Fr. 11.35–38 West). Tyrtaios describes the 'naked men' as scatted among the 'all-armoured', using their compatriots shields for protection as does Teukros in the *Iliad*. It is likely that this tactic was made possible by the new, larger shields carried by the *panoploi* from the early seventh century.[41]

Seventh-century poetry gives the impression that the light-armed remained prominent in combat, aside from the special case of the Euboians; they also remain so in dedications at the panhellenic sanctuaries.[42] Yet they begin to be far less prominent in the vase painting of the period, even in the Corinthian repertoire where we first see them crouched among the heavily-armed. Van Wees suggests that this is because they are of 'lower prestige' than the close-fighting *panoploi*;[43] if this is the case, what has happened between the prominent, high-status archers of the EIA and the 'naked men' of the late seventh century?

36 Greenhalgh (1973) 90–93; Jeffery (1976) 65.

37 Donlan (1970). Plutarch, who quotes the fragment of Archilochos in his *Theseus*, does not make the connection (*Vit. Thes.* 5.3), despite mentioning the war elsewhere (*Mor. Conv. sept. sap.* 153f, *Am. narr.* 760e–761b); nor does Strabo, who knew of Archilochos and the Lelantine War but does not cite this passage.

38 Forrest (1957) 163; Wheeler (1987).

39 Hall (2014) 1–8, quotation at 8.

40 Forrest (1957) 163–64.

41 *GWMR* 170.

42 Poetry: *GWMR* 169–70; dedications: Snodgrass (1964) 250; Baitinger (2001) 29–30.

43 *GWMR* 170.

It is first necessary to understand that it is not only the *gymnetes* but also the *panoploi* who appear to be a new feature in the seventh century. While the evidence for armour throughout the EIA is slim, it seems likely that the increase in Mediterranean connectivity in the eighth century resulted in the import and resumption in production of body armour.[44] In the eighth century, pictorial vase painting suggests that archers may have worn helmets, carried shields, and perhaps even ridden chariots and worn body armour. Going into the seventh century, and especially with the growing prominence of the double-grip *aspis*, which requires the shield-bearer's entire left arm, it became impractical for archers, who need both hands for their bow, to carry one – especially as the shield was capable of protecting more than one individual. Indeed, we may venture to suggest that such an arrangement may have been one of the reasons to adopt such a shield in the first place. However, we must also note that the distinction between the *panoploi* and the *gymnates* is not quite the same as the distinction between heavy- and light-armed that we make in our analysis. As noted for the Archilochos fragment above (Fr. 3 West), javelin-throwers are not distinguished alongside archers and slingers; it appears that the javelin remains the purview of the *panoploi*.

Another factor is the dual problem of affordability of armour and the need for bodies on the battlefield. As the seventh century continued, Corinthian vase-painting presents every warrior as fully encased in helmet and bell-cuirass, aside from the 'naked-men' with their bows; the full bronze panoply must have been more of a rarity than this represents.[45] If this is the case, then warriors who were not in a position to fully arm themselves would have been in the position of deciding how they might take part in combat without exposing themselves to the violence of close combat. The solution was to fight from a distance, behind the shields of wealthier allies.

For the wealthier men in their full armour, such a situation presents a problem. If all one needed to enter combat was a bow and an ally with a shield, the newly erected barrier to becoming a warrior was under threat.[46] As the Paros amphora indicates, archers and slingers could partake in battles over deceased enemies which might, as in the *Iliad*, result in the capture of the prestigious body armour – the *gymnetes* could become *panoploi*.[47] Such a change would be gradual – Teukros and Alexandros/Paris are examples of heroic archers still current in the seventh century and the bow is a skilled weapon that could not

44 Lloyd (2020) 511–12 and this volume.
45 Van Wees (2013); Van Wees and Fisher (2015) 12.
46 Lloyd, this volume.
47 On Homeric battles over the dead, Kucewicz (2021) 18–30.

simply be picked up without practice – but as the elite began to distinguish themselves through their armour,[48] and to be depicted as such, the distinction between the heavy- and light-armed would have begun to grow more distinct. As the norms of participation on the battlefield shifted, in order to gather larger armies the *panoploi* must allow those from lower social classes into their war band, but in order to maintain the social order, they must also change attitudes about what classifies nobility on the battlefield – that is, these people can fight as light-armed, and in order to prevent them from accruing the glory due to the nobly born, fighting as light-armed must become ignoble.

3 The Light-Armed and the Growth of Citizen Armies

The changing norms of participation between the heavily-armed, noble *panoploi* and the lightly-armed *gymnetes*, unable to afford hoplite equipment, meant that by the end of the seventh century missile warfare by infantry became exclusively associated with the lower social and economic classes.[49] As a result of this process, our knowledge about the light-armed in the sixth century is scant. Although the first written accounts of late Archaic conflicts begin to appear in Classical authors, in particular Herodotos, they give no attention to the role of light infantry in combat. This can be explained partly by the ideological agenda of the sources, written by and for upper-classes whose interest in the military activities of their social inferiors in the distant past would have been little at best. The marginalisation of the light-armed can be also witnessed in the artistic depictions of combat on sixth century pottery, arguably our richest source for the practice and ideology of warfare in the late Archaic era. Compared to the numerous depictions of archers in LG pictorial art, their relative rarity in black-figure representations of war and warriors, which focus almost exclusively on heavy-armed hoplites, is indeed striking. And while this might be seen as a historical reflection of the small numbers of light-armed in the Greek militias during the sixth century, their insignificance in the sources for the period should be rather interpreted as a general sign of

48 Elite warriors would also come to distinguish themselves by riding their horses to battle – see Greenhalgh (1973) 84–95; Brouwers (2007); Lloyd, this volume.

49 Regional variation, however, remained an important factor. The Akarnanians and Aitolians, for instance, based in the rugged area of north-west Greece, continued to rely on missile weapons, having 'little use for the heavy infantry phalanx': Wheeler (2007) 188. A comparative study of geopolitical factors in Archaic Greek warfare remains a *desideratum*.

the lower prestige associated with missile warfare.[50] The light-armed certainly did not disappear from the late Archaic battlefields.

That missile combat remained a feature of sixth-century warfare is indicated by the large quantities of arrowheads found in excavations. Long-range weapons were heavily employed by both the Lydian attackers and the defending Greeks during the siege of Smyrna in *ca.* 600, evidenced by multiple finds of bronze arrowheads.[51] A deposit of over eighty arrowheads dated no later than 525 has been found at Lindos.[52] By far the most numerous finds of arrowheads come from Archaic Greek sanctuaries; among the largest are the dedications from the Lakonian sanctuary of Apollo Maleatas, located on the Parnon mountain range in Kynouria. The excavations yielded a large amount of offensive weapons (e.g., javelin heads and spearheads, iron fragments of spears, and bronze axes) beginning in the seventh and continuing through the sixth and fifth centuries, which included no less than 461 iron and bronze arrowheads.[53] Although the high volume of such dedications would have been appropriate for the presiding deity at the sanctuary, Apollo – the god of bow and archery – Nicolette Pavlides has convincingly argued that the military character of the dedications reflects the geopolitical setting of the sanctuary, located in the vicinity of the much-disputed border area of the Thyreatis/Kynouria.[54]

The prominence of archers in Archaic Sparta can be also deduced from the lead miniature figurines dedicated in huge numbers in Spartan sanctuaries (esp. Artemis Orthia) from *ca.* 650 onward.[55] Among these, a sizeable number consists of kneeling archers, which, considering Tyrtaios' exhortations to the light-armed *gymnetes* 'crouching beneath a shield here and there' (Fr. 11.35), should be seen as 'a sort of iconographic shorthand for the ideal light-armed soldier', operating on the battlefields from the cover of the heavy-armed warriors.[56] The dedications of lead figurines of archers continue well into the Classical period, but in the fifth century most are standing up straight (Fig. 8.1), which may suggest that archers no longer mingled with the heavy-armed in combat (see below).

50 On the ideological distortion of sources, see n. 68 below.

51 Nicholls (1959) 128–34, who suspected based on the types of arrowheads that foreign auxiliary archers may have been used by both sides. See also Brouwers, this volume.

52 Snodgrass (1999) 81.

53 Pavlides (2018). On Archaic Spartan dedications of captured weapons and armour, see Kucewicz, this volume.

54 Pavlides (2018) 280–90, 300.

55 Dawkins (1929); Boss (2000).

56 Van Wees (2000, 152–53, *GWMR* 175, 296 nn. 37–38) provides drawings, further references and other examples of kneeling archers in Archaic art.

FIGURE 8.1 Lakonian lead figure of an archer (6th–5th century BCE; MMA 24.195.97). Gift of A.J.B. Wace, 1924

Light-armed warriors featured also in the militias of other Greek poleis in the sixth century. Although the evidence is scarce, we should imagine that their numbers varied depending on regional and political factors. According to Herodotos, Samos maintained a standing force of 1,000 archers under the tyrant Polykrates (Hdt. 3.39.3, 45.3). In a similar way, the Athenian tyrant Peisistratos is credited with hiring a force of Thracian peltasts and Scythian archers, who helped him secure victory in the battle of Pallene (*ca.* 546) and root his family's power in Athens until the last decade of the sixth century.[57] Whether contingents of foreign, light-armed troops remained in the service of the Peisistratids is a matter of debate (see below), but the sudden appearance of archers dressed in Scythian attire in Attic black-figure vase painting around

57 Hired Thracians: Arist. [*Ath. Pol.*] 15.2; Lavelle (2005) 133–34, 139–43.

FIGURE 8.2
Departure of a warrior and archer on an Attic black-figure terracotta neck-amphora in the Metropolitan Museum of Art, New York (last quarter of the 6th century BCE; MMA 41.162.171). Unattributed. Rogers Fund, 1941

530 may suggest that the Scythians did serve in Athenian armies of the late Archaic period (Figs 8.2–4).

According to Van Wees, archers appear on no fewer than 750 Attic black- and red-figure vases, the majority of which come from the last quarter of the sixth century. Nine out of ten are distinguished by pieces of Scythian clothing, most commonly the characteristic Phrygian cap.[58] Most of the images depict scenes of warriors departing from their homes, a commonplace of Athenian iconography of the period. The central figure is always the heavy-armed warrior, who in some images is accompanied by a single archer and his family (Fig. 8.2); in others he is attended by multiple archers and, occasionally, other heavy-armed warriors, as well as attendants who prepare his horses or his chariot – status symbols of the Athenian elites with clear connotations to the heroes of the Homeric epics (Figs. 8.3–4). Archers in Scythian dress appear also in scenes of

58 *GWMR* 175. Echeverría (2015) 53–54 provides more detailed breakdown. For a full study of Scythian archers in Attic vase painting, see Vos (1963).

FIGURES 8.3 AND 8.4 Departure of warriors on an Attic black-figure terracotta stamnos in the Metropolitan Museum of Art, New York (late 6th century BCE; MMA 2011.233). Attributed to the Painter of London B 343. Gift of Patricia Stickney, 2011

combat; a recent estimate counted a total of fifty-seven depictions of archers taking part in battle, usually in crouching position and aiming their bows horizontally from 'a safe distance behind other warriors'.[59]

The prominence of archers dressed in elements of Scythian attire in late Archaic Attic art raises two important questions: Did Athens employ a standing force of foreign, light-armed troops in the closing decades of the sixth century? And, if so, how did archers fight in the citizen armies of the Athenian, or indeed any other, contemporary Greek polis? Did they form a separate and tactically independent unit? Or did they mix in with the heavy-armed warriors in a loose and open formation, not unlike the one described in the elegies of Tyrtaios?

The answer to the first question depends largely on how we approach the evidence of late Archaic war iconography. Some scholars tend to dismiss the historical value of black-figure depictions of combat because of their supposed 'heroic' and fictitious nature, as artists deliberately portrayed episodes from myth (which included gods and monsters) and/or focused on themes and elements which appealed to their upper-class customers but had little or no relation to contemporary realities of war.[60] Images of Scythian archers were accordingly interpreted as products of artistic imagination; their 'exotic' and 'barbaric' appearance was set in contrast to the heavy-armed hoplites, symbolised as the Athenian warriors *par excellence*.[61] Other scholars, however, have rejected such claims, arguing that late Archaic art offers valuable insights into actual modes of fighting which must have been based on the actual experience of war of their viewers. The sudden appearance of Scythian archers in black-figure iconography, operating in consistently loose formations, should be, according to this view, seen as historical testimony to their employment in the citizen armies under the Peisistratids.[62]

How such light-armed units fought on sixth-century battlefields is a question no less difficult to answer. The images of archers in late Archaic art imply that they remained interspersed between the heavy-armed warriors, taking cover behind their shields and roaming freely among the front lines. This manner of fighting, if true, indicates that late Archaic armies did not yet employ tactical separation between units of heavy- and light-armed troops known

59 Echeverría (2015) 53. E.g., Beazley Archive 3818, 302101, 302750, 320014, 320227, 305643.

60 For recent discussions of Archaic iconography of war, see Viggiano and Van Wees (2013); Echeverría (2015); Kucewicz (2021) 81–86.

61 E.g., Lissarague (1990) 125–49; Osborne (2018) 95–100.

62 E.g., *GWMR* 175–76; Rawlings (2007) 56; Echeverría (2015) esp. 53–54; Wrightson (2019) 59.

from Classical battle accounts. The Greeks themselves believed that organisation into distinct units was first introduced by the Median king Kyaxares (*ca.* 625–585), who, according to Herodotos, 'first arrayed the men of Asia in companies and set each kind in bands apart, the spearmen and the archers and the horsemen: before this they were all blended alike confusedly together' (1.103.1). The lack of separation into distinct contingents of hoplites, light-armed, and horsemen in the Greek armies of the sixth century might reflect the largely amateur character of Archaic citizen militias, which lacked the necessary central organisation and discipline required for tactical separation of units and combined arms warfare.[63]

This reading supports the scholarly view of Archaic warfare as dominated by private war bands, led by wealthy individuals and organised along informal networks of friendship and dependence.[64] In this model, any light-armed troops would have consisted primarily of volunteers unable to afford hoplite equipment; or, more likely, attendants of the heavy-armed warriors, who escorted their masters to the battlefield and supported them in combat, which at times involved active participation in missile warfare.[65] This arrangement remained in place until the end of the Archaic period, which witnessed a series of public and institutional measures imposed by Greek poleis to exert control over the citizen militias, one of which included the tactical separation of units reflected in the literary sources.[66] Recent studies, nonetheless, tend to backdate the process of army institutionalisation to the sixth century, which suggests that the development from small and privately controlled bands to ordered masses operating in separate units occurred earlier than previously assumed.[67] Tracing the precise details of the latter remains a challenge and a subject for further study.

63 Wrightson (2019) 45–68.

64 For recent examples of this view, often traced back to Thucydides (1.5–6), see Pritchard (2010a), (2018a) 5–6.

65 *GWMR* 69. Van Wees rightly speculates that fewer (if any) light-armed troops took part in overseas expeditions in the Archaic era, 'when warships were small and their commanders presumably tried to fill them as far as possible with hoplite soldiers' (62). On the comparison of the cost of light- and heavy-armed equipment, see *GWMR* 62. Since archery required considerable skill, we should imagine that attendants and lower-class men fighting in the early armed forces confined themselves to throwing javelins and/or stones.

66 For unit division as a fifth-century phenomenon, *GWMR* 177; Krentz (2007) 75, 79–80; Rawlings (2007) 85; Wrightson (2019) 59–63.

67 E.g., Valdés Guía (2019); Kucewicz (2021) 135–76 and this volume.

4 The Fifth Century and Beyond

The nature of the surviving evidence allows us to write a different history of light-armed warfare from the early fifth century onward. Units of missile troops feature in the earliest narratives of historical battles; the appearance of discussions on military theory, whether in dedicated treatises or in the course of other literary works, gives us a sense of their place in tactics and strategy. Yet the problem of ideological distortion remains.[68] Light-armed warriors (typically referred to from this point on as ψιλοί, 'light ones') were still drawn from those of low social status: poor, foreign, unfree, or some combination of these categories. The picture of their nature and role that is painted by our elite sources is, therefore, patchy at best, and likely affected by negligence or misrepresentation.[69]

This is not to say that Greek authors were not interested in light-armed troops or their contributions to warfare. The presence of light infantry is duly recorded for nearly every major campaign and engagement in the Classical period; both Thucydides and Xenophon describe their tactics.[70] These and other authors were happy to count missile troops among the main military assets of a state (Thuc. 2.13.8; Andoc. 3.7; Lys. 34.4; Xen. *Hell.* 5.2.14, 6.1.9; Isoc. 8.118). But their actions are rarely the focus of war narratives and we almost never hear the names of light-armed warriors. Instead, these troops are typically presented as one of the challenges that armies of hoplites and cavalry had to overcome. Where the challenge was effectively met, missile troops quickly fade from view. Modern scholarship has tended to adopt a similar perspective. While there is no shortage of studies on aspects of light-armed warfare in ancient Greece,[71] the subject is poorly integrated into our understanding of warfare in general. Indeed, discussions of light-armed troops are traditionally accompanied by the remark that their uses were limited, if not irrelevant to the 'real' warfare of hoplites.[72] The purpose of this section is to suggest, however briefly, that this leap from a lack of focalisation to a declaration of irrelevance

68 Lissarrague (1990) 16–20; Van Wees (1995) 162–65; Hunt (1998) 42–46, 219–20; Trundle (2010) 142–46; Pritchard (2018b) 93–96; Lewis (forthcoming).

69 This could even extend to Persian infantry, whose archery is discounted by Herodotos in his account of Thermopylai: Van Wees (2018) 34–37.

70 Konijnendijk (2018) 96–102.

71 Lippelt (1910); Best (1969); Anderson (1970) 111–40; *GSW2* 117–25; Lissarrague (1990); Konecny (2001); *GWMR* 62–65; Williams (2004); Hunt (2007) 119–24; Trundle (2010); Kelly (2012); Davis (2013); Sekunda and Burliga (2014); Craven (2017); Pritchard (2018b); Lewis (forthcoming).

72 Konijnendijk (2018) 95 with nn. 1–2.

is reflective of a modern obsession with pitched battle rather than the reality of war in the Classical period.

From the beginning of this period, we can identify two distinct types of light-armed forces. The first is the light-armed element of mass levies, so armed either because its members could not afford heavier equipment or because service as missile troops was typical in their region of origin. The second is smaller units of light infantry specialists, often maintained as standing troops or hired as foreign mercenaries. The role and perception of these types are quite different, so it will be worthwhile to treat them separately here.

Mass levies of light-armed warriors are visible from the time of the Persian Wars. The light troops reported by Herodotos at Plataia are sometimes dismissed as mere servants and camp-followers, but the historian emphasises that the 35,000 helots brought as part of the Spartan levy were all armed for war (9.29.1).[73] We have no reason to believe the other men he identified as light troops were not there to fight. Even if their equipment and skill will have varied greatly, the sheer numbers of the sub-hoplite levy made it a significant presence in Greek conflicts. Where the sources give us an estimate, the light-armed levy outnumbers the hoplites (Thuc. 4.93.3). Elsewhere they appear as a vast, numberless mass (Thuc. 2.31.2, 4.94.1). The demographics of Greek states imply that the light-armed element would have formed the lion's share of any general levy[74] – in other words, that the average Greek warrior of the fifth century fought with missiles, not shield and spear.

Assessments of the effectiveness of these troops tend to be based on Thucydides' account of the battle outside Syracuse in 415. The main clash was preceded by a skirmish in the field between the hoplite lines, in which 'stone-throwers, slingers, and archers' attacked and routed each other, 'as is likely with light troops' (6.69.2). Scholars have assumed that, like other elements of this battle scene, the light-armed warriors' insignificant action is 'paradigmatic' and should be taken for granted in other battles where it is not made explicit.[75] Yet this battle is unique in its geographical constrains: the Athenians deliberately chose to anchor both flanks on impassable terrain to prevent any manoeuvres by the Syracusan cavalry, which inevitably reduced their light troops to a head-on collision as well.[76] As a result, their behaviour in this engagement may have been unusually ineffective. In any case, pitched

73 Hunt (1997) 129–33.
74 *GWMR* 65.
75 Hornblower (2008) 474–75.
76 Konijnendijk (2018) 99 and this volume.

battles were only a small part of war; their performance in this narrow context need not reflect their performance elsewhere.

Thucydides himself offers several examples of masses of light-armed warriors being used to great effect. He records the destruction, around 465, of a large Athenian army of prospective settlers at the hands of a Thracian levy – no doubt mostly consisting of peltasts armed with javelins.[77] He also describes how the light-armed element of the Athenian army invading Corinthian territory in the early 450s wiped out a part of the Corinthian levy that had fled into a field with no way out (1.106). He notes how, at the battle of Spartolos in 429, Chalkidian light troops and cavalry drove off their Athenian counterparts, then charged and scattered the Athenian heavy infantry, snatching victory from the jaws of defeat (2.79.3–7). He praises the Akarnanians collectively for their skill with the sling (2.81.8) and describes in detail how Demosthenes' expeditionary force based in Naupaktos was obliterated by a local levy of Aitolian javelin-throwers.[78] He stresses that the Athenian victory at Pylos and Sphakteria was largely the work of haphazardly armed naval crews, alongside smaller detachments of light-armed specialists and hoplites (Thuc. 4.9.3, 4.32.2). The Syracusan light infantry may have achieved little in the first battle outside their city, but Thucydides nevertheless treats them as a key Syracusan advantage over the Athenians in later engagements (7.5.3). Alongside the cavalry, they harried the Athenian army to destruction during the final retreat in 413. Xenophon records similar episodes and assessments. He also took for granted that any Rhodian could serve effectively as a slinger (*An.* 3.3.16–17) and had a speaker warn the Spartans that the levy of Thessalian peltasts could take on the whole world (*Hell.* 6.1.19).

In short, light-armed levies and improvised crowds of missile infantry could play important roles in the defence of territory and fortified positions, dealing with isolated enemies, and supporting hoplites in battle. In addition, arguably their principal offensive role was to ravage and plunder the enemy countryside, which was not merely a provocation to battle but an end in itself.[79] They were uniquely suited to such tasks precisely because of their large numbers and lack of cumbersome armour and equipment. In these situations, even non-specialist light infantry was acknowledged to be a significant asset – especially for communities or expeditionary forces whose hoplite numbers were low.

---

77 Thuc. 1.100.2; Best (1969) 20; Hunt (2007) 120.

78 Thuc. 3.94–98; Blome (2020) 29–49. Compare Roisman (1993) 24–27, who blames the defeat on Demosthenes' mistakes.

79 *GWMR* 122, 126.

That said, the irregular armament and lack of organisation of light-armed masses made them vulnerable and difficult to control. The sole example we have of a general trying to improve the cohesion and tactical capabilities of a large mass of missile troops speaks volumes about their usual limitations. When Demosthenes landed his largely light-armed force on Sphakteria, he grouped them into units of about 200 men, which allowed him to use them flexibly to occupy strong positions and coordinate attacks from all directions (Thuc. 4.32.3).[80] Without his reorganisation, the assault would most likely have dissolved into chaos as each warrior looked for his own path to the front and his own safety from Spartan counterattacks. Where enemy cavalry was present, the potential for grievous harm to the leaderless light-armed mob increased exponentially.[81]

These problems may explain the parallel use of the second type of light-armed infantry: specialist troops deployed in smaller, discrete units. The previous section covered their first emergence in the late sixth century. At the battle of Plataia, the Athenians already fielded a corps of archers that was separately drawn up and used to respond to emergencies. While the origins of this corps must lie in the trireme fleet's need of missile troops for deck-fighting – a necessity that stretches back through Polykrates' shipborne archers to the sailor-warriors of the post-palatial Bronze Age – the simultaneous operations of the fleet in the Aegean apparently did not prevent the detachment of a unit of these archers for the land campaign to Boiotia, where their usefulness was recognised by the other allies.[82] Throughout the Classical period, organised units of missile troops like the Athenian archers are regularly attested when forces are enumerated for a campaign or battle.[83] Sometimes their presence is specially noted alongside a much larger levy of irregular light-armed warriors (Thuc. 4.32.2, 93.3).

The numbers of these light-armed specialists may appear relatively modest next to the hoplite core of the armies in which they served, but they were often substantial in absolute terms. Gelon's army of 480 included 2,000 archers, 2,000 slingers, and another 2,000 light infantry to accompany the horsemen.[84] The Athenian archer corps grew in the course of the fifth century to a peak

80 It is sometimes assumed that Demosthenes learned his lesson in Aitolia and deliberately sought to master the art of light-armed warfare. But there is no reason to assume their potential was a revelation to any Greek of his day (Roisman 1993, 40) and no precedent in his Aitolian experience for his innovations on Sphakteria.

81 Konijnendijk, this volume.

82 Hdt. 9.21–22, 60.3; Pritchard (2018b) 87–91.

83 For example, Hdt. 7.158.4; Thuc. 3.98.1, 4.32.2, 6.43; Xen. *An.* 1.2.9; *Hell.* 4.2.16, 7.6.

84 Hdt. 7.158.4, with Hall, this volume.

strength of 1,600 (Thuc. 2.13.8). The combined force with which Demosthenes landed on Sphakteria included 800 archers and 800 peltasts. The Athenians sent 1,300 specialist missile troops to Sicily in 415 – mostly Rhodian slingers – and later planned to reinforce them with another 1,300 Thracian peltasts (Thuc. 6.43, 7.27.1). King Otys of Paphlagonia was able to furnish Agesilaos with 2,000 peltasts for the campaign of 395 (Xen. *Hell.* 4.1.3). In 389, the Athenians sent Iphikrates to the Hellespont with an army of 1,200 peltasts (Xen. *Hell.* 4.8.34). Since these troops were usually mercenaries or standing forces, the expense of maintaining large numbers of them would have been considerable. The readiness of larger Greek states to shoulder the cost suggests that these detachments were thought to be well worth their price.

It is worth noting that not all of these troops were mercenaries, hired from regions famous for light-armed warfare. There may already have been established channels to acquire peltasts from Thrace (Thuc. 2.29.5), archers from Crete, or slingers from Rhodes, and there may have been a recognised benefit in letting foreigners bear the brunt of military operations (4.28.4). Yet the Athenian archer corps included many citizens.[85] Where standing corps were established, as at Athens, and later at Sparta (Thuc. 4.55.2), local residents may well have jumped at the opportunity to earn a steady wage. Where forces were gathered only for the duration of a campaign, it is still difficult at times to tell the difference between local volunteers, mercenaries, and allied troops.[86] So, for instance, there is no definitive answer to the question whether the Iphikratean peltasts of the Corinthian War were Thracians, Athenians, or troops recruited from Pharnabazos' satrapy.[87] Light-armed specialists were expected to train regularly in the use of their weapons,[88] which would have reduced the importance of prior familiarity.

The value of light-armed specialists is apparent from the many roles they play in surviving historical accounts and military treatises. A full list would exceed the aims of this chapter. In pitched battle, they anchored and protected the flanks of the phalanx.[89] On the march, archers and slingers kept enemy light troops at a distance while peltasts could serve as a shock force to clear the way.[90] They could also be selected for service in mixed task forces,

85 Lys. 34.4; Pritchard (2018b) 91–93 draws on the epigraphic evidence.

86 The difference may be artificial, as argued by Rop (2019) 19–26 for Greeks serving in the East.

87 *GSW2* 118.

88 Craven (2017) 106–107; Konijnendijk (2018) 40–41.

89 Konijnendijk (2018) 113–14.

90 Our best evidence comes from Xenophon's *Anabasis*: see Perlman (1976–77) 272–75; Whitby (2004) 237–40.

whether to attack fortifications, carry out stealth manoeuvres, protect the marching column, or defend territory (Thuc. 3.107.3, 6.100.1; Xen. *An.* 3.4.42–43; Aen. Tact. 16.7). Archers, slingers, and peltasts led the way in siege assaults and the defence of city walls, attacks on enemy cavalry and light troops, pursuits, ambushes, and the defence of mountain passes. Xenophon records their use as a mobile reserve during marches through hostile terrain.[91] In short, these organised units of semi-professional light infantry were essential to most military operations. Their mobility and flexibility allowed them to excel in many situations where hoplites struggled due to their heavy equipment and the limited range of their weapons.

The scale of the use of light troops in this period – both levied masses and trained specialists – makes it difficult to account for the common view that their role was minor. The probable reason behind this view is the insistence that pitched battle was the central feature of Greek warfare.[92] This long-standing belief has narrowed the focus of many modern accounts to the type of military operation in which light infantry was least likely to have a decisive impact. Apparent confirmation from a handful of passages has been used to justify the dismissal of light-armed troops as inconsequential. Yet, as this section has shown, the surviving sources did not present light infantry as irrelevant. They only noted that it had different, complementary roles and proved its value in different situations. These alternative situations increasingly seized the spotlight in contemporary narratives of war. Therefore, at the other extreme of modern readings of the evidence, it has sometimes been suggested that the rise of the peltast, in particular, marked a fundamental change in the nature of Greek land warfare.

5 The Age of the Peltast?

At Lechaion in 390, the Athenian commander Iphikrates used a force of mercenary peltasts backed by an allied hoplite phalanx to destroy a *mora* of Spartan hoplites on open ground.[93] His spectacular victory has led some modern scholars to wonder whether the battle should be seen as a turning point in Greek military history. From this time onwards, mercenaries in general, and

91 Lee (2007) 160.

92 For discussion of this historiographical tendency, Konijnendijk (2018) 6–38, (2021).

93 Xen. *Hell.* 4.5.11–17; Konecny (2001). An English translation of this meticulous study forms the core of Sekunda and Burliga (2014). The *mora* represented one sixth of the Spartan levy of hoplites and cavalry.

mercenary peltasts in particular, appear to play an ever more prominent role in accounts of warfare on land; writing in 380, Isokrates complained that 'peltasts have seized the cities' in the aftermath of the King's Peace (4.115). Eventually, entire armies would come to be characterised as consisting of such troops, with no trace of a traditional citizen-hoplite core (Diod. Sic. 16.24.4; Dem. 9.49). Did the successes of peltasts render hoplites obsolete, shifting the centre of gravity of Greek warfare entirely to hired specialists in light-armed warfare?[94]

There can be no doubt that peltasts scored many successes in the fourth century, often in situations that do not have clear parallels in earlier periods. After his triumph at Lechaion, peltasts brought Iphikrates another victory when he ambushed the Spartans at Abydos (Xen. *Hell.* 4.8.35–39); Chabrias won a similar victory with a force of 800 peltasts on Aigina the following year (Xen. *Hell.* 5.1.10–12). During the Boiotian War, peltasts are seen successfully defending mountain passes against invading armies (Xen. *Hell.* 5.4.14, 59) and restricting the movements of armies in the field. If the mercenary army of Phokis was indeed built around a core of picked peltasts, as Diodoros suggests, then the Phokians owed their considerable success in the Third Sacred War to these troops. The majority of their thousands of mercenaries are likely to have been peltasts, too. By this time, units of these flexible troops seem to have been a common element of Greek armies.[95]

It is also true that, from Xenophon onwards, references to the mass levy of light infantry all but disappear. Where its presence could often be inferred in Thucydides even if it was not explicitly attested, the fourth-century sources give few indications that poorer citizens were still being mustered to fight as light-armed troops.[96] We might assume that this reflects only a deepening of our authors' prejudice against these low-status warriors, but it may also reflect a genuine absence. If so, the most likely explanation is that the scales had tipped in favour of the trained specialist. States no longer seem to have seen much value in putting large, cumbersome masses of irregular troops into the field when more organised and skilled troops were more effective in the same roles. The only major exception to the trend, already mentioned above, is a telling one: fourth-century Greeks were in awe of the might of the Thessalian

94 Parke (1933); Griffith (1935); Marinovič (1988) 47–77; for a historiographical approach, Konijnendijk (2014).

95 Note the presence of peltasts in the army of Arkadian Orchomenos in 370 (Xen. *Hell.* 6.5.17) and Sparta in 365 (7.4.22).

96 One possible exception is the unidentified Corinthian light troops who blocked a Theban attack on the city in 369 (Xen. *Hell.* 7.1.19). We may assume that such troops were still expected to serve in home defence.

levy commanded by Jason of Pherai, partly because its light-armed element consisted entirely of peltasts (Xen. *Hell.* 6.1.9, 19; Isoc. 8.118).

Yet, as several of the other contributions to this volume make clear, most scholars do not assume a fourth-century revolution in which peltasts eclipsed the citizen hoplite.[97] Greek states that could muster hoplite militias continued to field them; even Iphikrates would eventually command such an army (Xen. *Hell.* 6.5.49; Diod. Sic. 15.63.2). Communities also continued to invest heavily in their cavalry.[98] Where we receive enough detail to reconstruct the course of pitched battles, such as Leuktra and Mantineia, we see that light troops still played only a supporting role. The Athenian ephebes of the 330s may have been trained in several forms of light-armed fighting to prepare them for a variety of military actions, but the equipment they received at the end of the two-year programme was that of a hoplite ([Arist.] *Ath. Pol.* 42.3–4). If the many duties and operations of peltasts become more visible in the sources for this period and we hear complaints that they have supplanted the militia, it may be partly due to their greater spread throughout the Greek world – but also, most likely, to an ideological response to that development. The growing reliance of Greek states and armies on professional light-armed soldiers was as much a practical fact as it was an affront to moralists. Authors and orators used such sentiments to their advantage, decrying the debasement of the old ways even as they warned their fellow citizens against the belief that they could be brought back (Dem. 9.47–51).

This observation brings us back to the running theme of this chapter. Even in the fourth century, after the acknowledged merits of peltasts and other missile troops led to their inclusion in most expeditionary forces and campaign narratives, the notion persisted that their way of war was unmanly and unbefitting a free citizen. Thucydides' Spartans may have dismissed the methods of light-armed troops mainly to boost their own flagging morale (4.40.2, 126.4–6), but their attitude fit into a general ethic of standing one's ground and facing one's enemies. Plato reacted violently against the notion that there could be honour in the hit-and-run fighting style of amphibious and light-armed forces (*Leg.* 706b–d). Even Xenophon, whose works reveal the effectiveness of peltasts and their skirmishing tactics in the richest detail, still urged his own troops to bear in mind that 'it is not right for a good man to retreat from an enemy' (*An.* 6.5.17). In this moral environment, peltasts (as well as archers, slingers,

97 The notion of a set of 'Reforms of Iphikrates' creating a new hybrid soldier out of the mercenary peltast is now generally confined to its most probable context, the campaign in Egypt in the mid-370s: Anderson (1970) 129–32; Konijnendijk (2014); Sekunda (2014).

98 Konijnendijk, this volume.

and javelin-throwers) were never likely to cause a true tactical revolution, no matter how often and how effectively they played their part in the wars of the Greeks.

6 Conclusion

While it appears that the light-armed might have been high-status individuals in the EIA and early Archaic periods, their social position rapidly declined with the introduction of heavy armour in the late eighth century. Once among those given the most spectacular burials known in contemporary Greece and associated with some of the most prestigious figures of Greek myth, in combat archers were stripped of their status along with their clothes in the art of the seventh century, before largely disappearing in the sixth. Preference for the javelin seems to have slipped behind that of the thrusting spear. Slingers, meanwhile, seem never to have been ranked among the heroes and their presence is merely hinted at in our earliest evidence.

Although all social groups played a part in ancient Greek wars, most of the surviving testimony concerning ancient Greek battles, whether in the form of literary accounts or images on pots, was crafted by and for the social elites of the era. As a result, 'it was not the needs of warfare', as Van Wees pointed out, 'but the prevailing political and social order which determined who counted for something and who was dismissed as marginal'.[99] While our sources may disparage light-armed warriors, modern scholars should not simply accept the assessment of ancient writers at face value. Those who fought light-armed may not have written historical or tactical treatises, but they made themselves known through their dedications of their equipment at sanctuaries throughout the Greek world, and are made known even in the sources that belittle them. It is incumbent on us, in the present, to accept that ancient writers also operate from a position of bias, and to understand that bias when we come to assess the role of the light-armed.

When we do so, we can recognise that in the earliest centuries of the first millennium the light-armed mixed with heavily armed warriors, sharing their shields. We can see that, despite the neglect of writers such as Herodotos and Thucydides in describing their deeds, they made up a sizable percentage of those on the battlefield. On numerous occasions, these ancient authors show that the light-armed turned the tide of battle. While they were not always useful, a leader who knew how to deploy them could do well, especially against

99 *GWMR* 85.

an enemy who did not know how to defend against them. Eventually, they were recognised as being so important that Greek poleis turned principally to specialist mercenaries to fulfil their role. Although the evidence varies in quality and quantity, when we approach Greek warfare without preconceptions about the importance of elite, heavily-armed close-combat fighters over their lower-class, light-armed compatriots, we can develop a more balanced understanding that calls many commonplaces into question.

Bibliography

Ahlberg, G. (1971) *Fighting on Land and Sea in Greek Geometric Art*. Stockholm.

Anderson, J.K. (1970) *Military Theory and Practice in the Age of Xenophon*. Berkeley.

Baitinger, H. (2001) *Die Angriffswaffen aus Olympia. Olympische Forschungen* 29. Berlin.

Best, J.G.P. (1969) *Thracian Peltasts and their Influence on Greek Warfare*. Groningen.

Blome, D.A. (2020) *Greek Warfare Beyond the Polis: Defense, Strategy, and the Making of Ancient Federal States*. Ithaca.

Boss, M. (2000) *Lakonische Votivgaben aus Blei*. Würzburg.

Brouwers, J.J. (2007) 'From Horsemen to Hoplites: Some Remarks on Archaic Greek Warfare'. *BABesch* 82.2: 305–19.

Brouwers, J.J. (2013) *Henchmen of Ares. Warriors and Warfare in Early Greece*. Rotterdam.

Catling, H.W. (1985) 'The Arrangement of Some Grave Goods in the Dark Age Cemeteries of Lefkandi'. *ABSA* 80: 19–23.

Catling, H.W. (1993) 'The Bronze Amphora and Burial Urn', in Popham, Calligas, and Sackett (1993) 81–96.

Craven, S.P. (2017) 'The Mercenaries of Hellenistic Crete'. PhD Thesis. University of Texas, Austin.

Crielaard, J.P. (2016) 'Living Heroes: Metal Urn Cremations in Early Iron Age Greece, Cyprus and Italy', in Gallo (2016) 43–78.

D'Onofrio, A.M. (2011) 'Athenian Burials with Weapons: The Athenian Warrior Graves Revisited', in Mazarakis Ainian (2011) 645–73.

Dalsoglio, S. (2019) 'From Amphorae to Cauldrons: Urns at Athens in the Early Iron Age and in the Orientalizing Period', in Graml, Doronzio and Capozzoli (2019) 51–63.

Davis, T.A. (2013) 'Archery in Archaic Greece'. PhD Thesis. Columbia.

Dawkins, R. (ed) (1929) *The Sanctuary of Artemis Orthia at Sparta: Excavated and Described by Members of the British School at Athens, 1906–1910*. London.

Dickinson, O.T.P.K. (1994) *The Aegean Bronze Age*. Cambridge.

Dickinson, O.T.P.K. (2006) *The Aegean from Bronze Age to Iron Age: Continuity and Change Between the Twelfth and Eighth Centuries BC*. London.

Donlan, W. (1970) 'Archilochus, Strabo and the Lelantine War'. *TAPhA* 101: 131–42.

Echeverría, F. (2015) 'Heroic Fiction, Combat Scenes, and the Scholarly Reconstruction of Archaic Greek Warfare'. *BICS* 58.1: 33–60.

Fisher, N. and Van Wees, H. (eds) (1998) *Archaic Greece: New Approaches and New Evidence*. London.

Fisher, N. and Van Wees, H. (eds) (2015) *'Aristocracy' in Antiquity: Redefining Greek and Roman Elites*. Swansea.

Forrest, W.G. (1957) 'Colonisation and the Rise of Delphi'. *Historia* 6: 160–75.

Gallo, F. (ed) (2016) *Omero: quaestiones disputata*. Ambrosiana Graecolatina 5. Milan.

Graml, C., Doronzio, A. and Capozzoli, V. (eds) (2019) *Rethinking Athens Before the Persian Wars: Proceedings of the International Workshop at the Ludwig-Maximilians-Universität München (Munich, 23rd–24th February 2017)*. Munich.

Greenhalgh, P.A.L. (1973) *Early Greek Warfare: Horsemen and Chariots in the Homeric and Archaic Ages*. Cambridge.

Griffith, G.T. (1935) *The Mercenaries of the Hellenistic World*. Cambridge.

Hall, J.M. (2014) *A History of the Archaic Greek World ca. 1200–479 BCE*. 2nd edn. Oxford.

Hornblower, S. (2008) *A Commentary on Thucydides*. Vol. 3. Oxford.

Hunt, P. (1997) 'Helots at the Battle of Plataea'. *Historia* 46.2: 129–44.

Hunt, P. (1998) *Slaves, Warfare, and Ideology in the Greek Historians*. Cambridge.

Hunt, P. (2007) 'Military Forces', in *CHGRW* I: 108–146.

Jeffery, L.H. (1976) *Archaic Greece: The City-States c. 700–500 B.C.* London.

Kelly, A. (2012) 'The Cretan Slinger: A Weighty Exchange'. *ABSA* 107: 273–311.

Konecny, A. (2001) 'Κατέχοψεν την μόραν 'Ιφιχράτης. Das Gefecht bei Lechaion im Frühsommer 390 v. Chr.'. *Chiron* 31: 79–127.

Konijnendijk, R. (2014) 'Iphikrates the Innovator and the Historiography of Lechaion', in Sekunda and Burliga (2014) 84–94.

Konijnendijk, R. (2018) *Classical Greek Tactics: A Cultural History*. Leiden.

Konijnendijk, R. (2021) 'Playing Dice for the Polis: Pitched Battle in Greek Military Thought'. *TAPA* 151.1: 1–33.

Krentz, P. (2007) 'Warfare and Hoplites', in Shapiro (2007) 61–84.

Kucewicz, C. (2021) *The Treatment of the War Dead in Archaic Athens: An Ancestral Custom*. London.

Lagia, A. (2007) 'The Human Skeletal Remains', in Ruppenstein (2007) 273–81.

Lane Fox, R. (ed.) (2004) *The Long March: Xenophon and the Ten Thousand*. London.

Langdon, S. (2008) *Art and Identity in Dark Age Greece, 1100–700 BCE*. Cambridge.

Lavelle, B.M. (2005) *Fame, Money, and Power: The Rise of Peisistratos and "Democratic" Tyranny at Athens*. Ann Arbor.

Lee, J.W.I. (2007) *A Greek Army on the March: Soldiers and Survival in Xenophon's* Anabasis. Cambridge.

Lemos, I. (2000) 'Songs for Heroes: The Lack of Images in Early Greece', in Rutter and Sparkes (2000) 11–21.

Lemos, I. and Kotsonas, A. (eds) (2020) *A Companion to the Archaeology of Early Greece and the Mediterranean*. Hoboken, NJ.
Lewis, D. (forthcoming) 'The Cretan Way of War: Status, Violence, and Values from the Classical Period to the Roman Conquest'. *JHS* 143.
Lippelt, O. (1910) 'Die griechischen Leichtbewaffneten bis auf Alexander den Großen'. Dissertation. University of Jena.
Lissarrague, F. (1990) *L'Autre Guerrier: Archers, Peltastes, Cavaliers dans l'Imagerie Attique*. Paris.
Liston, M.A. and Papadopoulos, J.K. (2004) 'The "Rich Athenian Lady" Was Pregnant: The Anthropology of a Geometric Tomb Reconsidered'. *Hesperia* 73.1: 7–38.
Lloyd, M. (2014) 'The Archaeology of Greek Warriors and Warfare from the Eleventh to the Early Seventh Century BCE'. DPhil Thesis. Oxford University.
Lloyd, M. (2020) 'Weapons', in Lemos and Kotsonas (2020) 499–522.
Marinovič, L.V. (1988) *Le Mercenariat Grec au IVe Siècle avant Notre Ère et la Crise de la* Polis. Trans. J. Garlan and Y. Garlan. Paris.
Mazarakis Ainian, A. (ed) (2011) *The "Dark Ages" Revisited*. Volos.
Nicholls, R.V. (1958/1959) 'Old Smyrna: The Iron Age Fortifications and Associated Remains on the City Perimeter'. *ABSA* 53/54: 35–137.
Osborne, R. (2018) *The Transformation of Athens: Painted Pottery and the Creation of Classical Greece*. Princeton.
Parke, H.W. (1933) *Greek Mercenary Soldiers from the Earliest Times down to the Battle of Ipsus*. Oxford.
Pavlides, N. (2018) 'The Sanctuaries of Apollo Maleatas and Apollo Tyritas in Laconia: Religion in Spartan-Peioikic Relations'. *ABSA* 113: 279–305.
Perlman, S. (1976–77) 'The Ten Thousand: A Chapter in the Military, Social and Economic History of the Fourth Century'. *RSA* 6–7: 241–84.
Perrin, B. (1914) *Plutarch's Lives*. Vol. 1. Cambridge, MA.
Popham, M.R. and Lemos, I.S. (1996) *Lefkandi III: The Toumba Cemetery. The Excavations of 1981 to 1994 (Plates)*. Oxford.
Popham, M.R., Sackett, L.H. and Thelemis, P.G. (eds) (1980) *Lefkandi I: The Iron Age Settlement*. London.
Popham, M.R., Touloupa, E. and Sackett, L.H. (1982) 'The Hero of Lefkandi'. *Antiquity* 56: 169–74.
Popham, M.R., Calligas, P.G. and Sackett, L.H. (eds) (1993) *Lefkandi II: The Protogeometric Building at Toumba. Part 2: The Excavation, Architecture, and Finds*. Oxford.
Powell, A. (ed) (1995) *The Greek World*. London.
Pritchard, D.M. (2010a) 'The Symbiosis between Democracy and War: The Case of Ancient Athens', in Pritchard (2010b) 1–62.
Pritchard, D.M. (ed) (2010b) *War, Democracy and Culture in Classical Athens*. Cambridge.
Pritchard, D.M. (2018a) *Athenian Democracy at War*. Cambridge.

Pritchard, D.M. (2018b) 'The Archers of Classical Athens'. *G&R* 65: 86–102.

Rawlings, L. (2007) *The Ancient Greeks at War*. Manchester.

Ridgway, D. (1997) 'Nestor's Cup and the Etruscans'. *OJA* 16: 325–44.

Roisman, J. (1993) *The General Demosthenes and his Use of Military Surprise*. Stuttgart.

Rop, J. (2019) *Greek Military Service in the Ancient Near East, 401–330 BC*. Cambridge.

Ruppenstein, F. (ed) (2007) *Kerameikos XVIII: Die Submykenische Nekropole. Neufunde und Neubewertung*. Munich.

Rutter, N.K. and Sparkes, B.A. (eds) (2000) *Word and Image in Ancient Greece*. Edinburgh.

Rystedt, E. and Wells, B. (eds) (2006) *Pictorial Pursuits: Figurative Painting on Mycenaean and Geometric Pottery* (*Acta Instituti Atheniensis Regni Sueciae 4,53*). Stockholm.

Sekunda, N.V. (2014) 'The Chronology of the Iphicratean Peltast Reform', in Sekunda and Burliga (2014) 126–44.

Sekunda, N.V. and Burliga, B. (eds) (2014) *Iphicrates, Peltasts and Lechaeum*. Gdánsk.

Shapiro, H.A. (ed) (2007) *The Cambridge Companion to Archaic Greece*. Cambridge.

Snodgrass, A.M. (1964). *Early Greek Armour and Weapons: From the End of the Bronze Age to 600 BC*. Edinburgh.

Snodgrass, A.M. (1999) *Arms and Armor of the Greeks*. 1st ed. 1967. Baltimore.

Snodgrass, A.M. (2013) 'Setting the Frame Chronologically', in *MoB* 85–94.

Trundle, M. (2010) 'Light Troops in Classical Athens', in Pritchard (2010b) 139–60.

Valdés Guía, M. (2019a) 'War in Archaic Athens: Polis, Elites and Military Power'. *Historia* 68.2: 126–49.

Van Gils, L., De Jong, I. and Kroon, C. (eds) (2018) *Textual Strategies in Ancient War Narrative: Thermopylae, Cannae and Beyond*. Leiden.

Van Wees, H. (1994) 'The Homeric Way of War: The *Iliad* and the Hoplite Phalanx II'. *G&R* 41.2: 131–55.

Van Wees, H. (1995) 'Politics and the Battlefield: Ideology in Greek Warfare', in Powell (1995) 153–78.

Van Wees, H. (1998) 'Greeks Bearing Arms: The State, the Leisure Class, and the Display of Weapons in Archaic Greece', in Fisher and Van Wees (1998) 333–78.

Van Wees, H. (2000) 'The Development of the Hoplite Phalanx: Iconography and Reality in the Seventh Century', in *W&V* 125–66.

Van Wees, H. (2013) 'Farmers and Hoplites: Models of Historical Development', in *MoB* 222–55.

Van Wees, H. (2018) 'Thermopylae: Herodotus versus the Legend', in Van Gils, De Jong and Kroon (2018) 19–53.

Van Wees, H. and Fisher, N. (2015) 'The Trouble with "Aristocracy"', in Fisher and Van Wees (2015) 1–57.

Ventris, M. and Chadwick, J. (1973) *Documents in Mycenaean Greek*. London.

Viggiano, F.G. and Van Wees, H. (2013) 'The Arms, Armor, and Iconography of Early Greek Hoplite Warfare', in *MoB* 57–73.
Vlachou, V. (ed) (2015) *Pots, Workshops and Early Iron Age Society: Function and Role of Ceramics in Early Greece*. Brussels.
Vos, M.F. (1963) *Scythian Archers in Archaic Attic Vase-painting*. Groningen.
West, M. (1998) 'Grated Cheese Fit for Heroes'. *JHS* 118: 190–91.
Wheeler, E.L. (1987) 'Ephorus and the Prohibition of Missiles'. *TAPhA* 117: 157–82.
Wheeler, E.L. (2007) 'Battle: A. Land Battles', in *CHGRW* 186–223.
Whitby, M. (2004) 'Xenophon's Ten Thousand as a Fighting Force', in Lane Fox (2004) 215–42.
Williams, D. (2015) 'Ship, Horse, Battle: Some Attic Geometric Fragments from the Sanctuary of Aphaia, Aigina, and Attic Geometric Gold Jewellery', in Vlachou (2015): 253–66.
Williams, M.F. (2004) 'Philopoemen's Special Forces: Peltasts and a New Kind of Greek Light-Armed Warfare (Livy 35.27)'. *Historia* 53.3: 257–77.
Wrightson, G. (2019) *Combined Arms Warfare in Ancient Greece: From Homer to Alexander the Great and his Successors*. Abingdon.
Zaphiropoulou, P.N. (2006) 'Geometric Battle Scenes on Vases from Paros', in Rystedt and Wells (2006) 271–77.

CHAPTER 9

Assaults and Sieges

Rewriting the Other Side of Greek Land Warfare

Fernando Echeverría

In the summer of 431 the Athenians sent a great fleet of 100 ships carrying 1,000 hoplites and 400 archers round the Peloponnese (Thuc. 2.23.2). The fleet, joined later by fifty Korkyrian ships and other allies (Thuc. 2.25.1), made landings at different points on their route, ravaging the land, and then attacked the Lakonian town of Methone. The town was reinforced by a small Spartan contingent commanded by Brasidas, so the fleet weighed anchor and made another landing at Pheia, in Elis, where the Athenians ravaged the land and defeated in battle a unit of 300 picked Eleans. They managed to capture Pheia itself but abandoned it as soon as the Eleans gathered *en masse* to face the invasion. Continuing their cruise, they 'ravaged other places' and reached Acharnania, where they captured the Corinthian fort of Sollium and the town of Astakos, before sailing to Kephallenia and forcing it to join their alliance. They finally returned to Athens on time to join the massive Athenian invasion of the Megarid (Thuc. 2.25, 30–31). In the course of a few weeks, the Athenian fleet had performed a number of landings, devastated the territories of several regions, captured three towns and made an attempt against another, won a minor land battle, subdued a sizeable island, and participated in the invasion of a big neighbouring region as a supporting force.

This campaign has been interpreted as a series of raids or hit-and-run operations,[1] assuming that the objective was basically ravaging enemy territory, that the captured towns were immediately abandoned, and that the Athenians avoided open engagement with the enemy. But the Athenians had simultaneously sent another squadron, thirty ships this time, to Euboia and Lokris, with exactly the same pattern of selected landings, devastation, capture of enemy towns, and fighting against enemy forces (Thuc. 2.26), and the massive Peloponnesian army invading Attika was also following a similar strategy (Thuc. 2.19–23). In all those cases (and in fact in most cases during the Classical period) an army on the march was able to fulfil various objectives and perform different tasks during the same campaign. That included the attack and

1 Lazenby (2004) 35.

 | DOI:10.1163/9789004501751_010

capture of enemy settlements, what is commonly called 'siege warfare'. Sieges are usually regarded as a differentiated kind of warfare, characterised by specific aims and planning, tactics, troops, equipment, and logistics. In modern scholarship, and almost without exception, 'siege warfare' in the Archaic and Classical periods is treated separately from other theatres or aspects of war,[2] and particularly from 'land warfare' in general, which is usually thought to be focused on the pitched battle.

In the light of recent scholarship on the subject, this paper will primarily reconsider the strategic patterns of what we call 'siege warfare' from a more global perspective, within the broader framework of land and sea warfare in Classical Greece.[3] It will claim that the term 'siege' seems unsatisfactory to describe the nature and circumstances of the phenomenon, since it evokes just a particular phase of the process, a formal investment with construction works to reduce the target by starvation while storming the walls and gates with siege engines and artillery.[4] This restricted notion, however, predominates in modern uses of the term and I contend that, as a result, modern visions of ancient Greek poliorcetics seem to be conceptualised with the idea of the formal investment in mind. But in a culture such as ancient Greece between ca. 750–360, in which formal sieges were a minority, the term does not necessarily reflect the kind of practices and ideas that were the norm at that time. This, I believe, must be reconsidered, abandoning the formal investment as a paradigm and crafting new concepts to represent the reality of ancient Greek operations and campaigns.[5]

2 Adcock (1957); Ober (1991); *GWMR*; Rawlings (2007); Strauss (2007); Seaman (2013); Chaniotis (2013). Although I believe that the conclusions of this study can be applied to the Archaic period as well, the evidence used here belongs to the Classical period and the focus will be on Classical Greek warfare.

3 'Siege warfare' is the common label in English-speaking scholarship to refer to the whole process of attacking a fortification, with its different operations and practices: Kern (1999); Strauss (2007), Seaman (2013, 2020). French scholarship uses the term '*poliorcétique*', perhaps in better accordance with ancient Greek terminology: Garlan (1974); Will (1975); *DA* (1992); *DA* (1993); Rebuffat (2000).

4 See, for example, the treatment of the concept in the works listed in the previous note. Will (1975, 298) differentiates between the 'passive' nature of 'siege' and the 'active' nature of 'assault'. The term 'siege' seems to be presented, therefore, as more semantically restricted than 'poliorcetics'.

5 A note on the terminology employed in this paper: I will refer here as 'epistratic' (from *epistrateúo*; for an analysis of the term and its cognates, see Echeverría 2021) to campaigns of invasion of an enemy territory from a general perspective, including operations around or against one or more urban centres; this concept, apparent in the literature of the Classical period, represents a broader and more flexible approach to the dynamics of Classical Greek expeditions. On the other hand, 'poliorcetics' will refer to such operations centred on particular

The aim of this paper is to reinterpret Greek poliorcetics from a broader, holistic perspective, closer to how Greeks actually conceived and executed them according to the literary sources. For that purpose, it will address the basic questions of why, and how, a Greek army approached an enemy town, in the broader context of land operations in general. Military campaigns will be analysed from their inception, considering their aims and objectives, their context and targets, their planning and execution, and finally their development and consequences, to show that poliorcetics were not a separate branch of military operations but an integral part of the process. Classical Greek campaigns were scarcely planned, highly flexible invasions (*epistrateíai*) of enemy territory in which all operational possibilities (battles, assaults, devastations) remained always open.

I will then put forward three fundamental ideas: First, that poliorcetic actions were a consistent and integral part of Greek land operations and therefore must be understood in the general context of land warfare, not at all incompatible with the phalanx. Second, that enemy settlements were rather frequently the real and ultimate target of military campaigns, trying to accomplish the military goal of defeating (that is, imposing one's own will over) an enemy community. Third, that formal sieges were in fact a true minority against a broader horizon of more numerous and frequent assaults, captures, and destructions of urban centres in the Classical period, and that they must be conceived of as the final stage in an escalating process.

1 Academic Approaches to Greek Siege Warfare

Sieges in ancient Greece were for a long time associated with the study of fortifications and the evolution of construction techniques, and thus approached from the point of view of the defence of the community.[6] Considered a 'passive' side of war, the study of the capture and destruction of enemy fortifications attracted considerably less attention from military historians than the analysis of the 'active' side, the pitched battle and so-called 'hoplite' or

settlements. Finally, 'siege' will refer to a specific stage of poliorcetic operations, involving the formal investment of a target with the use of construction works and/or siege technology for a variable period of time.

6 Garlan (1974); Will (1975); Leriche and Tréziny (1986); *DA* (1992); *DA* (1993); Ober (1985, 1991); de Souza (2008); Frederikson (2016). Ancient sources, however, present the phenomenon consistently from the point of view of the attacker, as an aggression. For the methodological problems of studying ancient poliorcetics, see Echeverría (2019).

'phalanx' warfare. Even now, 'siege warfare' is frequently an appendix or addendum in general works and textbooks to a lengthier treatment of the nature and conditions of Greek land warfare.[7] This is partly a methodological choice to address the different aspects of Greek warfare in an orderly manner, but it is also to a great extent the result of sieges being conceived as a distinctive kind of warfare, characterised by practices and structures different from the rest of land operations. And land operations in Archaic and Classical Greece have been frequently assumed to be dominated by the hoplite phalanx.[8]

This created a divide between both dimensions of war, land warfare and 'siege warfare', excluding the ordered ranks of heavy-armed hoplites from poliorcetic action, but the implications for the scholarly reconstruction of Greek poliorcetics were even wider: operations around or against towns and fortifications were considered rare and secondary, often subordinated to open engagements between phalanxes,[9] and heavy infantry, allegedly adequate only for fighting in closed formations, was then unsuitable for assaults and sieges;[10] as a result, the Greeks were said to have shown no interest in sieges but focused instead on raids and agricultural attacks that did not affect urban centres;[11] Greek poliorcetics were consequently poorly developed and unsophisticated.[12] The face of Greek 'siege warfare', thus, was shaped according to the nature and conditions of 'phalanx warfare'.

The hoplite 'mirage', as it were, particularly developed in connection to poliorcetics after Garlan's seminal 1974 work. Garlan contributed decisively to the notion of Archaic and Classical Greek poliorcetics as the result of the ideological choice by 'hoplite-class' citizens to fight according to the chivalrous rules of the phalanx. In his view, Greek practices were static and passive, centred on blockading the enemy and forcing their surrender by starvation, and resulting from the will to devote their best troops, the hoplites, to phalanx warfare. This choice led in turn to an emphasis on open engagements over agricultural land as the best way to defend the community, taking war away from the

7 See above, n. 2.

8 See the introduction to this volume for a summary of the so-called 'hoplite debate'.

9 Hanson (*WWW*, 1999b, 2005). *Contra* Van Wees (2013) 243.

10 A hoplite outside the phalanx is useless: Bryant (1990) 498; Mitchell (1996) 89; Cartledge (1996) 712; Schwartz (2002) 33, 39–40; Hanson (2005) 19.

11 Anderson (1970) 8–9; Garlan (1974) 20–44; Will (1975); Kern (1999) 91, 93–94; Strauss (2007) 237.

12 Kern, for instance, talks of 'ineptitude' (1999, 93), 'inability' (97), and what he considers 'limitations' and 'failure' (112). See also Westlake (1983) 12–13; Hanson (1999a) 8, 93, (2013) 257–59; Campbell (2006) 31.

town and rendering poliorcetics unnecessary and thus underdeveloped. These views found ample support in later scholarship.[13]

Next, Hanson expanded the idea of agricultural devastation and the egalitarian nature of the middle-class hoplites integrating the polis and the phalanx,[14] and consequently developed a negative view of Greek poliorcetics: 'effective siegecraft' was 'in its infancy', determined by the 'heroic code of the citizen-soldier' that judged it 'repugnant' to 'shun decisive battle' and 'ominous' to 'divert capital and labor to fortify cities and passes'.[15] This encouraged in turn a line of study emphasising the 'rules' or 'protocols' of Greek warfare, allegedly imposing restrictions on the natural excesses of war and banning or despising military practices outside the phalanx.[16] That 'code' has been claimed to ensure that 'the "middling" hoplite class remained the gravitational center of Greek society',[17] thus linking it to the phalanx. If the goal of Greek armies was restricted to 'meet the enemy soldiers in an open battle'[18] and other strategies were ideologically despised, poliorcetics were then out of the question. These combined lines of phalanx warfare, agricultural devastation, and ritualised protocols ultimately supported the idea of Greek poliorcetics as faulty.[19]

Most of these assertions have been questioned in recent years, not only the debate around the introduction of the phalanx but also its multiple consequences. The idea of a gradual and slow evolution towards a closer formation by the end of the Archaic period has had the effect of liberating many lines of study from the chains of the phalanx and paved the way for interesting alternatives. A less formalised, less egalitarian, less 'hoplite' polis is reconstructed, in which the dominant ideology is not necessarily the solidarity and equality of the 'hoplite class' but the exclusiveness and agonistic ethos of a wealthy elite. In the military field, 'hoplite ideology' and 'hoplite warfare' have been reinterpreted as the ideal projections and aspirations of the citizen body of the polis, enabling the consideration of 'irregular' ways of war (particularly low-scale

13 Will (1975); Connor (1988) 12; Hanson (*WWW*, 1999a); Bryant (1990) 511–12 n. 8; Ober (1991, 1996b); Mitchell (1996) 92–94.

14 Hanson (*WWW*, 1991a, 1991b, 1999a, 1999b, 2005).

15 Hanson (1999a) 8.

16 Ober (1985, 1996b); Connor (1988); Hanson (*WWW*, 1999a); Mitchell (1996).

17 Ober (1996a) 54.

18 Ober (1996a) 61.

19 The notion of the late development of Greek poliorcetics is partly a consequence of the unbalanced state of our sources: Echeverría (2019) 324. In contrast, the pattern of consistent attacks against enemy settlements described here can be successfully detected during the Archaic period – Seaman (2020) – and the analysis of the vocabulary allows to identify it already in the Homeric epics, Echeverría (2021).

predatory warfare and amphibious operations) and the widespread use of surprise, deceit, and treachery in Greek warfare.[20]

In this line, in a crucial and liberating change of perspective, sophisticated siege techniques have been regarded as 'unnecessary':[21] ancient Greek poliorcetics were perfectly fine and adequate for the world of the Archaic and Classical polis, and, according to the narratives of the Classical historians, they were pretty effective too. The analysis of the vocabulary and narrative in ancient Greek literary sources, furthermore, reveals that the Greeks conceptualised poliorcetics as involving open-ended and flexible campaigns, escalating in several stages but indistinguishable from the rest of land and sea warfare and connected to the general ideas of power and hierarchy that governed interstate relations. All this seems to point at 'siege' as an inadequate concept to characterise ancient Greek poliorcetics.

The issue, however, still demands further treatment. Leaving aside purported ideological inhibitions and moral scruples on the part of the Archaic Greeks (that were clearly not an obstacle for the expansion of poliorcetics from the late fifth century onwards), a reasonable explanation for the alleged lack of sophistication at the beginning of the Classical period has been found in Thucydides' view (1.11) of the comparatively limited availability of material resources by Greek communities.[22] Operations around or against fortifications consumed most of the material and human resources of Greek armies, and the still unconsolidated political organisation of the late Archaic polis had a limited capacity to provide logistics and mobilise manpower in sufficient quantities for such a sustained effort. The consequence is clear: we need to interpret poliorcetics in the early Classical period not in connection with the phalanx, but in the context of Greek material culture and mentality.

The context for a revision of our reconstruction of poliorcetic operations has now been set. Effective ways to contribute to that revision would be, first, to analyse the vocabulary and conceptualisation of poliorcetics in the written sources, and second, to study Greek poliorcetic practices themselves. I will focus on the second task here,[23] offering some reflections on particular aspects of Greek poliorcetics in the Classical period and then integrating them into a more general view. The selection of aspects treated here responds to a personal judgement of the needs and conditions of current research on the issue: by

20 Krentz (2000, 2002); *GWMR* 115–50; Rawlings (2000), (2007) 63–68, (2013) 8–12; Seaman (2020).

21 Rawlings (2007) 131.

22 *GWMR* 144; Strauss (2007) 237.

23 The first has been explored in Echeverría (2021).

what means the attack on a settlement was justified and legitimised, which were the crucial factors that contributed to shaping a campaign, and what kind of process (the sequence of distinct operations) can be reconstructed. For that purpose, I will focus here on the Greek literature of the fifth and early fourth centuries, particularly historical and technical prose, not only because of the greater quality and quantity of the information, but also because the period is universally regarded as a turning point in Greek history and warfare. I will also focus on expeditions across borders in conflicts between enemy communities (as opposed to civil wars)[24] because poliorcetic actions were fundamentally carried out in enemy territory, and I hope the following pages will help to explain this particular choice. I conceive this analysis as a start, to be contrasted with complementary studies on other periods that will shed further light on the crucial question of why and how a Greek army approached an enemy town.

2 Justifying the Attack

Ancient Greek armies approached settlements on a regular basis, as we will argue later on. The practical reasons to do so were manifold, but they all responded to a combination of strategic, economic, political, and ideological concerns.[25] This cursory survey will show that the Greeks had many good reasons to attack a town, and that epistratic strategies served different but complementary purposes with the final aim to protect their own interests and damage the enemy's.

The Greeks sometimes presented reasons of an ideological nature to attack a settlement, such as liberation from slavery: it was invoked by Brasidas in his successful string of attacks against Athenian allies in the Chalkidike (Thuc. 4.86.1), and although it was more commonly associated with the declaration of wars,[26] particularly by the Spartans and Corinthians in 431 (Thuc. 1.124.3, 139.3) and in the Hellenic campaigns against the Persians in the Aegean (Hdt. 1.141, 5.49; Thuc. 1.96; Xen. *Hell.* 3.1.3), Thucydides openly regarded it as a pretext (Thuc. 1.23.6), albeit a powerful one. The other main ideological reason to attack a town (and frequently combined with the previous one) was

24 On invasions, see Whitehead (1990) 22–24; Krentz (2007); Rawlings (2013) 8–12.

25 Van Wees (*GWMR* 19–33) identifies ideological, economic, and political reasons for Greek wars; see also his treatment of the Greek discourse on the aetiology of war (19 n. 4). See also Lendon (2000, 1–3), who criticises the academic tendency to 'amend' the causes given to particular wars or campaigns by the ancient sources.

26 *GWMR* 21–22.

change of government, something popularised during the Peloponnesian War but intermittently practised throughout Greek history: the Spartans, for example, participated in the fall of the tyrants in Athens (Thuc. 1.18.1) and in an attempt against Polykrates of Samos (Hdt. 3.44–47); the Athenians attacked Samos again in 440 and established a democratic regime (Thuc. 1.115.2), and decades later the Athenian Thrasyboulos still considered it wise to change the constitution of a potential ally such as Byzantion (Xen. *Hell.* 4.8.27). This measure, however, seems more often the effect than the cause of the expedition, and other factors, such as old feuds and insults, play a part in it as well.

Economic factors were obviously a powerful motivation, fundamentally the collection of booty, not only as spoils of war but also as a way to provision the army. Booty could be presented as the main reason to attack a town: Thucydides describes how 'powerful men' in the old times used to fall upon unprotected towns to plunder them and thus obtain 'the source of their livelihood' in what he considers a general situation (Thuc. 1.5.1); that mirrors similar evidence present in the Homeric epics.[27] A desperate Spartan army attacked Chios in 406 to mitigate their lack of provisions (Xen. *Hell.* 2.1.1), and there are reasons to think that it was a widespread phenomenon: devastation of the land and plunder of settlements was a fairly regular stage of any epistratic action,[28] which means that invading armies always had an economic interest in warfare. Plunder and provision can be claimed to be the reasons behind the capture and destruction of an endless list of settlements by wandering armies for apparently no reason,[29] collateral victims of armies on the move: according to Xenophon, a Spartan fleet sent in 388 to Abydos turned to Tenedos to plunder the land and continued their journey once they collected the money they needed (*Hell.* 5.1.6); Iphikrates led a fleet of ninety ships in 372 against the Peloponnese but stopped first in Kephallenia to extract money from the towns there (Xen. *Hell.* 6.2.38). This need provides a likely pattern – armies provisioned in the field attacking random victims in their path, which also served to punish the enemy and bully them with acts of force.

Finally, there were strategic reasons connected to political and military interactions and relations of power between communities. One of them is expansion, the will to extend the territory under direct control, that

27 Van Wees (1992) 167–258; Jackson (1995); Rawlings (2007) 28–31; Trundle (2020) 21–25. See also Lloyd, this volume, on the role of booty in Early Iron Age and Archaic warfare.

28 Hanson (1999a), particularly 185–94, for the vocabulary of plunder and devastation.

29 Examples: Anaktorion (Thuc. 1.55.1), Therme and Pydna (Thuc. 1.61.2–3), Chalkidian towns (Thuc. 1.65.2), Naupaktos (Thuc. 1.103.3), Haliai (Thuc. 1.105.1), Boion, Kitinion and Erineus (Thuc. 1.107.2), Zakynthos (Thuc. 2.66), Spartolos (Thuc. 2.79), Acharnanian towns (Thuc. 2.102.1), Tanagra (Thuc. 3.91).

is presented at times as a cause for attack: the dispute between Samos and Miletos over Priene led to the blockade of Samos in 440 (Thuc. 1.115.2), while the will to control Plataia led to the Theban attacks of 431 (Thuc. 2.2.3); the Chalkidians denounced in 383 the aggressive policy of expansion by Olynthos, that led to the annexation of some towns and the capture of others (Xen. *Hell.* 5.2.12). Expansion, which is achieved at the expense of neighbouring territories, is just a dimension of a general will to subdue or subject other, sometimes distant, communities to your own leadership, ambitions described by Iason of Pherai as a process of irresistible submission to a powerful state (Xen. *Hell.* 6.1.7–12). The pursuit of hegemony explains Polykrates' expeditions in the late sixth century (Thuc. 1.13.6) and the Athenians' restless campaigning in the Aegean during the Pentekontaetia (Thuc. 1.89–118) that resulted in the capture and destruction of dozens of settlements. When the Corinthians were trying to persuade the Spartans of the hegemonic ambitions of Athens, they brought up the cases of two big towns, Korkyra and Potidaia, that were simultaneously being seized by the Athenians in two distant theatres (Thuc. 1.68.4), suggesting that the physical capture of urban centres is the best way to assert global leadership. The dynamics of hegemony can lead to aggressive strategies, punishing revolting allies (Thuc. 1.100.2, 114.1), which is the right of the hegemonic power (Thuc. 1.40.5), but also to pre-emptive strategies, increasing the influence in a certain region (Thuc. 2.33.2) and occupying cities to prevent them from revolting (Thuc. 1.57.3, 2.27.1, 3.3.1–2) or the enemy from capturing them (Thuc. 1.60.1, 4.73, 4.89).

Another group of strategic factors emphasises initiative and action on the side of the attacker. Some places are attacked because of their strategic value, such as Minoa by the Athenians (Thuc. 3.51.1–2) or Messina by the Syracusans (Thuc. 4.1). Strategic initiative is also implied in the technique called *epiteichismós*, literally to 'fortify a place against' the enemy, that involved the capture of a small settlement in enemy territory and its fortification to house an invading army.[30] There are multiple cases in Greek warfare,[31] and it is presented as a way to harm the enemy by hampering agricultural production, increasing the chances of social and political unrest, and facilitating the desertion of slaves (Thuc. 1.142, 7.27–28). The geography of Greece offered countless opportunities for *epiteichismós*: when in 425 Demosthenes insisted on fortifying Pylos, his colleagues, not totally convinced of the suitability of the region,

30 Garlan (1974) 33–38; Westlake (1983); Hanson (1999a) 28–29, 212; Krentz (2007) 177–78.

31 Examples: Pylos (Thuc. 4.3–4), Methana (Thuc. 4.45.2), Delion (Thuc. 4.76.5, 90), Dekeleia (Thuc. 6.91.6–7, 7.18.1, 18.4, 19.1), Kythera (Thuc. 7.26.2), Delphinion (Thuc. 8.38.2), Epieikeia (Xen. *Hell.* 4.4.13), Phoinikos (Xen. *Hell.* 4.8.8), Aigina (Xen. *Hell.* 5.1.2).

argued that there were 'plenty of desert hills in the Peloponnese to occupy' (Thuc. 4.3.3).

Finally, some strategic reasons were reactive: retaliation for a previous aggression could be the reason to attack a town, such as Leukas and Kyllene by the Korkyreans (Thuc. 1.30.2), Antissa by the Methymnians (Thuc. 3.18.2), Naupaktos by the Spartans (Thuc. 3.100), or the many settlements harassed by the Athenians in their campaigns round the Peloponnese (Thuc. 2.23.2, 56). Within this pattern of retaliation lays the phenomenon of old feuds and quarrels between communities that ignited intermittent wars on account of vengeance[32] and led to the capture of urban centres such as Amphilochian Argos (Thuc. 2.68), Kydonia (Thuc. 2.85.5–6), Rhegium (Thuc. 4.1.2) or Laodikeion (Thuc. 4.134). An interesting kind of reactive strategy is also the familiar pattern of a besieged community calling in a third party to invade the land of their invader in order to relieve pressure: several Spartan invasions of Attika are inspired by the request of communities threatened by the Athenians, such as the Thasians (Thuc. 1.101.1–2), Aiginetans (Thuc. 1.105.2–3), and the Mytileneans (Thuc. 3.13.3–4, 3.25).

In a sense, then, the practical reasons behind the attack of a town or fortification were a materialisation or application of the general causes of wars in ancient Greece, namely 'honour, fear, and profit', as Thucydides put it (1.76.2). This suggests that poliorcetic action was not a different kind of campaign from the rest of land and sea operations, but that war was conceived as an integrated phenomenon developed according to the changing circumstances of action. Those practical reasons were, moreover, embedded in an ideological framework that both explained and justified the fact of having urban centres as consistent and legitimate targets of military aggression: attacking and capturing towns was not only profitable or strategically wise, but also a potential source of power and prestige. Epistratic warfare is consistently connected in ancient Greek literature with a greater availability of economic resources. Thucydides particularly identifies economic growth with more intense piratical activities (Thuc. 1.2.2, 7), and he concludes that the failure at Troy was the result of lack of resources (Thuc. 1.11.1). In a speech by Perikles, he also presents the predominantly Athenian view of the Peloponnesian way of war as determined by poverty (Thuc. 1.41.3–7), later illustrated by the limited reach and length of the annual Peloponnesian invasions of Attika (Thuc. 2.23.2, 3.1), while wittingly making Archidamos recognise that the cause of the dreadful Athenian war making is their greater wealth and resources (Thuc. 1.80–81).

32 Lendon (2000).

Thucydides makes both politicians agree on the fact that war is a matter of money (Thuc. 1.83.2, 2.13.2–5).[33]

According to Thucydides, then, greater resources were the key to successful war making and enabled Greek communities to assault fortifications with better chances, but they were also the mark of powerful states.[34] He explicitly connects effective poliorcetics with reflections on power and strength: being the most powerful king in Greece is Agamemnon's legitimation for his claim to military leadership against Troy (Thuc. 1.9.1), allowing him to coerce other Greek leaders to support him (Thuc. 1.9.3–4); being powerful in ships enabled Polykrates to subdue part of Ionia (Thuc. 1.13.6), and Greek communities were able to wage greater wars when they started to cooperate in military leagues and alliances that increased their power (Thuc. 1.15.2). All these theoretical elaborations about the past reflect a fifth-century mentality about the relationship between power and war that is fundamentally confirmed by Thucydides' and Xenophon's descriptions of Classical wars. In this mentality, power is the supreme source of legitimacy for military action, because 'when it is possible to use force, it is not necessary to resort to law' (Thuc. 1.77.2), and the most advisable way to increase power is to rule over others.[35] Rule can be the result of a military alliance, because the allies are, in the words of the Athenians, 'the source of our strength' (Thuc. 1.43.5), but it is preferable to exert direct hegemony, because to rule over others (especially the weak) is in accord with human nature (Thuc. 4.61.5, 5.105.2), and aggressive policies befit mighty states (Thuc. 4.64.1). The vocabulary of poliorcetics, furthermore, emphasises this predominance of power and hierarchy through the notions of 'capture' and 'destruction', which refer to the rights and privileges of the conqueror as the result of victory.[36]

Van Wees argues that power was not, however, an end in itself but a means to a further end: to win respect and wealth.[37] In this context, then, the power and resources to wage war granted status and prestige, and assaulting and capturing enemy towns was an important part of that process. If taking goods from the enemy brought fame (*eúkleia*, Xen. *Hell.* 5.1.17), a campaign's success or failure could then be measured according to the number of places captured, like Agesilaos' expedition to Acharnania (Xen. *Hell.* 4.6.13). Thucydides identifies an old pattern in which powerful men plundered random towns and thus

---

33 Resources and Greek warfare: Anderson (1970) 8; Kern (1999) 89–90; *GWMR* 102–13, 235–39; Rawlings (2007) 158–69; Trundle (2020).

34 For power in Greek warfare and politics, see Raaflaub (1994); *GWMR* 24–25, 32–33. For power in Thucydides, see Mara (2009).

35 *GWMR* 24 n. 20 gathers the evidence on this point.

36 Echeverría (2021).

37 *GWMR* 32–33.

acquired reputation (*dóxes*, Thuc. 1.5.1), but although he implies that this has become a shameful practice in his own time, the numerous negotiations over captured towns in peace agreements indicate that Greek communities found a source of prestige in their capture and temporary control (and also a mark of power and a tool for diplomatic leverage, Thuc. 2.29.6, 30.1, 4.65.1, 118, 5.17.2, 18.5–8). This also helps to explain the efforts to hold certain towns with the prospect of future negotiations, like the Spartan officer who accepted the surrender of Plataia (Thuc. 3.52) or Brasidas' explicit strategy in the Chalkidike (Thuc. 4.81.2).

As a result, attacking enemy towns in a context that found it ideologically legitimate and prestigious was not a collateral or accidental action, but a fundamental part of Greek war strategy. Urban centres were in fact the intended objective of military expeditions on a fairly regular basis: when the strategic aims of the campaign are explicitly stated (which is not always the case), a human group, a certain region or even a specific urban centre is usually mentioned, so the campaign itself is planned around that primary objective.[38] General references to human groups or regions may not necessarily imply explicit poliorcetic operations, but indicate that the aggression is led against the enemy community as a whole, either in its human or its territorial dimension, so any kind of operation is in fact contemplated. Specific references to towns show more clearly that the aim is to threaten, capture, or even destroy the settlement itself. King Kleomenes, for example, 'launched an invasion (*esébale*) against Eleusis' (Hdt. 5.74.2) in 506, while the Hellenic alliance 'made an expedition (*estráteusan*) against Byzantion' (Thuc. 1.94.2) in 478, and the Athenians 'made an expedition (*estráteusan*) against Abydos' (Xen. *Hell.* 1.2.15) in 409.[39]

Towns could be the centre of a military campaign even when the strategic objectives are not explicitly stated, or even when they are but refer to other matters: in the expeditions of 431 described in the introduction, for example, not one but several urban centres were attacked with unequal results, either by the Athenians or by the Peloponnesians. This was in fact a fairly regular pattern:

38 Communities as targets: Hdt. 1.66.3 (Tegeans), 1.71.4 (Lydians), 1.166.1 (Phokaians), 5.15.2 (Paionians), 5.89.2 (Aiginetans); Thuc. 1.26.4 (Epidamnians), 1.30.3 (allies), 2.79.1 (Chalkidians), 3.54.2 (Plataians); Xen. *Hell.* 2.4.30 (Athenians), 2.4.43 (Eleusinians), 3.1.13 (Mysians and Pisidians), 6.5.36 (Lakedaimonians and Tegeans), 7.2.2 (Phliasians), 7.4.20 (Arkadians). Regions as targets: Hdt. 1.73.1 (Cappadocia), 7.9a.2 (Macedonia), 7.25.1 (Hellas); Thuc. 1.58.1 (Macedonia), 1.58.1 (Attika), 1.69.5 (Peloponnese), 1.107.2 (Doris), 2.30.2 (Kephallenia), 2.31.1 (Megarid); Xen. *Hell.* 1.1.37 (Sicily), 3.1.7 (Karia).

39 Other examples: Miletos and Smirna (Hdt. 1.15), Klazomenai (Hdt. 1.16), Sardis (Hdt. 1.27.3), Paros (Hdt. 5.117), Athens (Hdt. 7.138.1), Potidaia (Thuc. 1.58.1), Pharsalos (Thuc. 1.111.1), Miletos (Thuc. 1.115.5), Athens (Thuc. 2.11.6), Plataia (Thuc. 2.12.5), Kalydon (Xen. *Hell.* 4.6.1).

the Phokians attacked three towns in Doris around 457 and captured one of them (Thuc. 1.107.2); an Athenian fleet around 456 burnt a Spartan military post (Gytheion) and captured the Corinthian town of Chalkis, with a minor battle with the Sikyonians (Thuc. 1.108.5); the Korkyreans devastated the island of Leukas and burnt the town of Kyllene in 435 (Thuc. 1.30.2); the Athenians stormed three forts in Macedonia (Therme, Pydna, and Strepsa) in 432 before turning to Potidaia where they conducted a two-year siege (Thuc. 1.57.2, 59, 62–65). In his anecdote on Miltiades' failed campaign against Paros, Herodotos (6.132) seems to assume that targeting specific settlements or communities was a standard practice, because he finds it striking that Miltiades 'asked the Athenians for seventy ships, an army, and money, not revealing against what country (*chóren*) he would lead them (*epistraeúsetai*)'. He is not wondering about the purpose of the expedition, but about its target.

Furthermore, Greek generals had considerable freedom of action and a free hand to reconsider or rearrange the objectives of a campaign according to the circumstances. In the case of the Athenian force sent to Potidaia, Thucydides says (1.59) that, upon arrival and seeing the town already in revolt, the generals decided to stick to their original instructions and went to Macedon. The general Demosthenes led an Athenian fleet against Leukas in 426 but decided to attack the Aitolians instead, as he was persuaded by his Messenian allies, who argued that leading such a vast army constituted 'a fine opportunity' (Thuc. 3.94.3), but also because he had devised an entirely new objective by himself, namely, to attract all the peoples of central Greece to their side and thus have free hand to march against Boiotia, the real enemy, from the north (Thuc. 3.95.1).

This tendency to set open and flexible objectives seems to have had the effect of situating the invasion at the centre of Greek strategic planning in the Classical period. A standard campaign was conceived strategically as little more than an invasion of enemy territory, because, once there, Greek commanders enjoyed a high level of autonomy and initiative and, far from home, tended to improvise on the spot. The invasion started a military campaign of unpredictable course, which may include attacking a settlement. Attacking and destroying enemy towns was a legitimate way to conduct a war, serving multiple strategic purposes and fitting a militaristic mentality that privileged power and aggression on the one hand and wealth and resources on the other. Poliorcetic actions were part of the general Greek concept of war. Neither theoretical elaborations and reflections nor general conceptualisations across time ever treated them separately from other land or sea operations, but rather integrated in a consistent strategy of invasion (*epistrateía*), and they could even be the central piece of the planning of a campaign. This takes us to the question of the frequency of captures and assaults in Greek warfare.

3 Contextualising the Attack

Several aspects of the context of poliorcetic actions can help us to reconsider their role in Greek warfare. One of them is the relative frequency of captures and assaults of settlements in Archaic and early Classical warfare, commonly regarded as fairly low.[40] Analysis of the narratives of Herodotos, Thucydides and Xenophon, however, indicates that urban operations of different kinds (not only formal sieges) were in fact rather frequent, particularly in comparison to other operations such as pitched battles or ravaging expeditions. A systematic catalogue of epistratic campaigns and their targets in the Classical sources remains to be done, but some cautious estimates for particular periods can be offered. Seaman, for example, differentiates twenty-one operations or campaigns during the Pentekontaetia and more than eighty during the Peloponnesian War, which break down into at least thirty-one and ninety-five different targets attacked, respectively.[41] On the other hand, while different lists of pitched battles have been compiled in recent years by different specialists, most of them are either too general (only the most famous battles are included) or too specific (restricted according to the needs of a particular study).[42] Hanson's request for a detailed catalogue of land battles in Greek warfare is still to be answered.[43] Krentz estimates that some 101 'poliorcetic incidents' took place during the Peloponnesian War, 'plus numerous other invasions limited to ravaging', while only forty-seven battles were fought according to Thucydides' and Xenophon's narratives.[44] I can contribute a few statistics of my own, based on a personal reading of Thucydides and Xenophon, and compiled in Table 1.[45]

40 Low frequency of poliorcetics: Garlan (1974); Hanson (*WWW*, 1999a); Kern (1999) 89; Campbell (2000) 30–31. *Contra* Krentz (2007) 169–70; Van Wees (2013) 243; Chaniotis (2013) 440; Sears (2019) 96–97; Seaman (2020).

41 Seaman (2013) 653–55. His useful appendices illustrate the many troubles of statistical work with ancient evidence: all actions (captures, destructions, assaults, blockades) are listed as 'sieges'; operations involve periods of rather diverse length, from days to months or even years; some actions (such as the Athenian landing in Sphakteria) are included because the ancient sources label them as 'sieges'; others are omitted for different reasons (for example, operations with unnamed or even unnumbered targets).

42 Lists of battles: See *HCGBE* x; Krentz (2007) 169; Schwartz (2009) 235–92.

43 Hanson (1991c) 254.

44 Krentz (2007) 168–70.

45 For the sake of simplicity, three case studies with adequate literary coverage have been selected. To the figures presented in the table we must add Thucydides' five and Xenophon's eight references to the attack of an undetermined number of targets (towns, forts, or coastal places in general).

TABLE 9.1 Expeditions and their targets and battles in three case studies from the Classical period

Period	Expeditions	Targets approached (towns)	Individual towns	Individual regions	Individual islands	Captures	Battles
Pentekontaetia	55	42 (38)	32	–	4	22	21
Archidamian War	102	114 (95)	79	11	8	52	41
Corinthian War	42	24 (22)	16	–	2	9	16

These figures speak for themselves – if we combine the information from the three periods, Greek armies fought seventy-eight battles, but approached urban centres in 155 occasions and captured *a minimum* of eighty-three of them (many more if we include the thirteen times in which the number of targets is not recorded, or the eleven regions and fourteen islands, with their towns, that are attacked and devastated), in a total of 199 expeditions. The figures may be questionable on different grounds, but they provide a preliminary framework for study. What they indicate is that, far from restricting military operations to agricultural devastation or open engagements, Greek armies regularly targeted communities, territories, urban centres, and forts. Attacking a town was in fact more likely than fighting a pitched battle, at least twice as likely, as we can see.[46] So much for frequency.

In practice, any kind of settlement, from unfortified villages to sizeable cities, could be the target of an attack. The literary sources usually provide the name of the site, but they offer little information on its nature and characteristics, so we are left to speculate. Naturally, unwalled settlements such as Pallene or Klazomenai made easier targets,[47] but in most cases the sites were walled, even though they could be *astheneis*, 'ill defended' (Xen. *Hell.* 3.1.7). Literary sources incidentally refer to some sites as *polísmata* (towns), *choría* (places), *peripólia* (forts), or *kómai* (villages),[48] but these different labels do not seem to correspond clearly with different patterns or practices to capture

46 Konijnendijk (2020, 2021) argues that the Greeks actually distrusted pitched battles as an efficient way to settle conflicts and thus avoided them as much as possible.

47 Sites described as *ateíchiston*, 'without walls', in the literary sources: Pallene (Thuc. 1.64.1), Limneiai (Thuc. 2.80.8), Ionian towns (Thuc. 3.33.2), Aitolian towns (Thuc. 3.94.4), Solygeia (Thuc. 4.43.1), Klazomenai (Thuc. 8.31.3), Knidos (Thuc. 8.35.3), Kos (Thuc. 8.41.2), Kamiros (Thuc. 8.44.2), Samos (Thuc. 8.50.5), Lampsakos (Thuc. 8.62.2), Kyzikos (Thuc. 8.107.1), Elis (Xen. *Hell.* 3.2.27), Peiraieus (Xen. *Hell.* 3.5.16), Sparta (Xen. *Hell.* 6.5.28), Lakonian towns (Xen. *Hell.* 6.5.32).

48 See Hansen (1995).

them. Hansen offers details about the size of Greek settlements[49] and classifies them in different categories, from less than four hectares to more than 150 hectares, and he concludes that the great majority of urban centres had a size between ten and 100 hectares (most of them in fact between twenty and fifty hectares) so the average size was sixty-five hectares and the median forty hectares. This allows us to differentiate roughly a top category of really big cities (Athens, Corinth, Thebes, Syracuse), a secondary class of medium-sized cities (Potidaia, Sikyon, Korkyra), another group of smaller towns (Torone, Pydna, Mende, Methone), and finally a category of small forts and villages (Solygeia, Oinoe, Ithome, Limneiai).

The most recurrent target of Greek expeditions seems to be the third class of small towns. Fortified yet vulnerable to attacks, they are commonly captured or destroyed by wandering armies through direct assaults: a Phokian expedition against their Lokrian neighbours around 457 attacked the towns of Boion, Kitinion and Erineon, capturing one of them (Thuc. 1.107.2); an Athenian force sent to Chalkidike in 432 to prevent the defection of Potidaia captured the Macedonian town of Therme and assaulted Pydna and Strepsa unsuccessfully, before marching against Potidaia itself (Thuc. 1.61.2–4); the Spartans invaded the territory of Corinth in 392, capturing Sydos and Krommyon and occupying and fortifying Epieikeia (Xen. *Hell.* 4.4.7–13). During the Archidamian War, the Athenian expeditions round the Peloponnese tended to concentrate their attacks against small settlements in Lakonia or the Gulf of Corinth: Methone, Pheia, Sollion and Astakos in 431; Troizen, Halieis and Prasiai in 430; Oiniadai and a number of Lakonian *choría* in 428; Ellomenos, Potidania, Krokyle, Tichion and Segition in 426; and Skandeia, Kythera, Epidauros Limera, and Thyrea in 424. Peloponnesian and Athenian campaigns in other theatres such as Thrace and the Chalkidike in 431–22 also targeted primarily this kind of secondary centres that were easier to capture and offered a reasonable reward in booty at a tolerable risk.

Bigger cities, however, were not free from danger either. Between 431 and 427 the Athenians invaded the Megaris at least once every year and attacked Megara itself in 427 (Thuc. 3.51) and in 424 (Thuc. 4.67–69); Athens was approached by the Spartans in 411 (Thuc. 8.71) and occupied at the end of the Peloponnesian War after a long blockade (Xen. *Hell.* 2.1.5–23); Corinth was attacked several times by the Spartans between 392 and 390 (Xen. *Hell.* 4.4.7–19, 5.1–10); Syracuse was the centre of the Athenian operations in Sicily between 415 and 413 (Thuc. 6–7), and Sparta the target of Theban expeditions between 370 and 362 (Xen. *Hell.* 6.5.24–7.5.27). To capture them was not easy,

49 Hansen (2006) 73–76.

however: when Agis approached Athens in 411, he did it at night and relied on the political chaos in the city on account of the administration of the Four Hundred, but he failed and had to withdraw. Syracuse resisted after two years of intense fighting with a huge Athenian force deployed at the gates, and the attackers of Megara and Corinth managed only to capture the outer walls for a while.[50] Armies approaching big cities seem more cautious and tend to use indirect measures such as treason before launching a direct attack.

The size of the target seems to keep a correlation with another crucial aspect: the composition and numbers of the attacking army. Greek military expeditions almost universally involved troops of different kinds, at least light-armed and heavy-armed infantry in sufficient numbers, but we also hear of archers or cavalry as part of the force.[51] This can be deduced not only from the lists of troops that are sometimes offered in the literary sources at the beginning of the campaign, or before a relevant battle, but also from the occasional references to them during the account of a particular campaign: the allies broadly mobilised 'two thirds of their forces' (Thuc. 2.10) for the Peloponnesian invasion of Attika in 431, for example, but later we hear of an engagement between Athenian and Boiotian cavalry (Thuc. 2.22.2). It seems safe to assume that the rowers of the fleet could be used as light-armed troops if necessary, as recorded in the campaign of Sphakteria (Thuc. 4.32.2),[52] and in the wars of the first half of the fourth century it is common to see semi-professional, semi-mercenary armies with large numbers of light-armed forces of different kinds. This has been studied in detail,[53] so I will not dig again into the 'politics' of the battlefield[54] that gave preference to heavy infantry over other types of troops both in public memory and in historical accounts. Conversely, Rawlings demonstrated that heavy infantry was an adaptive force regularly involved in all kinds of campaigns, including amphibious operations, raids, and assaults.[55] I think it is worth emphasising that different kinds of troops were systematically mobilised in Greek expeditions, not necessarily depending on their strategic objectives, and that those multi-troop armies could be equally launched against

50 See also Diod. Sic. 14.14–16 on Dionysios of Syracuse, who attacked and captured his Sicilian enemies one by one using night attacks, treason, and trickery. I am grateful to the editors for pointing this out to me.

51 Archers: Thuc. 2.23.2, 3.107.1, 4.28.2, 129.2; Xen. *Hell.* 4.2.16. Cavalry: Thuc. 2.56.1, 79.1, 4.42.1, 53, 124.1, 5.2.1; Xen. *Hell.* 4.2.16–17. See further Echeverría (2011) 58–60 and the chapters by Konijnendijk and Kucewicz, Lloyd and Konijnendijk in this volume.

52 See *GWMR* 62–63 for discussion.

53 Anderson (1970) 111–38; *GSW5* 1–67; Hunt (1998, 2007); *GWMR* 61–76; Rawlings (2007) 85–88; Echeverría (2011) 58–60; Lee (2013) 147–51.

54 Van Wees (1995), *GWMR* 77–85.

55 Rawlings (2000).

an enemy territory, army, or town, according to the specific circumstances of the campaign.

Numbers, on the other hand, were in principle more determined by the expected aims of the campaign. Again, Classical historians do not always offer detailed information on numbers, and when they do, they usually focus on full citizens serving as heavy infantry, so it is hard to collect a full picture, but communities naturally mobilised their forces based on an estimate of the number of men required for the job.[56] Those armies were expected to accomplish different tasks, including the incidental capture of random towns in their path. Different examples will provide a sense of the different scales of the campaigns and the forces involved. The Peloponnesians, as already pointed out, invaded Attika with two thirds of their forces, and the Athenians invaded Megara in 431 and Boiotia in 424 'in full force' (*pandemeí*, Thuc. 2.31.1, 4.90.1); the Athenian campaigns round the Peloponnese were also huge fleets of 100 ships or more (Thuc. 2.23.2, 56, 3.16), conceived, as Thucydides says, as a display of military power (Thuc. 3.16.1). To secondary theatres, on the other hand, more modest forces were usually sent: thirty Athenian ships to Lokris in 431 (Thuc. 2.26), forty Corinthian ships and 1,500 hoplites to Acharnania in 431 (Thuc. 2.33), twenty Athenian ships to Naupaktos under Phormion in 430 (Thuc. 2.69), 2,000 Athenian hoplites to the Chalkidike in 429 (Thuc. 2.79), forty-seven Corinthian ships to Acharnania in 429 (Thuc. 2.80.1), or forty Athenian ships to prevent the revolt of Mytilene in 427 (Thuc. 3.3). Finally, smaller contingents were detached to secondary operations: Brasidas commanded 100 men that frustrated the Athenian attempt to capture Methone (Thuc. 2.25.2), the Athenians sent six ships to collect taxes from the allies (Thuc. 2.69), and the Spartans sent eight ships to help the Rhodian exiles in 391 (Xen. *Hell.* 4.8.20–23). A proof of that tendency towards moderate forces is that some expeditions had in fact to be reinforced: the Athenians sent thirty ships and 1,000 hoplites to prevent the defection of Potidaia in 432 (Thuc. 1.57.6) and forty ships to Sicily in 425 (Thuc. 4.2), but both asked for reinforcements when they found themselves unable to perform a specific operation, the blockade of Potidaia in one case and the defence of Pylos in the other.

What this quick survey indicates is that marching armies engaged in poliorcetic operations frequently in early Classical warfare, and that they had in fact more chances to attack a town than to face an enemy force in an open battle. It also shows that all kinds of urban centres could be the target of an attack, but that Greek armies tended to prefer secondary settlements that were easier to capture but still provided some profit when plundered; bigger

56 Mobilisation as an estimate: *GWMR* 102–104; Rawlings (2007) 69–71; Krentz (2007) 148–50.

cities entailed logistical problems and tactical risks that commanders were not always willing to face. Forces were mobilised according to the expected aims of the campaign, but the tendency was towards middle-sized armies incorporating different kinds of troops (a few dozen ships, a few hundred heavy infantry, and a supplement of light infantry, cavalry and allies or mercenaries); bigger armies were naturally harder to mobilise and therefore less frequently mustered, usually sent to primary theatres against direct and powerful enemies. All these point at flexible and diverse armies with a considerable freedom of action that were expected to take wise command decisions to adapt to the changing conditions of the campaign and to provide solutions for unexpected situations.

4 The Process

According to the literary sources, the ancient Greeks conceptualised poliorcetics as an escalating process that started with the invasion (*epistrateía*) of the enemy territory and developed in successive stages in a permanent interaction with the defenders.[57] Once the invading army had selected and approached a specific target, the attackers followed the general strategy of gradually reducing the mobility of the defenders in an increasingly tighter grip.[58] The different tactics and operations always took place in this context of escalating blockade, which is, I believe, the core of Archaic and Classical Greek poliorcetics. The strategy of the defenders, conversely, attempted to hamper the activities of the invaders in order to prevent them from tightening the blockade, first by fielding their forces to face the enemy, which depended on a perception of the balance of the respective forces: if the invaders had clear numerical advantage the defenders usually abandoned the fields and took shelter behind the walls, but if some balance was perceived they seemed to take the risk and face the attackers in an open engagement.[59] Greek poliorcetics, then, were a struggle for mobility, both sides contending for the control of the surrounding territory, the invaders trying to tighten the blockade and the defenders adapting their tactics to a context of increasingly reduced mobility.[60] Let us see this pattern in action.

Escalation begins with an initial attempt by the attackers to control the surrounding territory (Thuc. 2.68.9; *cf.* 8.25.4). This is sometimes explicitly

57 This section follows and complements the analysis presented in Echeverría (2021).
58 For a detailed description of the process of a campaign, see Krentz (2007).
59 Balance: Krentz (1997) 61, 65–70, (2000) 177.
60 See Seaman's application of this pattern to the Greek wars of the Archaic period (2020).

stated in the sources (Thuc. 1.111.1, 2.68.9, 3.6.2, 85.3, 115.3), and armies can also be described encircling the target (Thuc. 2.20.3, 101.2, 3.46.3, 5.7.2). This is a proper occupation of the enemy territory, by which the army is 'installed' or even 'seated' (*káthemai* and derivatives) in a position that allows it to continue the devastation of the land (that has started during the march itself) but also to watch over the target to check the reaction of the defenders. Archidamos, for example, 'seated' (*ekathézonto*) his army around Oinoe first in order to storm it (Thuc. 2.18.1), in the Eleusinian plain later (*kathezómenoi*) to devastate it (Thuc. 2.19.2), and in Acharnae finally (*kathezómenoi*) to launch their attacks from there (Thuc. 2.19.2). Kleon also approached Amphipolis in 422 with the intention to 'surround and capture the city by storm' (Thuc. 5.7.3). Occupation was then associated with the preliminaries of the assault, and it represented a transition towards a more direct focus on the target itself. The grip is not very tight yet, and individuals or even entire units can get into and out of the surrounded town relatively easily, like the Spartan Brasidas reinforcing the Lakonian town of Methone attacked by the Athenians in 431 (Thuc. 2.25.2); the Athenians themselves spent several weeks in the summer of 428 encamped in front of the walls of Mytilene without taking further measures (Thuc. 3.4–6), since the Mytileneans in fact 'commanded the whole country' (τῆς γῆς ἐκράτουν, Thuc. 3.6.2).

The reaction of the defenders depended, as we have seen, both on the perception of a tolerable balance of forces and on the tightness of the invaders' grip. These two aspects dictated whether to field the citizen army, to seek protection inside the town and make a series of sallies, or to stay put behind the walls and resist. Ainias Taktikos repeatedly recommends impeding the invasion and hampering the advance of the enemy by fielding the army (15.1–7, 16.1–13, 19–20), relying on the greater familiarity with the territory. The army was fielded usually at the initial stages of the invasion, when the enemy could still be stopped before causing any harm, while sallies were attempted to break the enemy's occupation of the territory or a blockade that was getting increasingly tighter. Both could escalate into a battle. Battles were then the result of the reaction of the defenders to the invasion, and they could be fought at any point of the campaign. According to our sources, the great battles of the Peloponnesian War (Potidaia, Delion, Mantinea, Anapos River) followed the pattern of fielding the citizen army to hamper an invasion, and the same can be said of later battles such as Nemea, Koroneia, or Leuktra, and of many other, 'nameless' battles.[61] Surprise sallies, too, could escalate into battles, like the

61 Kythera 424 (Thuc. 4.54.2), Aphrodisia 424 (Thuc. 4.56.1), Nisaia 424 (Thuc. 4.72.2–4), Sikyon 424 (Thuc. 4.101.4), Mende 423 (Thuc. 4.129.3–5), Torone 422 (Thuc. 5.3), Methymna 390 (Xen. *Hell.* 4.8.29).

Chalkidian victory over the Athenians attacking Spartolus in 429 (Thuc. 2.79), or the Mytilenean victory over the Athenians in 428 (Thuc. 3.5.2).[62] From this perspective, battles were not independent operations or a strategic end in themselves, but a stage within the pattern of epistratic warfare, an episode in the process of the invasion; for the defenders, victory could mean the withdrawal of the attackers and the end of the invasion, but defeat meant that the enemy controlled the territory unmolested and was now ready to focus on the target.

Then began the next stage. The attackers stormed the fortification in an action that is described in the sources as an 'attempt' (*peiráo*), but more commonly as a head-on assault, literally 'throwing' the troops against the walls (*eisbállo*). It carries connotations of quickness ('rush', 'hasten', 'hurry') in order to reduce losses and cause havoc and surprise in the defenders, as Demosthenes planned in his failed attack against Syracuse in 413 (Thuc. 7.42). The vocabulary seems to imply quick and violent attacks that could be repeated several times, sometimes both by land and sea (Thuc. 4.8.4, 9.1, 11.2, 25.10) and possibly on various fronts. Assaults are specifically mentioned rather often in the Classical historians, like at Strepsa (Thuc. 1.61.4), Oinoe (Thuc. 2.18.1), Methone (Thuc. 2.25.1), Haliartos (Xen. *Hell.* 3.5.18), Sidous and Krommyon (Xen. *Hell.* 4.4.13), and we may assume that they are implied as well when a target is said to have been 'captured by force' (*bía*/*krátos*, Thuc. 1.102.2; Xen. *Hell.* 2.1.19). The assault could drag on for days on end, like the Spartan attacks against the Athenian fort in Pylos (Thuc. 4.13.1), and it could be combined with more permanent strategies, such as the formal siege: in 430, the Athenians sent a new army to reinforce the contingent already blockading Potidaia, and upon arrival the generals conducted a full-scale assault against the walls with the use of machines (Thuc. 2.58.1), exactly the same as Demosthenes in 413.

The assault required speed and mobility, so its first wave consisted of light-armed troops, either men equipped with light weapons and missiles, like the Plataians assaulting the Peloponnesian circumvallation wall in 427 (Thuc. 3.22), or specific units of light infantry detached to spearhead the attack. In the Athenian attempt against Megara in 424, the first troops to engage were Demosthenes' units of Plataian light troops (*psiloí*) and Athenian *perípoloi*, in fact a multi-task force (Thuc. 4.67); the Boiotians recruited 'javelin throwers and slingers from the Malian gulf' to storm the Athenian fort at Delion

62 Samos 440 (naval, Thuc. 1.117.1), Antissa 428 (land, Thuc. 3.18.2), Naxos 425 (land, Thuc. 4.25.9), Messina 425 (land, Thuc. 4.25.11), Amphipolis 422 (land, Thuc. 5.10), Lechaion 390 (land, Xen. *Hell.* 4.5.14–18).

(Thuc. 4.100.1), while Brasidas sent a unit of 100 peltasts in advance during the assault of Torone (Thuc. 4.111.1).[63]

Descriptions confirm the verdict of vocabulary – assaulting troops tried first to climb or jump over the walls. In Torone, Brasidas' troops climbed the walls using the structures of timber that the defenders had placed to conduct repairs (Thuc. 4.112.2), and to climb (*anabántas*) the Syracusan defences was also Demosthenes' plan in 413 (Thuc. 7.43.2). Brasidas captured the walls of Lekythos by taking advantage of the Athenian confusion and havoc caused by the collapse of one of the towers, and he even offered a prize of thirty minae to the first man on the wall (Thuc. 4.116.1–2). Ladders were the most basic tool for the assault, explicitly mentioned in a number of operations,[64] as well as 'machines' (generally assumed to be rams or platforms to protect the troops from enemy fire, at least until the innovations of Dionysios I in the early fourth century), but they are rarely described:[65] at Plataia (Thuc. 2.76.4) they seem to have been rams, while those used by the Spartans in the assault of Pylos were specifically devised to 'capture the wall' (Thuc. 4.13.1). They were mobile devices intended to approach the walls, and Xenophon mentions a smart Athenian engineer who planted large rocks on the roads to Athens to hamper the approach of Thrasyboulos' engines from the Peiraieus (*Hell.* 2.4.27). A particular case is represented by the Boiotian 'flame-thrower' at Delion (Thuc. 4.100), and a 'machine' used a few weeks later in the assault of Lekythos by Brasidas devised to 'throw fire upon the wooden defences' (Thuc. 4.115.2).

In many cases, particularly those involving secondary settlements, the assault was enough to break the defences and capture the town, but it sometimes turned out unsuccessful against particularly motivated defenders, so the attackers had to decide whether the target was worth the effort or not. If it was not, they simply withdrew, possibly looking for an easier target, but if it was, the process escalated once more towards a tighter blockade. The blockade is clearly implied by the vocabulary of the Classical historians, who use terms such as *katakleío*, 'shut in, enclose', *eírgo*, 'shut in', and more clearly *ephorméo*, 'blockade', and Garlan persuasively suggested that the meaning of *poliorkéo* was in fact metaphorical and should be interpreted as 'blockade' as well.[66] The lack of systematic details in the sources produces an incomplete picture, but

---

63 Assaults and light troops: Kern (1999) 106–107; Kucewicz, Lloyd and Konijnendijk in this volume.

64 Ladders (*klímakes*): Thuc. 3.20–23, 4.135, 5.56.5; Xen. *Hell.* 4.4.11, 7.2.7–9.

65 Machines (*mechanaí*) in the context of assaults: Thuc. 2.18.1, 58.1, 76.4, 4.13.1, 100.1, 115.2, 5.7.5, 7.43.1, 8.100.5; Xen. *Hell.* 2.4.27. For the introduction and meaning of siege machinery in Greece, see Trundle (2019).

66 Garlan (1974) 5.

what can be deduced is that the blockade was achieved primarily by the distribution of the attacking troops (including sometimes ships for a complete blockade by land and sea) in permanent positions around the target. In some cases, circumvallating walls were used to strengthen the blockade, probably temporary palisades (the sophisticated Peloponnesian wall around Plataia in Thuc. 3.21 seems more the exception than the norm).[67] Finally, formal siege devices such as ramps, mines, or sabotage were attempted at times,[68] probably combined with the slower but less risky practice of letting the target starve to surrender: some blockades are reported to have lasted for months or even years, like those of Thasos (Thuc. 1.100–101), Plataia (Thuc. 3.52), and Potidaia (Thuc. 2.70). In effect, blockades combined the strategies of launching repeated assaults and holding positions for long periods to starve the defenders in a context in which the attackers almost completely controlled the territory. In some cases, when logistically and financially possible, walls could be erected to facilitate both strategies and to tighten the grip around the target, and finally some siege techniques could be applied. However, those cases remained a minority and their techniques were isolated practices that never entered the common pattern of Classical poliorcetics, as far as our sources indicate.

The impression we get from the Classical historians, then, is that the operations of invading armies tended to escalate according to the circumstances and the availability of resources. It is true that the literary sources do not systematically mention every step, but they sometimes emphasise certain stages. Thucydides, for example, seems to imply that the Athenians started to build a circumvallation wall around Potidaia right away, without any indication of the other stages (Thuc. 1.64), and similar examples may be adduced.[69] But

---

67 Examples of blockades with circumvallating wall: Potidaia (Thuc. 1.64), Samos (Thuc. 1.116.2), Plataia (Thuc. 2.75.1–2, 77.1, 78, 3.21), Mytilene (Thuc. 3.18.4–5, 27), Nisaia (Thuc. 4.69.2), Mende (Thuc. 4.130.7), Skione (Thuc. 4.131), Melos (Thuc. 5.114.1), Syracuse (Thuc. 6.97–103, 7.4–7), Miletos (Thuc. 8.25.5), Kalchedon (Xen. *Hell.* 1.3.4), Byzantion (Xen. *Hell.* 1.3.14), Mantineia (Xen. *Hell.* 5.2.4–5), Phleious (Xen. *Hell.* 5.3.16), Kromnos (Xen. *Hell.* 7.4.21–25).

68 Machines: Thuc. 2.18.1, 58.1, 76.4, 3.51.3, 4.13.1, 115.2, 6.102.2, 7.43.1, 8.100.5; Xen. *Hell.* 2.4.27. Mining and counter-mining: Hdt. 4.200; Thuc. 2.76; Aen. Tact. 37. Water sabotage: Thuc. 6.100.1; Xen. *Hell.* 3.1.7, 1.18.

69 Particularly if we consider the cases in which the literary sources merely referred to the poliorcetic operations with simple words such as 'captured', 'destroyed', or 'besieged', in a narrative technique we may call 'referential', see Echeverría (2021); examples: Periandros 'sent an army against Prokles, his father-in-law, [...] and he took (*heíle*) Epidauros, captured Prokles, and imprisoned him' (Hdt. 3.52.7); the Athenians 'sailing to Prasiai, a maritime town in Lakonia, ravaged part of its territory, and took and sacked (*heílon kai*

the blockade of Potidaia was carried out after the Athenians had defeated the enemy in a pitched battle and became masters of the land, so we may conclude that the emphasis on certain stages is perhaps a narrative choice. The cumulative impression of descriptions in the Classical historians is that this strategy of increasing pressure on the target was fully active, and it tended to escalate from informal occupation to static blockade. This is confirmed by the fact that some operations were fairly common, particularly at the beginning of the process (devastation, occupation, and assault), while some others (blockade, siege technology, mines, and construction works) were rare and attempted only in specific conditions.

Escalation was not a rigid and fixed process, one stage invariably after the other, but the result of this strategy of increasing pressure. Greek commanders naturally had flexibility to decide the most suitable operation on the spot and according to the circumstances and to the characteristics of the target and of their own army. In practice, then, the specific order or intensity of the different stages could be slightly modified, and an assault could be attempted without a proper occupation of the territory, for example, while devastation could be carried out at any time throughout the process. But the invaders reacted to the response of the defenders, and it is this interactive dynamic that explains Classical Greek poliorcetics as an escalation.

5 *Epistrateia*: Approaching Greek Poliocetics from a General Perspective

I have claimed so far that settlements were frequent targets for marching armies in the Archaic and Classical periods, and that they could even be the main objective of an expedition. Yvon Garlan claimed long ago that this pattern of poliorcetic actions was not different from the rest of Greek land warfare: he saw a 'certain continuity' between the invasion of a territory and the investment of a settlement, which were different stages of the same process.[70] I believe it is necessary to recover that approach: the capture and destruction of towns was in fact a crucial part of the land strategy in the Classical period, and from a broader perspective, the general objectives of a campaign

---

epórthesan) the place itself' (Thuc. 2.56.6); the Athenians and Achaians 'marched against and laid siege (*estráteusan kai epoliórkoun*) to Oiniadai in Akarnania, but failing to take (*heílon*) it, they returned home' (Thuc. 1.111.3).

70 'Il existe une continuité certaine, du point de vue stratégique, entre les incursions dans un territoire et l'investissement d'une place forte, qui sont deux formes d'agression analogues, mais de degré different', Garlan (1974) 5.

were permanently renegotiated to adapt to changing circumstances. On the eve of the Peloponnesian War, the Greeks seemed to think that the nature of war was completely unpredictable: Thucydides makes the Corinthians state in the pre-war debates that 'war least of all conforms to fixed rules, but itself in most cases has to form its plans to suit the occasion as its own resources allow' (Thuc. 1.122.1). Archidamos agrees when he admits that 'the events of war cannot be foreseen, and attacks are generally sudden and furious' (Thuc. 2.11.4; trans. C. Forster Smith). The patterns of military aggression that can be reconstructed from the literary sources corroborate this view of war as a changing and adapting reality.

If we are left to judge by our sources, most campaigns were launched with only general and unspecific objectives: an Athenian fleet is sent against Macedonia in 432 (Thuc. 1.57.6), another to reinforce Phormion in the Gulf of Corinth in 429 (2.85.5), another in a raid around the Peloponnese in 428 (Thuc. 3.3.1–2), another to Sicily in 425 (Thuc. 4.8.20), and a Spartan fleet in 391 to help the cause of the Rhodian exiles (Xen. *Hell.* 4.8.20).[71] These may just be the historian's summary of the real, potentially more detailed, objectives, and numerous hints indicate that sometimes there were other additional aims. Thucydides mentions 'other objectives' in an unsuccessful Athenian campaign to Thessaly around 455 (Thuc. 1.111.1), and we occasionally hear of complementary instructions that are added impromptu to the original ones: the Athenian fleet sent to reinforce Phormion in 429, for example, is instructed to intervene in Crete upon a *proxenos*' request (Thuc. 2.85.5). A fleet that had been mustered to raid the Peloponnese in 428 was sent to prevent the defection of Lesbos instead (Thuc. 3.3.1–2), and the fleet to Sicily in 425 was also ordered to look to their Korkyrean allies in their internal civil war (Thuc. 4.2.3). The Athenian fleet to Macedon in 432 also received complementary instructions, namely, to punish Potidaia for revolting (Thuc. 1.57.6), and the generals apparently were free to prioritise and decide which objective was easier to achieve (Thuc. 1.59).

71 Other examples: Thuc. 1.114.1 (discipline Euboia), 2.10 (invade Attika), 2.20 (attack Athens), 2.26 (devastate Lokris), 2.33.1 (reinstate a tyrant in power), 2.69.1 (collect taxes), 2.95.1–2 (intervene in Macedon), 3.3.1–2 (raid the Peloponnese), 3.15 (help the Mytileneans), 3.16 (display military might), 3.19.1 (collect taxes), 3.26.1 (help the Mytileneans), 3.26.1 (divert the attention of the Athenians from the Peloponnesian relief fleet to Lesbos), 3.51.1–2 (occupy a strategic place), 3.80.2 (reinforce a previous fleet), 3.86 (intervene in Sicily, explore a future invasion, intercept corn imports to the Peloponnese), 3.91.2 (subdue Melos), 3.100 (attack Naupaktos), 4.2.2 (fleet to Sicily); Xen. *Hell.* 4.6.3 (help the Achaian allies).

What this shows, in any case, is that expeditions were both strategically and tactically flexible, and an open and changing list of objectives is just the beginning. In practice, expeditions seemed prepared for any eventuality, and they were ready to change or adapt their objectives on the spot. The Corinthians sent to protect Potidaia from the Athenians in 432, for example, abandoned the place after a defeat and decided to campaign in the Chalkidike for a while before returning home (Thuc. 1.65.2). Thucydides tells of a Peloponnesian fleet sent to attack the Peiraeus after a naval defeat in Naupaktos that ended up attacking Salamis instead (Thuc. 2.93). Another Peloponnesian fleet sent to Mytilene had to reconsider its aims after learning of the capture of the town by the Athenians and finally set sail to Korkyra (Thuc. 3.31, 69). A failed Athenian expedition against Melos in 426 ended up landing in Boiotia (Thuc. 3.91), and the Athenian Demosthenes, sent that year in a campaign round the Peloponnese, decided to invade Aitolia 'seeing that so large an army was collected' among his Acharnanian allies (Thuc. 3.94–95).[72] Greek commanders, therefore, enjoyed great strategic flexibility: Antalkidas is shown improvising his moves in the Aegean in 388 (Xen. *Hell.* 5.1.25–28), and Agesilaos displays a remarkable mobility that leads to a permanent change of objectives (Xen. *Hell.* 4.1.1–2).

The pattern of military aggression started with the invasion of enemy territory, after which multiple possibilities were at hand according not only to the objectives and conditions of the invading army but also to the reactions and conditions of the defenders. Devastating and controlling the enemy lands and farms was a first and almost universal step. Armies could engage in a series of raids looking for multiple targets in a row, as in the Athenian campaign against the territory of Corinth in 425, when a defeat in a land battle

72 Other examples: Thuc. 1.94.2 (the Hellenic coalition against Cyprus and Byzantion), 1.108.5 (Athenian campaign against the Peloponnese), 1.111.1 (Athenian campaign in Thessaly), 1.111.2–3 (Athenian campaign in Sikyon), 1.112.2–4 (Athenian campaign in Cyprus), 1.114 (Athenian campaign against Euboia and Megara), 1.116.1 (Athenian campaign in Samos), 2.5 (Theban attack against Plataia), 2.18–20 (Peloponnesian invasion of Attika), 2.25 (and 2.30, 31.1, the Athenian fleet round the Peloponnese joins the land invasion of the Megaris), 2.33 (Corinthian campaign in Acharnania), 3.33–34 (an Athenian fleet in the Aegean), 3.100–102 (an expedition against Naupaktos is diverted against Amphilochian Argos), 4.8.2 (a Peloponnesian fleet to Korkyra is called to help in Pylos), 4.46 (the Athenian fleet going to Sicily takes part in the events at Pylos and later intervenes in Korkyra), 4.70.1 (Brasidas' expedition to Thrace goes to the rescue of Megara), 4.75 (Athenian expedition to the Hellespont); Xen. *Hell.* 4.28.3 (a Spartan fleet to Rhodes ends up raiding Ionia), 5.1.6 (a Spartan fleet to Abydos stops at Tenedos to plunder and collect money), 5.1.10 (an Athenian fleet to Cyprus stops at Aigina to intervene in the fighting there).

after a landing in Solygeia did not interrupt the campaign, but the Athenian ships continued their cruise and made other surprise landings (Thuc. 4.45). If the defenders decided to face the invaders, land or naval battles were fought, such as at Sybota (Thuc. 1.46–47), Potidaia (Thuc. 1.62), Thasos (Thuc. 1.100.2, 101.1), Sikyon (Thuc. 1.111.2), Samos (Thuc. 1.116.1), Mylai (Thuc. 3.90.2), Tanagra (Thuc. 3.91.5), or Solygeia (Thuc. 4.42–44). Settlements of different kinds, such as Anaktorion (Thuc. 1.55.1), Therme and Pydna (Thuc. 1.61.2–3), Oinoe (Thuc. 2.18–19), Kolophon (Thuc. 3.34), and many others, were approached, attacked, and sometimes captured, the specific means depending on the logistics, conditions, and resources of the invaders. At any time, the army could turn its back and look for a different target, or move to another theatre, or explore an unexpected alternative, or even return home. Mobility, flexibility, and improvisation were the keys to Greek warfare.

Wars during the Classical period had different purposes, whether territorial conquest, predatory raids, or border disputes, but they were fought through a number of individual campaigns that followed a consistent pattern of invasion (*epistrateia*) of enemy territory. The Athenian expedition round the Peloponnese in 431 did not end the war, but it represented the standard military action carried out by both sides during the conflict. It shows that the Greek approach to strategy was broad, open, and flexible, because in the course of those campaigns anything, from battles to assaults, could happen. Pitched battle was in fact not more relevant or significant than other operations, and conversely, 'siege warfare' was not a separate branch of war. Battles and sieges were in fact part of the same epistratic dynamic, complementary mechanisms to defeat the enemy, and fought by the same troops in the course of the same actions.[73]

Bibliography

Adcock, F.E. (1957) *The Greek and Macedonian Art of War*. Berkeley, CA.

Anderson, J.K. (1970) *Military Theory and Practice in the Age of Xenophon*. Berkeley, CA.

Bryant, J.M. (1990) 'Military Technology and Socio-Cultural Change in the Ancient Greek City'. *Sociological Review* 38.3: 484–516.

73 This paper is part of a research project funded by the Spanish Ministry of Innovation under two connected programs (RYC-2011-08161 and HAR2015-65456-P). Its original draft, written in 2015, benefitted a great deal from the several revisions prompted by the careful reading and detailed assessment of the editors of the volume and Brill's readers, to whom I thank for their interesting suggestions and comments. All remaining mistakes are naturally my own.

Campbell, D. (2006) *Besieged. Siege Warfare in the Ancient World*. Oxford.
Cartledge, P. (1996) 'La nascita degli opliti e l'organizzazione militare', in Settis (1996) 681–714.
Chaniotis, A. (2013) 'Greeks Under Siege: Challenges, Experiences, and Emotions', in *WCW* 438–56.
Connor, W.R. (1988) 'Early Greek Land Warfare as Symbolic Expression'. *P&P* 119: 3–29.
DA (1992) *Les Fortifications Grecques de Mycènes à Alexandre, Dossiers d'Archéologie* 172.
DA (1993) *À la Découverte des Forteresses Grecques, Dossiers d'Archéologie* 179.
De Souza, P. (2008) 'Greek Warfare and Fortification', in Oleson (2008) 673–90.
Echeverría, F. (2011) '*Taktikè Technè*: The Neglected Element in Classical "Hoplite" Battles'. *AncSoc* 41: 45–82.
Echeverría, F. (2019) 'Epilogue: Ancient Sieges and Modern Perspectives', in *Sieges* 321–41.
Echeverría, F. (2021) 'Notes on the Conceptualization of Poliorcetics in Archaic and Classical Greek Literature'. *DHA* 47.1: 71–95.
Euben, J.P., Wallach, J.R., and Ober, J. (eds) (1994) *Athenian Political Thought and the Reconstruction of American Democracy*. London.
Fontaine, S. and Helas, S. (eds) (2016) *Le fortificazioni arcaiche del Latium vetus e dell'Etrurie meridionale (IX–VI sec. a.C.). Stratigrafia, cronologia e urbanizzazione*. Rome.
Frederikson, R. (2016) 'Fortifications and the Archaic City in the Greek World', in Fontaine and Helas (2016) 252–66.
Garlan, Y. (1968) 'Fortifications et histoire grecque', in Vernant (1968) 245–60.
Garlan, Y. (1974) *Recherches de Poliorcétique Grecque*. Paris.
Hamilton, C.D. and Krentz, P. (eds) (1997) *Polis and Polemos: Essays on Politics, War, and History in Ancient Greece in Honor of Donald Kagan*. Claremont, CA.
Hansen, M.H. (1995) '*Kome*. A Study in How the Greeks Designated and Classified Settlements which were not Poleis', in Hansen and Raaflaub (1995) 45–82.
Hansen, M.H. (2006) *Polis. An Introduction to the Ancient Greek City-State*. Oxford.
Hansen, M.H. and Raaflaub, K.A. (eds) (1995) *Studies in the Ancient Greek Polis*. Stuttgart
Hanson, V.D. (1991a) 'The Ideology of Hoplite Battle, Ancient and Modern', in *HCGBE* 3–11.
Hanson, V.D. (1991b) 'Hoplite Technology in Phalanx Battle', in *HCGBE* 63–84.
Hanson, V.D. (1991c) 'The Future of Greek Military History', in *HCGBE* 253–56.
Hanson, V.D. (1999a) *Warfare and Agriculture in Classical Greece*. Berkeley, CA.
Hanson, V.D. (1999b) *The Other Greeks: The Family Farm and the Agrarian Roots of Western Civilization*. Berkeley, CA.
Hanson, V.D. (2000) 'Hoplite Battle as Ancient Greek Warfare: When, Where, and Why?' in *W&V* 201–32.

Hanson, V.D. (2005) 'Genesis of the Infantry 600–350 BC', in Parker (2005) 15–29.
Hanson, V.D. (2013) 'The Hoplite Narrative', in MoB 256–75.
Hunt, P. (1998) *Slavery, Warfare and Ideology in the Greek Historians*. Cambridge.
Hunt, P. (2007) 'Military Forces', in *CHGRW* 1:108–46.
Jackson, A. (1995) 'War and Raids for Booty in the World of Odysseus', in *W&S* 64–76.
Kern, P.B. (1999) *Ancient Siege Warfare*. Bloomington, IN.
Konijnendijk, R. (2020) 'Risk, Chance and Danger in Classical Greek Writing on Battle'. *JAH* 8.2: 175–86.
Konijnendijk, R. (2021) 'Playing Dice for the Polis: Pitched Battle in Greek Military Thought'. *TAPhA* 151.1: 1–33.
Krentz, P. (1997) 'The Strategic Culture of Periclean Athens', in Hamilton and Krentz (1997) 55–72.
Krentz, P. (2000) 'Deception in Archaic and Classical Greek Warfare', in *W&V* 167–200.
Krentz, P. (2002) 'Fighting by the Rules: The Invention of the Hoplite Agôn'. *Hesperia* 71: 23–39.
Krentz, P. (2007) 'War', in *CHGRW* 1:147–85.
Lazenby, J.F. (2004) *The Peloponnesian War: A Military Study*. London.
Lee, J.W.I. (2013) 'The Classical Greek Experience', in *WCW* 143–61.
Lendon, J.E. (2000) 'Homeric Vengeance and the Outbreak of Greek Wars', in *W&V* 1–30.
Leriche, P. and Tréziny, H. (eds) (1986) *La Fortification dans l'Histoire du Monde Grec*. Paris.
Lloyd, A.B. (ed) (1996) *Battle in Antiquity*. London.
Mara, G. (2009) 'Thucydides and Political Thought', in Salkever (2009) 96–125.
Mitchell, S. (1996) 'Hoplite Warfare in Ancient Greece', in Lloyd (1996) 87–107.
Ober, J. (1985) *Fortress Attica: Defense of the Athenian Land Frontier, 404–322 BC*. Leiden.
Ober, J. (1991) 'Hoplites and Obstacles', in *HCGBE* 173–96.
Ober, J. (1996a [1994]) 'The Rules of War in Classical Greece', in Ober (1996c) 53–71.
Ober, J. (1996b [1985]) 'Thucydides, Pericles, and the Strategy of Defense', in Ober (1996c) 72–85.
Ober, J. (ed) (1996c) *The Athenian Revolution*. Princeton.
Oleson, J.P. (ed) (2008) *The Oxford Handbook of Engineering and Technology in the Classical World*. Oxford.
Parker, G. (ed) (2005) *The Cambridge History of Warfare*. Cambridge.
Powell, A. (ed) (1995) *The Greek World*. London.
Raaflaub, K.A. (1994) 'Democracy, Power and Imperialism in Fifth-Century Athens', in Euben, Wallach and Ober (1994) 103–46.
Rawlings, L. (2000) 'Alternative Agonies: Hoplite Martial and Combat Experiences Beyond the Phalanx', in *W&V* 233–59.
Rawlings, L. (2007) *The Ancient Greeks at War*. Manchester.

Rawlings, L. (2013) 'War and Warfare in Ancient Greece', in *WCW* 3–28.
Rebuffat, F. (2000) *Guerre et Société dans le Monde Grec (490–322 av. J.-C.)*. Paris.
Salkever, S. (ed) (2009) *The Cambridge Companion to Ancient Greek Political Thought*. Cambridge.
Schwartz, A. (2002) 'The Early Hoplite Phalanx: Order or Disarray?'. *C&M* 53: 31–64.
Schwartz, A. (2009) *Reinstating the Hoplite: Arms, Armour and Phalanx Fighting in Archaic and Classical Greece*. Stuttgart.
Seaman, M. (2013) 'The Peloponnesian War and its Sieges', in *WCW* 642–56.
Seaman, M. (2020) 'Early Greek Siege Warfare', in *NAGRW* 29–38.
Sears, M.A. (2019) *Understanding Greek Warfare*. London.
Settis, S. (ed) (1996) *I Greci: Storia, Cultura, Arte, Societá. Vol. II.I, Formazione*. Torino.
Strauss, B. (2007) 'Battle. Naval Battles and Sieges', in *CHGRW* 1:223–47.
Trundle, M. (2019) 'The Introduction of Siege Technology into Classical Greece', in *Sieges* 135–49.
Trundle, M. (2020) 'Wealth and the Logistics of Greek Warfare', in *NAGRW* 17–27.
Van Wees, H. (1992) *Status Warriors. War, Violence and Society in Homer and History*. Leiden.
Van Wees, H. (1995) 'Politics and the Battlefield: Ideology in Greek Warfare', in Powell (1995) 153–78.
Van Wees, H. (2013) 'Farmers and Hoplites: Models of Historical Development', in *MoB* 222–55.
Vernant, J.-P. (ed) (1968) *Problèmes de la Guerre en Grèce Ancienne*. Paris.
Westlake, H.D. (1983) 'The Progress of *Epiteichismós*'. *CQ* 33.1: 12–24.
Whitehead, D. (1990) *Aineias the Tactician: How to Survive Under Siege*. Oxford.
Will, E. (1975) 'Le territoire, la ville et la poliorcetique grecque'. *RH* 235: 297–318.

CHAPTER 10

The Western Greeks and the 'Greek Warfare' Narrative

Joshua R. Hall

Scholars often talk about 'the Greek world' as if it was a monocultural entity. This can be seen easily in the titles of survey-type monographs, such as Peter J. Rhodes' *A History of the Classical Greek World* or Simon Hornblower's *The Greek World*.[1] This is also true in general approaches to 'Greek warfare'. These works tend to eschew the subtleties of Greek culture throughout the Mediterranean world.[2] Even the best modern discussions of warfare in the Classical period do not afford the inhabitants of various regions much individual consideration, often discussing them alongside their Athenian or Spartan cousins.[3] Specialist studies have filled some of the gaps left by these broader analyses, but there has been a dearth of scholarly attention paid to the regional differences in Greek warfare. This has led to scholars concluding – or simply asserting – that similar approaches to war were taken by Hellenes in Sicily as by the Athenians. For example, Arthur G. Woodhead asserted without discussion of the evidence that 'in earlier days the pattern of Western [Greek] warfare had resembled that in Greece'.[4] Jeff Champion, in a recent discussion of the Sicilian tyrants, tells his readers that '[hoplites] are the troops that would be the backbone of the armies of the Sicilian Greeks'.[5] These views are grounded in an understanding of Greek warfare which presents hoplites as the only meaningful type of soldier in Hellenic armies. This is a now generally discredited approach, grounded in a perspective centred almost entirely in the Aegean. However, with the rise of heterodox views of Aegean Greek warfare, it is past time that we deconstruct this narrative even further and look at how Greeks waged war at a regional level.

This chapter functions as a survey of evidence for how the inhabitants of Magna Graecia practised warfare. It generally focuses on the sources for this

1 Rhodes (2010); Hornblower (2011).
2 For example, *WWW*; *GWMR*; Rawlings (2007); Sears (2019).
3 Konijnendijk (2018).
4 Woodhead (1962) 133.
5 Champion (2010).

 | DOI:10.1163/9789004501751_011

region, which are admittedly weaker than for the Aegean. However, by recentring the discussion to Sicily and Southern Italy, we are able to develop a better understanding of both the practices in this region and how they fit into a wider 'Greek world'. The importance of studying how the Western Greeks fought their wars goes beyond creating a more nuanced understanding of Hellenic warfare. It forces us to rethink the meaning of evidence for war making in Italy in non-Greek communities. Proximity and contact between Etruscan and Italic peoples and their Greek neighbours have often been pointed to as vectors through which their practice of warfare changed.[6] Considerable archaeological evidence exists which shows a connection, such as Corinthian helmets and other elements of the hoplite panoply.[7] Although scholars have questioned how Etruscans and Romans (in particular) could adapt Greek practices of warfare to their societies for quite some time, the use of hoplite tactics has only recently been dismissed outright.[8] Thus, exploring how the Western Greeks went to war will impact two different fields of ancient military studies.

My aims in this chapter are to present the practice of warfare in Magna Graecia on its own terms. When analogies with the Greeks of the Aegean are warranted, I employ them. However, my intention is to create a new narrative which puts Sicily and Southern Italy at its centre. By avoiding engagement in now outdated arguments over hoplite practices, I give the Siceliotes and Italiotes a new identity. The extent to which this approach compromises their broader identity as Greeks is best left to others. But I hope that this discussion sparks that conversation and an interest in the regionality of Greek warfare. My contribution to this conversation begins with a discussion of recruitment and the composition of armies in Magna Graecia and then moves on to look at patterns of warfare in the region.

1 Recruitment and Composition of Armies

One of the most important means of understanding how war was practised by a historical culture is looking at who did the fighting. In the armies of

6 For example, Stary (1979) 193; Snodgrass (1986) 51–52.

7 A telling example of a Corinthian helmet used by an Etruscan was dedicated by Hiero of Syracuse at Olympia: Kunze (1967) 106–7; *SEG* 23.253. More examples of these, and of *aspides*, can be found in Saulnier (1980); Stary (1981).

8 This topic has a rather deep bibliography: Momigliano (1963) 119; Snodgrass (1965) 116–17; Saulnier (1980) 115–20; Jannot (1985) 135–38, (1986) 111–19; Ridgway (1988) 665; Adam and Rouveret (1990); D'Agostino (1990); Ampolo (1999); Smith (2006) 286–90; Rosenstein (2010); Armstrong (2016) 111–26. However, see the recent comments by Taylor (2020).

hierarchical societies, this often included only a portion of the adult (typically male) population. Which portions of this served as warriors or soldiers is telling of how that society perceived war. Unfortunately, our evidence for the recruitment of Hellenic armies in the central Mediterranean is not extensive. The earliest recruitment for which we have any kind of detail is from the last quarter of the sixth century, although it is preserved in a first-century source. In 524 or 523, the city of Cumae was threatened by a supposed allied army of barbarians: Etruscans, Umbrians, and Daunians (Dion. Hal. *Ant. Rom.* 7.4).[9] The response from the Cumaeans was to send 4,500 infantry and 600 cavalry into the field, and to hold two similarly sized groups in reserve to defend the city (Dion. Hal. *Ant. Rom.* 7.3.4).[10] What is interesting about this episode is that Dionysios seems to imply that the recruits were only youths (ἀκμή). If this reading is correct, it would mean that only the youngest men of the community were expected to serve in the field, with the rest of the population of military age exempt from this, perhaps only required to defend Cumae itself. Regardless of who exactly served in the army, they defeated the barbarians.

Twenty years after the Cumaeans were victorious over the combined army of several Italic peoples, they were called to give aid to their ally, the Latin town of Aricia. The son of Lars Porsenna, Arruns, had been sent to attack the Latins after the assault on, and possible capture of, Rome (Livy 2.14; Dion. Hal. *Ant. Rom.* 5.36.1). Aristodemos was chosen to lead the allied forces of Cumae, but the leading men of the city refused to raise a proper, or perhaps typical, army. Instead, in a plot to get Aristodemos killed, they recruited only the poor (ἄπορος) and the cowardly (πονηρός).[11] If we can believe this story, rather than

9 That the ethnicities provided by Dionysios reflect the reality of an allied army is not certain. Greek authors tended to use broad ethnic labels to describe large parts of Italy (Bradley 2000, 22). Modern authors have generally been sympathetic to the historicity of this episode suggesting that the tradition preserved was based on a local Cumaean source: e.g., Alföldi (1965) 56–57; Bradley (2000) 112 n. 34. Gallia (2009), though, has challenged the idea of a 'Cumaean Chronicle' and has called into question how this episode, and indeed all of the Cumaean digression in Dionysios, made its way into the Roman historical tradition. His arguments are persuasive enough to make us question the level of detail that would have been available to Dionysios in the first century.

10 I am unsure of Mitchell's (1990, 43 n. 30) reading of 'τούτων ἱππεῖς μὲν ἦσαν ἑξακόσιοι, πεζοὶ δὲ τετρακισχίλιοι καὶ πεντάκόσιοι' as meaning the entire body of recruits, rather than the last mentioned, which was the group sent into the field. If his reading is correct, it would mean that the city could only muster 4,500 infantry and 600 cavalry, which would have been a comparatively small army, though perhaps not for a single polis.

11 Alternative translations could read the 'unmanageable' and the 'worthless', amongst a number of other choices. As the leading men of the city were choosing who should serve, it seems probable that many of those being pressed into service were from the (or a) lower class.

ascribing it to the category of propaganda, it indicates that Cumae did not have a legally binding recruitment system based on equity of service, but rather a system based on the whims of a powerful aristocracy or oligarchy. Whether or not this practice was more widespread amongst Greek cities in Italy, we cannot say.

We also know a bit about the recruitment practices of Syracuse. Hiero attempted to kill off his brother Polyzelos through the use of a similar strategy to that of the Cumaeans with Aristodemos, by sending him to war with a mediocre army (Diod. Sic. 11.48.4–5).[12] Later in the fifth century, Nikias' describes the Syracusan army, as presented by Thucydides, as 'a mass levy, not picked troops like ourselves [the Athenians], and a levy of Sicilians, too, who may look down on us but will not stand up to us, since they are more rash than experienced as soldiers' (Thuc. 6.68).[13] While there is little reason to describe the recruitment of Athenian armies as anything but a 'levy', clearly Nikias is implying a significant difference between the two systems. The implication could be that rather than a commanding general(s) selecting the troops to serve under him, in Syracuse all available men were put under arms.[14] This would mean that Syracusan commanders would have to deal with poorly trained and disciplined men, alongside those who made better soldiers. This seems like a different system to that of Cumae, where, as we have just seen, the elites were able to handpick those who made up their armies, for better or worse.

It is important to note in this passage, however, that Nikias is also accusing *all* Sicilian Greeks of being poor soldiers. This means that his criticism is not only that the Syracusans levy *en masse*, taking into their army even those who are ill-disciplined, but that there are simply *no* good soldiers amongst them. It is possible that this was a ploy to enthuse the Athenians about attacking Sicily, by demeaning the troops of the greatest city of the island. Taken at face value, however, it is evidence of a Syracusan militia which lacked something that the Athenians had, whether that was an effective system of calling soldiers or simply a lack of skilled fighters generally. Alkibiades had earlier referred to the Sicilian armies pejoratively as a mob (ὅμιλος) that was unlikely to organise a resistance to the Athenians (Thuc. 6.17.4). He also noted that they lacked armour, which could be reflective of this part of the army consisting of people

12 Diodoros does not explicitly state that the army was recruited only from groups which typically made bad soldiers, but it is safe to assume this was the case, otherwise Hiero should have had little reason to think that his plan had any chance of success.

13 Trans. Warner (1972).

14 On Athenian recruitment, see Christ (2001); *GWMR* 102–104; Crowley (2012) 22–39; Pritchard (2018) 46–52.

from a lower economic class (Thuc. 6.17.3).[15] This assertion may be confirmed by the need of the Syracusan state to supply arms and armour to their troops. But we must keep in mind that the speeches delivered in Athens were primarily propagandistic, so how much weight we should put in them as historical evidence for the Syracusan way-of-war is questionable.

There is, though, evidence that military-aged men in a number of Sicilian Greek cities did not always bear their own arms, as 'hoplites' in the Aegean are often claimed to have. When a Carthaginian army invaded Sicily in 406, the people of Akragas were forced to raise an army. We are told that they levied all persons of military age and armed (καθοπλίζω) them (Diod. Sic. 13.85.3). The implication is that the state provided the arms for this army, rather than the soldiers themselves, as was common in mainland Greece.[16] Similarly, when Dionysios I was preparing his forces to attack the Puno-Phoenician settlement of Motya, he provided the arms for his soldiers, supposedly to the number of 140,000 shields (ἀσπίδες), daggers (ἐγχειρίδια), and helmets (περικεφαλαῖαι) (Diod. Sic. 14.43.2).[17] Polyainos notes that he confiscated the arms from the citizens, returning them only whilst on campaign, and taking them back once the army reached the gates of Syracuse (*Strat.* 5.2.14; cf. Diod. Sic. 14.10.4). This strategy kept his position safe, but it also necessitated a system of arming and disarming citizen troops during wartime. It is interesting that in the context of Greek warfare he did not commission the creation of any spears.[18] It is possible that this was an item possessed by many of his soldiers; but the general picture is that they did not possess any arms of their own. Spears were the key weapon for Greek hoplites and their absence here implies that Dionysios' soldiers did not adhere to the modern image of a Greek soldier. It is true that his army contained non-Hellenic mercenaries, but there was also a considerable number of Greeks, whether Syracusan, Sicilian subjects of Dionysios, or

---

15 This may be a rather glib conclusion, as not all hoplites in mainland Greece wore armour into battle: see *GWMR* 47–57.

16 The dramatic take on this by Hanson (*WWW* 29) may be outdated but it is still a stark contrast with the situation being discussed. See the more controlled discussion by Van Wees (*GWMR* 52–54).

17 Dionysios also commissioned 14,000 cuirasses which were distributed only to the cavalry, his mercenary bodyguard, and to the commanders of the infantry. If this is to be trusted, it implies that the groups of soldiers we would expect to have been best armed already, and in the case of the cavalry and infantry commanders wealthier in general, did not regularly possess body armour.

18 McKechnie (1989) 83–84, did not seem to notice the absence of spears from the list when he claimed that it 'also implies that the bulk of it [the arms manufacturing] was concerned with producing Hellenic arms ... the infantry officers were evidently to have full Greek panoply, while their followers were to be lightly armed'.

Peloponnesian mercenaries (Diod. Sic. 14.44.2).[19] Thus, Syracuse's armies may have in fact been different than their Athenian contemporaries, as Nikias had implied decades earlier. In fact, as far back as the 480s, Syracuse's army looked different than those of many of their Aegean cousins.

Gelon's army at the time of the Persian Wars consisted of 20,000 hoplites, but also 2,000 each of cavalry, archers, slingers, and ἱπποδρόμοι ψιλοί (Hdt. 7.158).[20] The last component of this army is not as easy to identify as one might think on a cursory reading. The component parts most commonly mean 'race-track' and 'bare', respectively. The latter, ψιλοί, translates to 'light troops' as a substantive in military contexts, and often indicates infantry. The former, however, is found only in this passage in the context of a type of soldier. How and Wells argued that we should read this as 'light infantry who fought interspersed among the cavalry'.[21] Translators have often followed their reasoning. This was the translation followed by Alfred D. Godley in his 1920 Loeb edition. For Liddell and Scott, however, this phrase should be translated as 'light horsemen'.[22] David Asheri echoed this by describing them in the same way.[23] This is, in part, followed by Robin Waterfield in his translation for the Oxford World Classics series, but at the expense of the other cavalry mentioned.[24] Similarly, Thomas J. Dunbabin ignores one of these, although it is the ἱπποδρόμοι ψιλοί, only listing '2,000 horse' in his description of Gelon's army.[25] Van Wees elides the 'ἱπποδρόμος' and describes this group simply as 'other light-armed'.[26] The view that these were infantry is tacitly accepted without discussion by others, as well.[27] Which of these interpretations we should believe is unclear. This is due in part to both words being used in the accusative. If ἱπποδρόμοι was in the genitive, perhaps, the meaning of 'light infantry who run with the cavalry' would be more certain, as the phrase would then be read close to 'the light armed who race with the horses'. In fact, the Greeks had a word for warriors fighting in this way, ἅμιπποι, which in a military context

19 Caven (1990) 96–97. The system of keeping the citizens unarmed continued under Dionysios' son and successor, see Diod. Sic. 16.10.3; Plut. *Vit. Dion* 27.3.

20 These figures supposedly came from Gelon himself in a speech given to ambassadors from the Aegean Greeks seeking his assistance in the coming war with Xerxes.

21 How and Wells (1912) s.v. 7.158.

22 *LSJ* 834 s.v. ἱπποβόσιον (ἱπποδρόμος).

23 Asheri (1988) 772.

24 He excludes the first instance of cavalry: Waterfield (1998) 460.

25 Dunbabin (1948) 419. His description, '20,000 hoplites, 2,000 horse, 2,000 bowmen, and 2,000 slingers', completely ignores this unit.

26 Van Wees (2002) 341. This is the same approach taken by Ameling (1993) 16.

27 For example, De Angelis (2016) 104; Konijnendijk (2018) 111.

translates to infantry which are mixed in with the cavalry.[28] Herodotos does not use this term, but his near contemporary, Thucydides, does (5.57). We may expect the former to have had knowledge of the word, but that is no guarantee that he would have used it.

The translation of ἱπποδρόμοι ψιλοί as 'light cavalry' is also not straightforward, and requires us to read the phrase as something similar to 'light-armed race horses (horsemen?)'. This could have been some sort of specialised force of scout cavalry, whose speed would have been their most distinguishing characteristic. In the days before the battle of Himera, Gelon's cavalry is noted to have effectively dealt with Punic foraging parties whom they are said to have caught by surprise (Diod. Sic. 11.21.2). His horsemen were quick enough to catch a messenger sent from Hamilcar, the Carthaginian commander, to his Greek allies (Diod. Sic. 11.21.4). Both of these stories may indicate that the Syracusan cavalry was superior in speed to what the Carthaginians expected or could field, and may support the postulation that the ἱπποδρόμοι ψιλοί are to be read as exceptionally quick horsemen. If this is correct, the total number of cavalry available to Gelon when the Greek embassy arrived would have been 4,000, closer to the 5,000 attributed to his army during the Himera campaign (Diod. Sic. 11.21.1). Regardless of which interpretation we believe is correct, it is important to note again that this is the only instance of this particular phrase in the Greek corpus and thus may represent a unit type that was unique to the Western Greeks, or perhaps to Syracuse.[29]

One of the most striking aspects of army recruitment amongst the Western Greeks is their proclivity towards using mercenaries.[30] This could be due to the tendency towards tyranny in this part of the Greek diaspora, which is often cited as being supported, in part, by soldiers of fortune.[31] Gelon's army of some 28,000 soldiers, discussed above, had as many as 10,000 mercenaries in it, whom he settled in Syracuse as citizens (Diod. Sic. 11.72.3). When Hiero became tyrant of Syracuse, following Gelon's death, his first deed was to enlist an army of foreign soldiers and surround himself with a force of mercenaries (Diod. Sic. 11.48.3). When the population of Syracuse rose against Thrasyboulos,

---

28 *LSJ* 83 s.v. ἅμιππος (e.g., Thuc. 5.57; Xen. *Hell.* 7.5.23).

29 Herodotos or his sources may not have known how to describe these troops, or maybe they misunderstood Gelon, if he did indeed deliver an offer of assistance as a speech to the Aegean Greek delegation. If so, searching for any 'true' interpretation of this term would be folly.

30 Tagliamonte (1994); Péré-Noguès (2000).

31 For example, Luraghi (1994) 119–85 and 231–373; Quesada Sanz (1999) 12; Trundle (2004) 5, (2006).

he was forced to enlist an army of mercenaries to oppose the citizen force.[32] Although the 'real' citizens were successful in eventually expelling those who would suppress their freedom, we are told that they were inferior to their opponents in their experience of war (Diod. Sic. 11.73.3). While it is probable that some of the citizens already had arms at hand, this revolt is not proof positive that the vast majority of the population was regularly armed. After all, tools typically meant for other activities, such as forestry or farming, could be turned into deadly weapons when they needed to be and were readily found in Greek cities (Xen. *Hell.* 3.3.7).

Other Sicilian poleis also used mercenaries. When the Carthaginians marched on Akragas in 406, the defenders called into service Dexippos, a Spartan *condottiero* who had been resident at Gela, and asked him to raise a force of mercenaries in order to support the city's army (Diod. Sic. 13.85.3–4). Dexippos has been described as a *chef mercenaire* of the Greeks.[33] As opposed to the mercenaries who supported the tyrants of Syracuse, he was sent back to the Aegean by Dionysios I, the ascendant Syracusan leader, for fear that he would restore liberty to the city (Diod. Sic. 13.96.1). This is perhaps evidence of contemporary suspicion towards the loyalty of mercenary groups. A fear that they would switch sides, or turn on their employers at some point, is well founded. Some Campanian mercenaries were hired by Carthage for their invasion of Sicily at the end of the fifth century, but ended up fighting alongside the Greeks after the initial offensive, as they had been upset at the Carthaginians for not giving them what they viewed as their fair share of the spoils of the campaign (Diod. Sic. 13.62.5, 80.4).

It is also important to note a minor point of language which relates to the composition of Western Greek armies. Diodoros, for all his supposed failings, is perhaps our best reflection of the Western Greek historians of the Classical period, none of whose works are extant.[34] Thus, his use of certain words and phrases is important to the present discussion. In this light, it is significant that Diodoros used the term 'hoplite' (ὁπλίτης) only once in reference to a Western Greek unit (16.67.3). This is not because he was uncomfortable using, or unfamiliar with, the term; it can be found in thirty-five other instances throughout his extant work down to the time of Alexander the Great (e.g., 11.2.5, 84.3–6; 12.40.4, 60.3). When Diodoros mentions Sicilian troops, he uses more general

32 Diodoros' δύναμις is broad and we should not read this, necessarily, as an organised 'militia'.

33 Péré-Noguès (1998).

34 On Diodoros' use of earlier historians and his methods of constructing his own work, see Drews (1962); Sacks (1990); Dudziński (2013). Cf. Giovannelli-Jouanna (2011); Robert (2011).

terms, primarily στρατιῶται, which is found 283 times over the same period. We also hear of 'select' troops (λογάδες and ἐπίλεκτοι) performing certain functions, though we hear of no other details regarding their armament or possible status as hoplites.[35] The absence of 'hoplites' suggests that the Western Greeks, or at least their historians, viewed warfare in a fundamentally different way to their Aegean cousins.[36]

We do not possess many descriptions of Western Greek armies. Their recruitment may have been different from what is expected based on Aegean parallels, such as at Cumae where anyone could be called up for service based on the desires of the elite. But ultimately their composition is only occasionally revealed. Based on what we know of these armies, we can deduce that cavalry was prominent amongst their troop types, and it appears that it was an important arm in battles. Ranged troops and picked units were also present, but notably, hoplites are in many ways absent. The heavy infantry divisions of Western Greek armies are not emphasised in our sources, with Diodoros generally referring to the soldiers within them in the most basic and general terms. These may have been people who bore arms in war, but were otherwise dispossessed of these in peacetime. And the absence of spears in the armament prepared by Dionysios I is a lingering problem of interpretation. Ultimately, the picture of the typical soldier in Magna Graecia is different from that of the Aegean Greeks. Perhaps this is a problem of evidence, of which we have relatively little, but overriding what little we have with universals based on Aegean evidence is inappropriate.

2 Patterns of Warfare amongst the Western Greeks

No continual narrative of the Archaic and Classical periods exists for the central Mediterranean Greeks; thus, we do not have comprehensive accounts of the military activities of this region. What evidence we do have seems to be biased towards certain types of events, such as those which were celebrated in Panhellenic sanctuaries (e.g., the Battle of Cumae). Nevertheless, by examining the patterns of warfare in this region we learn more about how the inhabitants of Magna Graecia perceived and practised conflict.

35 Vinci (2010).

36 In contrast to earlier writers such as Herodotos and Xenophon, Diodoros uses 'hoplite' in a more technical sense. On the use of the term 'hoplite', see Echeverría (2012) 300–303 and the introduction to this volume.

The earliest armed conflicts that we hear of amongst the Western Greeks are those fought by the Hellenic immigrants against native peoples. The settlers of Syracuse, originally from Corinth and Tenea (Strabo 8.6.22), drove a Sikel population from what would become their city (Thuc. 6.3.2). When Naxos sent a group of colonists to settle what would become Leontinoi, they forcibly displaced a group of Sikels (Thuc. 6.3.3). And it was through military strength that the inhabitants of Massalia took control of the land around their settlement (Strabo 4.1.4–5). This wave of foundations occurred over the course of the late eighth and seventh centuries, precisely during the period traditionally viewed as that in which something that looked like 'hoplite warfare' came to be.[37] Thus, in the Archaic and Classical periods, when the Greek poleis of the Aegean were fighting against one another and developing a unique form of warfare, the Greeks moving to the west were fighting local populations whose armament and tactics could scarcely have resembled those of the east.[38] With relatively scant evidence for violent confrontation during the settling of Greek cities on Sicily, some scholars have challenged the significance of military conquest in this narrative, suggesting that it was only in the seventh and sixth centuries that warfare was extensively practised by the Greek settlements in this region.[39] Other commentators, however, have postulated that it was commonly practised against native populations.[40] That early encounters between Greeks and Sicilian natives were violent, or led to some sort of violence, is perhaps corroborated by a passage in the *Odyssey* which features an enslaved woman originally from the island (Hom. *Od.* 24.211–12, 386–89).[41]

37 On the chronology of Greek foundations in Sicily, see Dunbabin (1948) 435–71; Leighton (1999) 222–23; De Angelis (2016) 66–73. On developments in Greek warfare in Greece in the eighth and seventh century, see Lloyd, this volume.

38 Evidence of native grave goods includes spears and short swords, with the occasional inclusion of horse bits, axes, and greaves. Large defensive elements, such as shields, are absent, although this of course could be a matter of ritual or survival of the items, rather than concrete evidence for their absence from native Sicilian warfare. Leighton (1999, 198) suggests that the type of warfare practised by these peoples was 'a loose style of fighting or skirmishing that required ease of movement, reminiscent of Homeric combat'. The Homeric comparison should be questioned, though, as the importance of massed infantry therein has recently been asserted: Van Wees (1995) 166–70.

39 Malkin (1998) 1–61, (2002) 254, (2011); De Angelis (2010).

40 For example, Boardman (1999) 189.

41 The historical 'usefulness' as well as the dating of the Homeric poems are perennial problems. Snodgrass' seminal 1974 article has convinced many that we should not read a 'unified' Homeric society out of the epics, but rather the stories therein coalesced piecemeal over a wide period of time. This paper was a reaction to Finley (1954) which argued for an historical Homeric society of the tenth to ninth centuries. Since Snodgrass' publication, the debate has ballooned with countless approaches to these issues, with dates proposed

Although this woman could have been enslaved through a number of different means, there was a tendency in Greek warfare to subject the defeated to this fate, which may have been the woman's pathway into servitude.[42] Regarding Greek settlement in Southern Italy, aggression and violence have been pointed to as common based on the archaeological material.[43] The violent reactions that met the expeditions of Pentathlos and Dorieus are historical examples of the possible reactions endured by earlier attempts at settlement. Pentathlos led a group of settlers between 580–576 from Knidos and Rhodes to Sicily, but they were driven off by native Elymians and Phoenicians (Diod. Sic. 5.9.2–3; Paus. 10.11.3).[44] Dorieus, a Spartan prince, set out to found a city in North Africa, on the Kinyps river, but he was driven from there by the Makai, a local tribe, and the Carthaginians. He, or his companions, then tried to do the same on Sicily, near Mount Eryx, but this attempt was thwarted by the Elymians and Phoenicians, perhaps supported by Carthage (Hdt. 5.42–46; Diod. Sic. 4.23.3; Paus. 3.16.4–5).[45] Although these are later Hellenic attempts at founding cities, it is folly to think that similar responses were not experienced by earlier expeditions, even if they were not universal.

The military superiority of the Greek migrants is often assumed in modern scholarship.[46] Snodgrass proposed that early Hellenic victories over natives in Sicily and Southern Italy helped to spread the practice of hoplite warfare throughout the Mediterranean, through peer polity interaction.[47] As was mentioned in the introduction to this chapter, Greek styles of arms and armour are commonly found throughout Italy, so there has been relatively little challenge to this hypothesis. There are, however, reasons to doubt its sweeping

for the writing, or at least the society reflected by the poems, between those of Finley (tenth century) and the seventh century. Raaflaub (1998) provides a discussion and bibliography of the major studies prior to this century and is still an important synthesis of the discussion. Even if we are to accept a mid to late-seventh century date for the *Odyssey*, as has been proposed by Carpenter (1946, 90–102 and 111), S. West (1988, 33–4), and M. West (2003, 13), then this still provides early evidence for a Sicilian slave. I am rather convinced of the evidence and argument of M. West (2005) on the aggregate nature of the poem and the influences which led to its contents as we know them, which as he argues certainly suggest a seventh-century date. On the significance of the Sicilian slave in Homer, see Lepore (2000) 55; Lane Fox (2008) 121–23.

42 On the enslavement of defeated enemies, see Kern (1999) 135–62; Van Wees (2003, 2011).

43 Lomas (1993) 27–28.

44 On the expedition, see Dunbabin (1948) 328–32; Krings (1998) 1–32. For a traditional explanation of Puno-Phoenician involvement in a campaign against Pentahlus, see Warmington (1993) 40–41.

45 Dunbabin (1948) 349–52; Krings (1998) 161–215. Cf. Miles (2010) 98–100.

46 For example, De Angelis (2016) 58–59.

47 Snodgrass (1986).

conclusion. Defeats of Greek armies by native populations considerably later than the phase of colonisation are important in this light. Tarentum found variable luck in their confrontations with the local indigenous population. At the beginning of the fifth century, the Tarantines won a victory over the Messapians and dedicated bronze sculptures of horses and captured women at Delphi (Paus. 10.10.6). During the same period, they were victorious over an army of Iapygians and Peuketioi, led by the Iapygian king, Opis, and dedicated statues of πεζοί (infantry) and ἱππεῖς (cavalrymen) (Paus. 10.13.10).[48] However, by 473 whatever military superiority the Tarantines may have had was gone, and they, along with their allies from Rhegium, suffered a crippling defeat at the hands of the Iapygians (Hdt. 7.170; Diod. Sic. 11.52.3–4). 3,000 Rhegians were killed, while an uncountable number of the Tarantines fell; according to Herodotos, this was the largest slaying of Greeks on record. So many of the aristocrats from Tarentum were killed that it caused, or spurred on, a change in governance to a more radical form of democracy (Arist. *Pol.* 1303a). Whatever technological superiority the Greeks may have had was undone over the course of two decades. If there was a superiority at some point, it was not what Snodgrass had in mind. We also should not *a priori* assume the Tarantines to have fought as a hoplite phalanx and that this is what secured their earlier victories. In fact, the evidence of their victory dedications shows that cavalry was important in both of their triumphs, but infantrymen only make an appearance in the second. Cavalry superiority, then, is what we should read into these battles.

2.1 *Maritime War Related to Land-Based Warfare*

Our understanding of the conflict between Greek migrants and the indigenous populations of Italy depends not only on land warfare, but on the sea. The earliest warfare that appears in more concrete narrative in the central Mediterranean was naval warfare. It was here that Greeks came into conflict with local 'piratical' forces, which set the tone for many of their conflicts on land. In the snippet view we have of the history of the sixth century, an overarching theme is the struggle between Greeks on one side and Carthaginians[49] and Etruscans on the other for the dominance of the western Mediterranean. The first episode of this supposed struggle occurred in the early sixth century when Phokaian and Carthaginian fleets met in combat around the time that the former were establishing the city of Massilia in southern France (Thuc. 1.13). The second was a naval battle between different Phokaian colonists and

48 On the chronology of these two battles, see Dunbabin (1948) 149.

49 It may be better to describe them as Phoenicians rather than Carthaginians, though this period is obscure to us. See the discussion in Pilkington (2019); López-Ruiz (Forthcoming).

a combined fleet of Carthaginians and Etruscans off the coast of Alalia, on Corsica, around 540. The Phokaians won a 'Cadmean victory' in this instance but were eventually forced to abandon their settlement (Hdt. 1.166).[50] The latter battle was supposedly provoked due to Phokaian piracy, itself a form of small-scale warfare.[51]

As noted above, piracy is a common theme in the literary evidence for the Western Greeks. Chalkidian pirates were remembered as having settled the Sicilian town of Zankle/Messene (Thuc. 6.4). The Liparians were also notable Greek pirates in the central Mediterranean. Originally attempting to settle on Sicily, they were driven to the islands of Lipari and founded a rather successful town (Diod. Sic. 5.9; Paus. 10.11.3–4). Their notoriety comes from an event at the beginning of the fourth century; after the Roman sack of Veii, a dedication was sent to Delphi, but was stopped by a Liparian pirate fleet led by Timasitheos. When they found out the cargo and destination of the Romans, however, they escorted them to the sanctuary (Livy 5.28; Diod. Sic. 14.93.3; Plut. *Vit. Cam.* 8.8).[52] Something of a piratical rivalry may have developed between the inhabitants of the islands and a group of Etruscans, against whom the Liparians resisted frequent incursions (Strabo 6.2.10).[53]

---

50 Justin 18.7.11, 43.5.2; Strabo 6.1.1; Paus. 10.8.6, 18.7. We do not hear of a land engagement in relation to this episode, although archaeological remains from Phoenician Cagliari, on Sardinia, are described by Bartoloni (1997, 160–61) as being 'suitable for light infantry'. Unfortunately, the author did not document the sources which led to his conclusions on this point and I have not had the opportunity to review the archaeological corpus from Cagliari. It is possible, however, that Phoenicians and Greeks came to blows on land as well as at sea on nearby Corsica, but without secure literary notices any inference is dangerous.

51 A general association of Phokaians with piracy is common in the literature. We hear of a certain Dionysios who settled in Sicily and preyed on Etruscan and Punic ships as a pirate after the expulsion of the Ionian Greeks from the coast of Anatolia (Hdt. 6.17), L. Scott (2005) 112. Justin (43.3) tells us that the Phokaians of Massilia supported themselves above all by piracy, perhaps what provoked the first of the conflicts noted above. In the same passage, Justin preserves a story in which the founders of Massilia sailed up the Tiber and entered into an alliance with the Romans under a member of the Tarquin dynasty (*ca.* sixth century). It is possible that this is evidence of piratical practices similar to the Viking river-raids of much later times. According to a notice in Strabo 4.1.10, the founders of Massilia fortified the Stoechades Islands in order to fend off pirates.

52 This event is generally considered historical, thanks in part to the testimony of Appian who cites the dedication's base and that the escort of the Roman ship was remembered when the Romans annexed the islands in 252 (App. *Ital.* fr. 8.3).

53 This rivalry may be represented by what scholars have seen as a competitive display of dedications at Delphi, see Colonna (1984); M. Scott (2010) 91–93. Cf. Torelli (1996) 570.

Etruscan piracy is something of a *topos* in Greek literature.[54] Etruscan pirates may have challenged the early Greek settlers in eastern Sicily and were remembered as having infested the seas in the region before the coming of the Greeks (Strabo 6.2.2). It is coincidental, but intriguing, that one of the earliest known pieces of painted pottery produced in a Western Greek workshop, the so-called Aristonothos *krater*, found in an Etruscan setting, depicts a naval battle which is often postulated to be between Greeks and Etruscans (Fig. 10.1).[55] Depictions of armed ships also appear in native Etruscan productions.[56] Piracy was likely widespread throughout the central Mediterranean, as attested by the first two treaties between Rome and Carthage (Polyb. 3.22–24). Although there are a number of economic points in these treaties, much of the behaviour being regulated seems to relate to piracy.[57] Raiding of land-based wealth was a major concern in the clauses relating to Latium, the alluring results of which we have some knowledge.[58] Dionysios I of Syracuse led a seaborne raid of the Caeretan port of Pyrgoi in 384 and supposedly took away 500 talents

54 This assumes the identification of 'Tyrsenians' with Etruscans, which is not entirely secure. Thucydides (4.109) speaks of Tyrsenians in the Chalcidian peninsula, who used to inhabit Lemnos and Athens. A recent thorough discussion of the setting of this story and the unlikely connection to the Etruscans of Italy has questioned the uniform association of these two groups of people: Paleothodoros (2012). The situation is complicated by the notices we have of Tyrrhenoi raiding Attika (Suid. Κώλτας) and plundering the statue of Hera from Samos (Ath. 15.671e–674a = Menodot. *FGrH* 541 F 1). Further evidence is needed to swing the argument in either direction, but see the discussions of Gras (1976) and Bruni (2013, 760) who both link the various Greek terms to the Etruscans of Italy.

55 Izzet (2004). For the original publication of the vase, see Förster (1869). On the krater, in general, see Schweitzer (1955).

56 One of the most spectacular of these images is on an Etruscan black-figure *hydria* from Vulci which depicts a ship armed with archers, two figures wearing linen corselets, and a figure armed with what looks like an *aspis* and possibly a Corinthian helmet (or a helmet of the Etrusco-Corinthian type): Walters (1893) 69 B60; Stary (1981) Taf. 22.2. Cherici (2006, 355–66) has assembled an impressive catalogue of depictions of armed ships from early Italy.

57 The economic clauses are often the focus of treatments of these treaties. It is clear, however, that seaborne raiding was being conducted against Latium by the Carthaginians and the purpose of these regulations was in part to mitigate this particular brand of piracy. On the historicity of these treaties see Walbank (1957) 337–43, who provides a discussion of the prominent theories and approaches of the early scholars. Most modern authors accept a sixth-century date for the first treaty, although an exact date within the first consular year of the Roman republic is problematic. For the modern discussion after Walbank, see Scullard (1989); Palmer (1997) 16–22; Scardigli (1991); Cornell (1995) 211; Serrati (2006). For a more comprehensive discussion of my opinions on the historicity of the earliest treaty, see Hall (2016) 21 n. 88.

58 In the first treaty, the fourth, fifth, sixth, and seventh clauses concern Punic raiding of Latium, while the second clause of the second treaty deals with the same. The fourth

FIGURE 10.1 Aristonothos Krater, second quarter of the 7th cent. BCE. Currently in the Musei Capitolini, Rome
PHOTO BY DAN DIFFENDALE, USED WITH PERMISSION

worth of booty from this one action alone (Diod. Sic. 15.14.3–4).[59] With this kind of 'money to be made' it is no surprise that Greek fleets are cited as preying on the coast around Antium, Laurentum, and the mouth of the Tiber in 349–348 (Livy 7.25). This raiding was so severe that it spurred the Romans to appoint a dictator for the year (Livy 7.26).[60]

clause of the second treaty allowed for ships from both states to take water from one another's territory but not to harm those persons belonging to that state.

59 We are given two reasons for this raid taking place. The first, and official, reason was to suppress pirates originating from this port. The second, and 'actual', reason for the raid which we hear of was simply plunder; Dionysios needed money to fund his other projects: Caven (1990) 190–92. Regarding the 'official' reason given, it is interesting that Strabo (5.2.3) noted that Pyrgoi was renowned amongst the Greeks because its inhabitants did *not* practise piracy.

60 Livy notes that the Greeks refused to fight a land battle, and that the Romans would not fight on the sea. He and his sources were ignorant of the origin of these particular Hellenes, but he postulates that they were forces belonging to Sicilian tyrants, although we are not given the names of any particular tyrants.

Similar practices were used on land. Hermokrates, a Syracusan, had been the leader of a fleet sent in the late fifth century to aid the Peloponnesians in the Aegean. Upon his return, he saw that his political position in Syracuse was untenable and was forced into exile. He first went to Messene where he had five triremes built and recruited 1,000 mercenaries (Diod. Sic. 13.63.2). He had originally intended on returning to Syracuse with these men, plus 1,000 Himerans who had been displaced by the Punic destruction of their polis, but this was not possible. Hermokrates resolved, instead, to take his troops to the ruins of Selinous, where he fortified a small portion of what remained, and attracted an additional 4,000 men from the countryside (Diod. Sic. 13.63.3–4). From here, he and his forces pillaged the territories of both Motya and Panormos, amassing a considerable amount of booty (Diod. Sic. 13.63.4–5). This behaviour was admirable enough amongst his countrymen that he was consequently welcomed back to Syracuse (Diod. Sic. 13.63.5–6). This shows that piracy and brigandage were both reputable approaches to armed conflict.

Sea battles were no less important than piracy to the practice of warfare in Western Greece and our understanding of the particularities of Western Greek warfare. Pindar made the Battle of Cumae, 474, the most famous of these early battles (*Pyth.* 1.71–78).[61] Hiero, the Deinomenid tyrant of Syracuse, defeated an Etruscan fleet off the Campanian coast. This battle was compared by Pindaros to the Battle of Salamis against the Persians. Indeed, the poet also compared Gelon's victory at Himera to that of Plataia, also during the Persian Wars, a comparison which is discussed below. Although war at sea was pursued by the Greeks of the Aegean, what makes it more interesting in the west is how frequently naval engagements appear to be the crux of a particular war and how much investment seems to have gone into innovations in naval technology. During the Athenian siege of Syracuse, it was the Syracusan naval victory in the Great Harbour which led to the end of the campaign (Diod. Sic. 13.13–18, esp. 13.14.1).[62] Naval battles were so prevalent that they led to something of an 'arms race' between Carthage and Syracuse, resulting, we are told, in the creation of the quinquereme, if not larger ships (Diod. Sic. 16.70; Ael. *VH* 6.12). These were bigger, stronger, and more powerful vessels than those previously used by the major naval powers of the Mediterranean.

---

61 This is the same victory for which Hiero dedicated three helmets at Olympia, Kunze (1967) 106–107; *SEG* 23.253.

62 Thucydides is less overt in his description of the significance of the battle, but does note that afterward the Syracusans were confident in their naval superiority and that they raised trophies for the naval actions of the past few days (7.41).

2.2 *Cavalry*

The Western Greeks were renowned for their cavalry. Modern commentators often make passing comments regarding the viability of raising horses on Sicilian and Italian plains and the impact that this had on war-making in these regions.[63] The basis of this discussion is Nikias' speech against the Sicilian Expedition in which he repeatedly warns the Athenians of the threat of the superior Sicilian cavalry (Thuc. 6.21–22). The victories of Sicilian tyrants in equine events in various Panhellenic games are a testament to the reputation's validity.[64] However, just because they can raise numerous horses does not mean they can field a large cavalry force, or that it would be an important force in battle. A cavalry force takes time and resources to train, and is generally thought to be predicated on the presence of a leisured class.[65]

In the narrative tradition, cavalry plays a prominent role in early battles. In the highly mythologised narrative of the First Battle of Himera, in 480, there is little that can be said about the tactics employed. The focus of the narrative is on a stratagem employed by Gelon in which he had a group of cavalrymen infiltrate the Punic camp and kill the Carthaginian commander, Hamilcar, thus throwing the Carthaginian army into disarray (Diod. Sic. 11.21). While it may be folly to read too much into this story, it must be recognised that the central event revolved around a cavalry action, whether it was the stratagem described by Diodoros or simply that the cavalry was the essential element of the Greek army during this battle. Any acceptance of Diodoros' reckoning is complicated by the version of events recorded by Herodotos (7.167). According to this account, Hamilcar killed himself by jumping into a sacrificial pyre when he saw that his army was being routed. Herodotos' account may have been based on the version of events which travelled via hearsay or by some other mechanism across the Mediterranean, while Diodoros' was certainly based on the version recorded by at least one of the Western Greek historians.

Another event which highlights the importance of horsemen to the Greek practice of warfare in the central Mediterranean occurred in 524 or 523. In one of these years, a combined force of Etruscans, Umbrians, and Daunians attacked Cumae and a great battle ensued outside of the city walls. It was the cavalry which was remembered as having won the day (Dion. Hal. *Ant. Rom.* 7.4.3). That the cavalry victory was indeed spectacular is supported by the

63 For example, Worley (1994) 100.

64 For instance, Hiero's victory in the chariot races (Pind. *Pyth.* 1, 2). See Cummins (2010) 1 n. 1.

65 Drews (2004). For further discussion of the importance of wealth/the leisure class in sustaining cavalry forces, see Konijnendijk in this volume.

aristocracy's desire to give special honours to Hippomedon, the commander of the horse (Dion. Hal. *Ant. Rom.* 7.4.4).[66] It is possible that the victories of Tarentum over their neighbours mentioned earlier were also due to their cavalry. This would explain the dedication of statues of horses after their victory over the Messapians (Paus. 10.10.6) and of both infantry and cavalry after their defeat of the Iapygians (Paus. 10.13.10).[67]

Further evidence of the importance of cavalry to the Western Greek 'way of war' was the realisation of Nikias' fears during the Sicilian Expedition. The first encounter that the Athenian expeditionary force had with the Syracusan horse was during an amphibious raid on the hinterland of the city, during which a number of light infantry were ridden down (Thuc. 6.52). The Athenians were further hindered by the Syracusan cavalry when, after having routed the Sicilian infantry, they were prevented from running them down and slaughtering them by the horsemen (Thuc. 6.70; cf. Polyaenus *Strat.* 1.39.2). As the war progressed, the Syracusans continued to experience defeats, one of which their commander blamed on the terrain of the battlefield which did not allow for their cavalry and light infantry to take part (Thuc. 7.5). In the next battle, Gylippos, Spartan commander of the Syracusan forces, deployed so as to allow his cavalry to fight. Because of this, the horsemen were able to rout the Athenian left flank which led to the retreat of the entire army (Thuc. 7.6). In the end, the Athenian inferiority in cavalry contributed to the defeat of their expedition.[68]

2.3 *Poliorcetics*

While other types of battles were fought by the Western Greeks, sieges are a prominent feature in the literary sources. This is at odds with the 'rule' at one

66 In the same passage (7.4.4) we hear that the people clamoured for these honours to be bestowed upon Aristodemos, the future tyrant, for his bravery in the battle. He supposedly killed the enemy general, a feat repeated by Aristodemos outside the walls of Aricia (Dion. Hal. *Ant. Rom.* 7.6.2). It is possible that this means he fought as a cavalryman, as Etruscan and early Roman commanders were often mounted (e.g., Livy 2.6–7; Hall 2016, 67–68).

67 There are, of course, other possible explanations. The dedication of horse statues after the first victory could be due to the capture of a considerable amount of the animals in the battle; this explanation makes sense as the other statues dedicated were of captured women. It would make the victory dedication one of the plunder taken in the aftermath of the victory. The second dedication, as it included statues of infantrymen, as well, could be a testament to the triumph of a 'combined arms' force. Unfortunately, without more details of either battle we are unable to know conclusively.

68 Brice (2013). For an account of the campaign with an emphasis on cavalry, see Worley (1994) 100–19.

time favoured by historians that Greek armies did not regularly attack settlements or civilians.[69] Poleis were often the target in warfare during the sixth and fifth centuries on Sicily. The destruction of communities through depopulation, especially the relocation of a population, was common.[70]

In the wars between the Sicilian Greeks and the Carthaginians, sieges were the typical scenes of combat. Hannibal Mago's expedition against the Sicilian cities, beginning in 410, opened with the siege and destruction of Selinous (Diod. Sic. 13.54.6–7) and then of Himera (Diod. Sic. 13.59.6–8). The war continued in a similar way until the Carthaginians were forced to retreat outside the walls of Syracuse. This conflict saw one of the first uses of siege towers against a Greek settlement (Diod. Sic. 13.54.7). The Greek response to these campaigns was to attack settlements loyal to Carthage. Dionysios I, the tyrant of Syracuse, first laid siege to Aigesta and Entella (Diod. Sic. 14.48.5). The crowning moment of his campaign, however, was the spectacular siege and sack of Motya in 397. For this battle, he employed complicated siege techniques, including a mole and towers (Diod. Sic. 14.49–51).

During sieges, fighting between the besiegers and the besieged regularly occurred.[71] Most of these amounted to little more than skirmishes, without the entire army of either force being engaged. When the Carthaginians laid siege to Himera in 409, for example, the settlement's defenders sallied out and fought a small battle with some of the attackers (Diod. Sic. 13.60). One of the most striking of these skirmishes, however, was fought outside of Gela in 405. The Carthaginians attacked the town after wintering in Akragas, following an impressive victory there the year before. The Greek army, led by Dionysios I of Syracuse, was split into three forces, one of which would advance between the city's walls and the coast, the second would march through the town, and the third would swing around it on the inland side, while his ships would sail directly against the Punic camp (Diod. 13.109.4). The assault failed, in part because the three groups of infantry did not arrive on the battlefield at the same time (Diod. Sic. 13.110). Dionysios would later use a similar plan of attack when assaulting the Punic siege works outside of Syracuse in 395, where it met with more success.[72] While battles such as this are similar to pitched battles in their conduct, the circumstances are different. Dionysios had not sought to

69 See Ober (1996). This 'rule' has been rightly refuted in recent scholarship, see Krentz (2002); Van Wees (2011); Hall (2018); Seaman (2020); Echeverría in this volume.

70 Full discussion of these occurrences can be found in Moggi (2006) 77–80.

71 Hall (2018) 187–89.

72 Diod. Sic. 14.72–75. Discussion in Roisman (2017) 256–59.

meet his foe in the field before they could attack Gela; instead, he used the city as part of his battle plan.[73]

The frequency of sieges on Sicily helped to spur on the development of technologies used to fight these kinds of battles. The most significant of these inventions was the catapult.[74] As part of his massive preparations to bring war to the Carthaginian allies of western Sicily, Dionysios I assembled a team of craftsmen to create a large cache of arms and armour, as well as to build great engines of war (Diod. Sic. 14.41–42). One of the machines specifically mentioned, and said to have been first constructed at this time, was the catapult (καταπελτικὸν) (Diod. Sic. 14.42.1, 50.4; Ath. Mech. 10.1).[75] Although rams were known to the Greeks since at least the period of the Persian Wars, wheeled towers had not been used by a Hellenic army (as far as we know) until Dionysios' siege of Motya (Diod. Sic. 14.51.1). These towers were six stories tall and wheeled so that they could be brought up against the walls of the town. It is likely that these were copies of Carthaginian examples used against the Sicilian Greeks in earlier wars.[76]

Taken together, a naval focus, cavalry combat, and sieges form the typical practice of warfare in the west. This is not to say that pitched battles did not occur (see, for example, Diod. Sic. 13.87); however, when even important battles occurred, they do not receive the attention that we would expect from a 'mainland' Greek historian. The Battle of Kabala was a major victory for Dionysios I over the Carthaginians while the Battle of Kronion was a major defeat. Despite these being described as the two 'important and famous pitched battles which occurred during this period', Diodoros only devoted a couple of lines to the battles and their aftermath (15.15–16).

3 Conclusions

This chapter has examined the evidence for warfare amongst the Western Greeks. It has been approached on its own terms, eschewing as far as possible preconceptions of what Greek warfare should be. We have seen that in this part of the Greek world, patterns of conflict do not easily reconcile with older

73 Caven (1990 66) has argued that Dionysios' strategy abandoned the 'principle of the phalanx that had dominated Greek military thinking since the mid-seventh century'. As the principle of the phalanx was not a dominant element of Greek thinking since the mid-seventh century, though, this cannot stand as an acceptable description.

74 *GWMR* 142.

75 Rawlings (2007) 138–39. Cf. Schellenberg (2006).

76 *GWMR* 142.

views of Greek warfare.[77] Wars against local inhabitants were not likely fought in a manner resembling those in the Aegean, especially between Greek poleis. It is unlikely that much beyond the arms and armour of the Aegean was found in Italian and Sicilian warfare. As we have seen above, the evidence we have for land battles won by southern Italian Greeks were victories attributed to cavalry, not the heavily armed infantry with whom many authors, ancient and modern, are preoccupied. On Sicily, we have seen evidence for early examples of what could be called 'mixed arms' tactics, whether in the variety of troop-types enrolled in Gelon's army or the combination of land and naval forces used by Dionysios I at Gela and Syracuse. The wars we hear of in the literary sources were typically fought at sea or through sieges.

These patterns of warfare do not conform to the orthodox view of Greek warfare. Indeed, they do not entirely line up with more modern assumptions, either. The Hellenicity, arms, and armour of the Greeks of the central Mediterranean did not dictate that wars were to be decided in pitched battle between citizen hoplites. More often, and seemingly more important to our sources, was combat at sea and the siege and sacking of settlements. Often found on these battlefields were mercenaries, fighting alongside citizens and their allies, another feature which stands in opposition to certain views on how the Greeks of the Archaic and Classical eras fought their wars. All of the evidence points to a style of warfare which supports a heterodox view; it was pragmatic, bloody, and destructive, not an agonal struggle between equals.

Magna Graecia should be treated on its own terms, and *a priori* assumptions about how war was waged in this part of the world should be left behind. The diversity of the Greek world means that this axiom should also be extended to other regions. Assumptions as to the superiority of 'hoplite warfare' as a reason for Greek success abroad and as the basis for the transmission of Hellenic patterns of arms and armour to regions such as Etruria must be abandoned. More interesting patterns of warfare, and cultural phenomena, contributed to these occurrences and deserve to be studied without the constraining force of the preconceptions attached to earlier conclusions on Greek warfare. The most secure path forward in the study of warfare amongst the ancient Greeks is through jettisoning universalisms and engaging in detailed regional studies. These will produce new generalisations, though hopefully ones which are closer to the experiences of ancient people in those places.

77 See the Introduction to this volume.

Bibliography

Adam, A.-M. and Rouveret, A. (1990) 'Les cites étrusques et la guerre au Ve siècle avant notre ère'. *Crise et transformations des societies archaïques de l'Italie antique au Ve av. J.-C. Actes de la table-ronde organise par l'École Française de Rome et l'Unité de recherches étrusco-italiques associée au CNRS*: 327–56.

Alföldi, A. (1965) *Early Rome and the Latins*. Ann Arbor, MI.

Ameling, W. (1993) *Karthago: Studien zu Militär, Staat und Gesellschaft*. Munich.

Ampolo, C. (1999) 'La città riformata e l'organizzazione centuriata. Lo spazio, il tempo, il sacro nella nuova realtà urbana', in Giardina and Schiavone (1999) 49–85.

Armstrong, J. (2016) *War and Society in Early Rome: From Warlords to Generals*. Cambridge.

Asheri, D. (1988) 'Carthaginians and Greeks', in *CAH* 4: 739–80.

Bakker, E.J, De Jong, I.J.F. and Van Wees, H. (eds) (2002) *Brill's Companion to Herodotus*. Leiden.

Bartoloni, P. (1997) 'Army, Navy and Warfare', in Moscati (1997) 160–66.

Bekker-Nielsen, T. and Hannestad, L. (eds) (2001) *War as a Cultural and Social Force: Essays on Warfare in Antiquity*. Copenhagen.

Bellelli, V. (ed) (2012) *Le origini degli etruschi: storia archeologia antropologia*. Rome.

Boardman, J. (1999) *The Greeks Overseas: Their Early Colonies and Trade*. 4th edn. London.

Bradley, G.J. (2000) *Ancient Umbria: State, Culture, and Identity in Central Italy from the Iron Age to the Augustan Era*. Oxford.

Brice, L.L. (2013) 'The Athenian Expedition to Sicily', in *WCW* 623–41.

Bruni, S. (2013) 'Seafaring: Shipbuilding, Harbors, the Issue of Piracy', in Turfa (2013) 759–77.

Carpenter, R. (1946) *Folk Tale, Fiction, and Saga in the Homeric Epics*. Berkeley, CA.

Carratelli, G.P. (ed) (1996) *The Western Greeks*. Milan.

Caven, B. (1990) *Dionysius I: War-Lord of Sicily*. New Haven.

Champion, J. (2010) *The Tyrants of Syracuse: War in Ancient Sicily. Volume I: 480–367 BC*. Barnsley, UK.

Cherici, A. (2006) 'Talassocrazia: aspetti tecnici, economici, politici con un brevissimo cenno a novilara, nesazio e ai feaci'. *Annali della Fondazione per il Museo "Claudio Faina"* 13: 323–66.

Christ, M.R. (2001) 'Conscription of Hoplites in Classical Athens'. *CQ* 51.2: 398–422.

Collin-Bouffier, S. (ed) (2011) *Diodore d'Agyrion et l'histoire de la Sicile*. Dialogue d'histoire ancienne, supplément no. 6. Besançon.

Colonna, G. (1984) 'Apollon, les Étrusques et Lipara'. *MEFRA* 96: 557–78.

Cornell, T.J. (1995) *The Beginnings of Rome: Italy and Rome from the Bronze Age to the Punic Wars (c. 1000–264 BC)*. London.

Crowley, J. (2012) *The Psychology of the Athenian Hoplite: The Culture of Combat in Classical Athens*. Cambridge.

Cummins, M.F. (2010) 'Sicilian Tyrants and their Victorious Brothers II: The Deinomenids'. *CJ* 106.1: 1–20.

D'Agostino, B. (1990) 'Military Organization and Social Structure in Archaic Etruria', in Murray and Price (1990) 59–82.

De Angelis, F. (2010) 'Re-Assessing the Earliest Social and Economic Developments in Greek Sicily'. *Römische Mitteilungen* 116: 21–53.

De Angelis, F. (2016) *Archaic and Classical Greek Sicily: A Social and Economic History*. Oxford.

Dowdall, A. and Horne, J. (eds) (2018) *Civilians Under Siege, From Sarajevo to Troy*. Basingstoke.

Drews, R. (1962) 'Diodorus and his sources'. *AJPh* 83: 383–92.

Drews, R. (2004) *Early Riders: The Beginnings of Mounted Warfare in Asia and Europe*. London.

Dudziński, A. (2013) 'The Bull of Phalaris and the Historical Method of Diodorus Siculus'. *Histos* 7: 70–87.

Dunbabin, T.J. (1948) *The Western Greeks: The History of Sicily and South Italy from the Foundation of the Greek Colonies to 480 B.C.* Oxford.

Echeverría, F. (2012) 'Hoplite and Phalanx in Archaic and Classical Greece: A Reassessment'. *CPh* 107.4: 291–318.

Fagan, G.G. and Trundle, M. (eds) (2010) *New Perspectives on Ancient Warfare*. Leiden.

Finley, M.I. (1954) *The World of Odysseus*. London.

Fisher, N. and van Wees, H. (eds) (1998) *Archaic Greece: New Approaches and New Evidence*. Swansea.

Förster, R. (1869) 'Vaso Ceretano con rapresentazione dell'Accamento di Polifemo'. *Annali dell'Instituto di Corrispondenza Archeologica* 4: 157–72.

Gallia, A.B. (2009) 'Reassessing the 'Cumaean Chronicle': Greek Chronology and Roman History in Dionysius of Halicarnassus'. *JRS* 97: 50–67.

Giardina, A. and Schiavone, A. (eds) (1999) *Storia di Roma*. Turin.

Giovannelli-Jouanna, P. (2011) 'Sicile mythique, Sicile historique: la place de la Sicile dans l'histoire universelle de Diodore', in Collin-Bouffier (2011) 21–41.

Gras, M. (1976) 'La piraterie tyrrhénienne en mer Egée: mythe ou réalité?'. *L'Italie préromaine et la Rome républicaine. I. Mélanges offerts à Jacques Heurgon*: 341–70.

Hall, J.R. (2016) 'The Tyrrhenian Way of War: War, Social Power, and the State in Central Italy (c. 900–343 BC)'. PhD Thesis. Cardiff University.

Hall, J.R. (2018) 'Gods Under Siege: Civilians and Sacrilege in Greek Warfare', in Dowdall and Horne (2018) 185–206.

Heubeck, A., West, S., and Hainsworth, J.B. (eds) (1998) *A Commentary on Homer's Odyssey Volume I: Introduction and Books I–VIII*. Oxford.

Hornblower, S. (2011) *The Greek World 479–323 B.C.* London.

How, W.W. and Wells, J. (1912) *A Commentary on Herodotus*. Oxford.

Izzet, V. (2004) ‘Purloined Letters: The Aristonothos Inscription and Krater’, in Lomas (2004) 191–210.

Jannot, J.-R. (1985) ‘Les cites étrusques et la guerre. Remarques sur la fonction militaire dans la cite étrusque’. *Ktema* 18: 127–41.

Jannot, J.-R. (1986) ‘Les cavaliers étrusques: armament, mode de combat, fonction. VIIème au IVème siècle’. *Römische Mitteilungen* 93: 109–33.

Kern, P.B. (1999) *Ancient Siege Warfare*. Bloomington, IN.

Konijnendijk, R. (2018) *Classical Greek Tactics: A Cultural History*. Leiden.

Krentz, P. (2002) ‘Fighting by the Rules: The Invention of the Hoplite Agôn’. *Hesperia* 71: 23–39.

Krings, V. (1998) *Carthage et les Grecs c. 580–480 av. J.-C.* Leiden.

Kunze, E. (1967) *Bericht über die Ausgrabungen in Olympia VIII*. Berlin.

Lane Fox, R. (2008) *Travelling Heroes: Greeks and their Myths in the Epic Age of Homer*. London.

Leighton, R. (1999) *Sicily Before History: An Archaeological Survey from the Palaeolithic to the Iron Age*. Ithaca.

Lepore, E. (2000) *La Grande Grèce: aspects et problems d’une « colonisation » ancienne. Quatre conferences au Collège de France (Paris 1982)*. Naples.

Lewis, S. (ed) (2006) *Ancient Tyranny*. Edinburgh.

Linder, M. and Tausend, S. (eds) (2011) *Boser Krieg: exzessive Gewalt in der Antiken Kriegsfuhrung und Strategien zu deren Vermeidung*. Graz.

Lomas, K. (1993) *Rome and the Western Greeks 350 BC–AD 200*. London.

Lomas, K. (ed) (2004) *Greek Identity in the Western Mediterranean: Papers in Honour of Brian Shefton*. Leiden.

López-Ruiz, C. (Forthcoming) *Phoenicians and the Making of the Mediterranean*. Cambridge, MA.

Luraghi, N. (1994) *Tirannidi arcaiche in Sicilia e Magna Grecia da Panezio di Leontini alla caduta dei Dinomenidi*. Florence.

Luraghi, N. and Alcock, S.E. (eds) (2003) *Helots and their Masters in Laconia and Messenia: Histories, Ideologies, Structures*. Cambridge, MA.

Lyons, C.L. and Papadopoulos, J.K. (eds) (2002) *The Archaeology of Colonialism*. Los Angeles.

Malkin, I. (1998) *The Return of Odysseus: Colonization and Ethnicity*. Berkeley, CA.

Malkin, I. (2002) ‘A Colonial Middle Ground: Greek, Etruscan, and Local Elites in the Bay of Naples’, in Lyons and Papadopoulos (2002) 151–81.

Malkin, I. (2011) *A Small Greek World: Networks in the Ancient Mediterranean*. Oxford.

McKechnie, P. (1989) *Outsiders in the Greek Cities in the Fourth Century BC*. London.

Miles, R. (2010) *Carthage Must Be Destroyed: The Rise and Fall of an Ancient Civilization*. London.
Mitchell, R.E. (1990) *Patricians and Plebians: The Origin of the Roman State*. Ithaca.
Moggi, M. (2006) 'Peculiarità della guerra in Sicilia?', in Vaggioli (2006) 67–89.
Momigliano, A. (1963) 'An Interim Report on the Origins of Rome'. *JRS* 53: 95–121.
Moscati, S. (ed) (1997) *The Phoenicians*. London.
Murray, O. and Price, S. (eds) (1990) *The Greek City: From Homer to Alexander*. Oxford.
Paleothodoros, D. (2012) 'Dionysus and the Tyrrhenian Pirates', in Bellelli (2012) 455–85.
Palmer, R.E.A. (1997) *Rome and Carthage at Peace*. Stuttgart.
Péré-Noguès, S. (1998) 'Un mercenaire grec en Sicile (406–405): Dexippe le Lacédémonien'. *DHA* 24.2: 7–24.
Péré-Noguès, S. (2000) *Mercenaires et mercenariat en Occident de la fin du Ve siècle avant J.-C.* Toulouse.
Pilkington, N. (2019) *The Carthaginian Empire: 550–202 BCE*. Lanham.
Powell, A. (ed) (1995) *The Greek World*. London.
Pritchard, D.M. (2018) *Athenian Democracy at War*. Cambridge.
Quesada Sanz, F. (1999) 'Soldada, moneda, tropas ciudadanas y mercenaries profesionales en la antiguo mediterráneo: el caso de Grecia'. *Moneda y exèrcits: III Curs d'Historia monetaria d'Hispania*: 9–38.
Raaflaub, K.A. (1998) 'A Historian's Headache: How to Read "Homeric Society"?', in Fisher and Van Wees (1998) 169–93.
Rawlings, L. (2007) *The Ancient Greeks at War*. Manchester.
Renfrew, C. and Cherry, J.F. (eds) (1986) *Peer Polity Interaction and Socio-Political Change*. Cambridge.
Rhodes, P.J. (2010) *A History of the Classical Greek World, 478–323 B.C.* Oxford.
Ridgway, D. (1988) 'The Etruscans', in *CAH*^{2} 4: 634–75.
Robert, R. (2011) 'Diodore et la patrimoine mythico-historique de la Sicile', in Collin-Bouffier (2011) 43–68.
Roisman, J. (2017) *The Classical Art of Command: Eight Greek Generals Who Shaped the History of Warfare*. Oxford.
Rosenstein, N. (2010) 'Phalanges in Rome?', in Fagan and Trundle (2010) 289–303.
Sacks, K. (1990) *Diodorus Siculus and the First Century*. Princeton.
Saulnier, C. (1980) *L'armée et la guerre dans le monde étrusco-romain (VIII*^{e}*–IV*^{e} s.). Paris.
Scardigli, B. (1991) *I Trattati Romano-Cartaginesi*. Pisa.
Schellenberg, H.M. (2006) 'Diodor von Sizilien 14,42,1 und die Erfindung der Artillerie im Mittelmeerraum'. *Frankfurter elektronische Rundschau zur Altertumskunde* 3: 14–23.
Schweitzer, B. (1955) 'Zum Krater des Aristonothos'. *Römische Mitteilungen* 62: 78–160.
Scott, L. (2005) *Historical Commentary on Herodotus Book 6*. Leiden.

Scott, M. (2010) *Delphi and Olympia: The Spatial Politics of Panhellenism in the Archaic and Classical Periods*. Cambridge.

Scullard, H.H. (1989) ‘Carthage and Rome’, in *CAH* 7.2: 486–569.

Seaman, M. (2020) ‘Early Greek Siege Warfare’, in *NAGRW* 29–38.

Sears, M.A. (2019) *Understanding Greek Warfare*. New York.

Serrati, J. (2006) ‘Neptune’s Altars: The Treaties between Rome and Carthage (509–226 B.C.)’. *CQ* 56.1: 113–34.

Smith, C. (2006) *The Roman Clan: The Gens from Ancient Ideology to Modern Anthropology*. Cambridge.

Snodgrass, A. (1965) ‘The hoplite reform and history’. *JHS* 85:110–22.

Snodgrass, A. (1974) ‘An Historical Homeric Society?’. *JHS* 94: 114–25.

Snodgrass, A. (1986) ‘Interaction by Design: The Greek City State’, in Renfrew and Cherry (1986) 47–58.

Stary, P.F. (1979) ‘Foreign Elements in Etruscan Arms and Armour: 8th to 3rd Centuries B.C.’. *Proceedings of the Prehistoric Society* 45: 179–206.

Stary, P.F. (1981) *Zur eisenzeitlichen Bewaffnung und Kampfesweise in Mittelitalien*. Mainz.

Taylor, M.J. (2020) ‘The Evolution of the Manipular Legion in the Early Republic’. *Historia* 69.1: 38–56.

Tagliamonte, G. (1994) *I figli di Marte: mobilità, mercenari, mercenariato italici in Magna Grecia e Sicilia*. Rome.

Torelli, M. (1996) ‘The Encounter with the Etruscans’, in Carratelli (1996) 567–76.

Trundle, M. (2004) *Greek Mercenaries: From the Late Archaic Period to Alexander*. London.

Trundle, M. (2006) ‘Money and the Great Man in the Fourth Century BC: Military Power, Aristocratic Connections and Mercenary Service’, in Lewis (2006) 65–76.

Turfa, J.M. (ed) (2013) *The Etruscan World*. London.

Vaggioli, M.A. (ed) (2006) *Guerra e pace in Sicilia e nel Mediterraneo antico (VIII–III sec. a.C.): arte, prassi e teoria della pace e della guerra, vol. I*. Pisa.

Van Wees, H. (1995) ‘Politics and the Battlefield: Ideology in Greek Warfare’, in Powell (1995) 153–78.

Van Wees, H. (2001) ‘The Myth of the Middle-Class Army: Military and Social Status in Ancient Athens’, in Bekker-Nielsen and Hannestad (2001) 45–71.

Van Wees, H. (2002) ‘Herodotus and the Past’, in Bakker, De Jong and Van Wees (2002) 321–49.

Van Wees, H. (2003) ‘Conquerors and Serfs: Wars of Conquest and Forced Labour in Archaic Greece’, in Luraghi and Alcock (2003) 33–80.

Van Wees, H. (2011) ‘Defeat and Destruction: The Ethics of Ancient Greek Warfare’, in Linder and Tausend (2011) 69–110.

Vinci, M. (2010) 'Reclutamento di truppe scelte a Siracusa in età classica'. *Hormos* 2: 55–66.
Walbank, F.W. (1957) *A Historical Commentary on Polybius. Volume I. Commentary on Books I–VI*. Oxford.
Walters, H.B. (1893) *Catalogue of the Greek and Etruscan Vases in the British Museum. Volume II, Black-Figured Vases*. London.
Warmington, B.H. (1993) *Carthage*. Rev. edn. New York.
Warner, R. (1972) *Thucydides: History of the Peloponnesian War. Translated by Rex Warner with an Introduction and Notes by M. I. Finley*. London.
Waterfield, R. (1998) *Herodotus: The Histories. A New Translation*. Oxford.
West, M.L. (2003) '*Iliad* and *Aethiopis*'. *CQ* 53.1: 1–14.
West, M.L. (2005) '*Odyssey* and *Argonautica*'. *CQ* 55.1: 39–64.
West, S. (1988) 'The Transmission of the Text', Heubeck, West and Hainsworth (1998) 33–49.
Woodhead, A.G. (1962) *The Greeks in the West*. London.
Worley, L.J. (1994) *Hippeis: The Cavalry of Ancient Greece*. Boulder, CO.

CHAPTER 11

The First Greek Soldiers in Egypt

Myths and Realities

Hans van Wees

The Greeks liked to think that their heavy infantry had no parallel or peer in the world, and some modern scholars warmly endorse that self-image.

> Like the German soldier of 1940–41, the Greek hoplite of the classical period drew an almost smug assurance from the natural superiority of his own weapons over any other in the Mediterranean world.[1]

> They began to surpass all other dwellers around the Mediterranean in sheer physical strength and toughness.[2]

Claims such as these are based on two kinds of evidence: Greek successes against the Persian empire in the Classical period, and the demand for Greek soldiers in the Middle East, especially Egypt, in the Archaic age. This chapter revisits in detail the role played by Greek troops in Egypt, 664–525, and argues that our Greek accounts paint a deeply misleading picture. Demand for hoplites had little to do with their supposed intrinsic advantages in equipment, skill, or courage. Instead, the great powers of the Middle East sometimes needed forces that could serve for longer terms than they could mobilise their own armies, and the Greek world was just one recruiting ground among many, including Karia, Phoenicia, and Judea, for such additional manpower.

The fullest and best-known story about Greek hoplites in Egypt is told in Herodotos' *Histories*. According to this tale, an oracle had promised the exiled Pharaoh Psammetichos that his 'revenge would come from the sea when men of bronze appeared', and this was fulfilled when Ionian and Karian raiders in hoplite armour arrived and were mistaken by an awe-struck native for 'men of bronze'. Psammetichos hired them, defeated his rivals (2.152.3–5), and kept his new 'allies' (*epikouroi*) in Egypt (2.154). Psammetichos' great-grandson Apries employed as many as 30,000 Ionian and Karian soldiers (2.163.1) who defended him bravely against an uprising by an army of hundreds of thousands of

1 *WWW* 56.
2 Hale (2013) 190.

 | DOI:10.1163/9789004501751_012

professional Egyptian soldiers, 'but they were far fewer in number, and that is the reason they were defeated' (2.169.1). The victorious rebel Amasis in turn hired them to serve as his personal guard in Memphis (2.154.3), and they are last seen proving their loyalty to Amasis' successor 'Psammenitos' by fighting an overwhelmingly large Persian army and fleet at Pelousion in 525,[3] and drinking the blood of the sons of the sole deserter in their midst (3.11.1–3).

Contemporary Egyptian and Assyrian records tell a different story from Herodotos and other Greek accounts, especially about the career of Psammetichos, although they do show that some elements of the Greek and Karian traditions reflected real events. We shall examine the history of Psammetichos I and his Saite dynasty to see what roles Greek and Karian soldiers played in reality, try to explain the origin and purpose of the Greeks' largely legendary versions of events, and reassess the status of Greek hoplites as hired troops in the Archaic Eastern Mediterranean world.

1 Psammetichos I's Ascent to Power: What Really Happened

In the eighth century, Egypt was a patchwork of small territories ruled by a variety of kings, princes and chiefs, each with his own army, fleet, and fortified city.[4] A common assumption in Egyptological scholarship is that this fragmentation continued well into the seventh century, and that Psammetichos forged a unified Egypt under his control almost from scratch.[5] In this context, the use of foreign soldiers gave him a vital edge.[6] This picture is heavily influenced by

3 I use the conventional date of the Persian invasion, but see Quack (2011), arguing for 526.

4 See, e.g., O'Connor (1983); Kitchen (1986). Fortified cities: Lloyd (1983) 318–25. Early versions of Assurbanipal's annals refer to support from 'the kings of Egypt ... together with their boats [and] their forces' in 667: Prism E_1 fr. 8.16–21; E_2 fr. 3.10–12; LET 31'–32'. In what follows, Assurbanipal's inscriptions are cited in the edition and translation of RINAP 5.1 (with additional reference to paragraph numbers in the edition of Borger 1996). The key texts are: RINAP 5.1 nos. 1 (Prism E_1, *ca.* 666–665), 2 (Prism E_2, *ca.* 665–664), 3 (Prism B, 649–648; no. 4 of 648 is identical to B in the relevant parts and not cited separately below), 6 (Prism C, 647; nos. 7 [Prism Kh or CKalach] and 8 [Prism G] of 646 are identical to C in the relevant parts and not usually cited separately below), 9 (Prism F, 645), 11 (Prism A, *ca.* 644–642: for the date, see RINAP 5.1, 32–33), 12 (Prism H, 639), 21 (Inscription from the Ishtar Temple [IIT], *ca.* 638). For fr. 82–5–22, 10, see Onasch (1994) 91–92; for the Large Egyptian Tablets (LET, *ca.* 661), see Novotny (2014) no. 20; cf. Onasch (1994) 104–15.

5 See, e.g., Pope (2014) 257–74; Agut-Labordère (2013) 974–85; Perdu (2010) 141, (2004) 107; Török (2009) 351, 357; Kitchen (1986) 395–402; Chevereau (1985) 305, 310–11; Spalinger (1978b) 15–16, (1976) 133, 135–36; Lloyd (1975) 15–16, (1988) 134–35.

6 E.g., Chevereau (1985) 310–11: 'a turning point in the country's military history'; O'Connor (1983) 246: 'tipped the balance'; Lloyd (1983) 284: 'cutting edge of his army', (1975) 15; Spalinger

Herodotos' story that Psammetichos was initially just one of twelve kings of Egypt and had to defeat eleven rivals in war to make himself sole ruler. If we set aside this story for the moment and consider only the contemporary evidence, however, we find that the unity of Egypt had been largely restored well before his accession in 664, and that Psammetichos' claim to be the legitimate ruler of Egypt was widely accepted. Moreover, he had substantial military resources at his disposal and the backing of the Assyrian Empire, but he relied mainly on diplomacy and consensus to govern. Until 644, as we shall see, Psammetichos had no use for Karian or Greek troops.

Psammetichos' home city, Sais, was not just one of many. By *ca.* 730, under the leadership of Tefnakht, 'Great Chief of the West', it had in effect become the capital of the western Delta and taken control of Memphis, provoking a military intervention by Piye, ruler of the expansionist kingdom of Kush (Nubia/Sudan).[7] Tefnakht's son Bocchoris proclaimed himself king of Egypt, but was killed *ca.* 715, when the new king of Kush, Shabaqa, conquered the entire Delta, united Egypt under his imperial authority, and made Memphis his capital city.[8] Around 685, another Tefnakht of Sais proclaimed himself king of Egypt, thereby declaring independence from Kush. By 671 one of the most powerful men in the central Delta accepted Tefnakht's successor Necho (I) of Sais (672–664), not the king of Kush at Memphis, as legitimate ruler of Egypt.[9] Despite the relative independence of local rulers, then, by the time Psammetichos came to the throne, Egypt had been unified for two generations under the kings of Kush, and the rulers of Sais had a long record of challenging for the kingship of the whole of Egypt.

The Assyrian Empire mounted a series of invasions of Egypt under Esarhaddon and Assurbanipal in 673, 671, 669, 667 and 664. This is surprising because Assyria had previously made no attempt to conquer Egypt but been

(1978b) 19–20: 'the backbone of his infantry', (1976) 137. More cautiously Kitchen (1986) 402: 'a small nucleus of Greek mercenaries'; Perdu (2010) 143: 'an elite corps'; Agut-Labordère (2013) 989.

7 According to the Victory Stela of Piye, translated and discussed in Lichtheim (2006) 66–84; Eide (1994) 62–117; Grimal (1981); cf. Onasch (1994) 7–11; Morkot (2013) 957–60.

8 The chronology of the 'Nubian' 25th dynasty, the kings of Kush, is controversial, but this date (established by Kitchen 1986, esp. 550–59, 583–87) is quite widely accepted; a different view in, e.g., Morkot (2000) 200–203; ibid., 205, 217–21 for Memphis as new capital, and 207–208 for Shabaqa's quick acceptance as king in both Sais and the eastern Delta; cf. Török (2009) 351–52.

9 Perdu (2002) 1221–33, convincingly argues that a stela of a king Tefnakht (Athens, NM 32) refers to Manetho's 'Stephinates' of Sais who ruled *ca.* 685–678. Stela of Akanosh (B) recognising Necho I as king: Perdu (2004), (2002) 1235–39. Kushite activity at Memphis: Onasch (1994) 10–11.

content to have its border in Palestine. The explanation for this sudden flurry of activity, given that the rulers of Sais and their supporters in the Delta were at this time bidding to take over the kingship from Kush, is surely that Tefnakht and Necho asked for Assyrian military intervention on their side. Fragmentary records of Esarhaddon's campaign of 671 and early versions of Assurbanipal's annals stress that Memphis was besieged and sacked, 'Egypt' (i.e., the Delta) incorporated into Assyria, its kings regarded as Assyrian citizens and perhaps given Assyrian names, its cities certainly renamed, 'taxes and tribute' imposed.[10] Esarhaddon's claim that he appointed a whole new administration was exaggerated; he left most or all local rulers in place. Already by the time he prepared to send another army to Egypt in 669, he feared that Necho and Šarru-lū-dari of Pelousion were no longer loyal.[11] An attempt by Taharqa of Kush (690–664), to recover Memphis in 667 brought a new Assyrian intervention, supported by the rulers of the Delta with their own armies and fleets, but soon afterwards Necho and others were accused of plotting to massacre the Assyrian garrison in Memphis, apparently imposed after the renewed capture and sack of the city in 667.[12]

The Assyrian investigation into this conspiracy ended with a compromise. The garrison stayed at Memphis, but Necho was 'shown greater favour' than before, formally recognised as king of Memphis as well as Sais, and given symbols of kingship. His son, presumably Psammetichos but referred to by an Assyrian name (Nabu-šezibanni), was made ruler of Athribis in the southern Delta, a position traditionally held by the crown prince.[13] Another sign that the Assyrians were willing to make concessions is that later versions of

10 Esarhaddon's texts include a list of renamed cities, each with the name of either a renamed king or an Assyrian official: RINAP 4 no. 9 col. Ii.1'–11'; Onasch (1994) 23–25. Cf. Assurbanipal's Prism E_1 fr. 5.6–21 (incorporation of Egypt), fr. 6 (kings of Egypt referred to as 'Assyrians'; ditto in LET 10'–12'); E_1 and LET use Assyrian names of cities alongside original Egyptian names.

11 Kahn (2006) 252–57; in n. 15, he observes that Esarhaddon's records imply that the alleged reorganisation took less than a month. I would take this as evidence that Esarhaddon merely confirmed existing conditions. Suspicion of Egyptian rulers: *SAA* 04 88.

12 In most versions, the vague target of their 'conspiracy' is 'the troops of Assur', but Prism A adds 'that I had stationed (there) to help them' (col. i.127; §12), i.e., a garrison at Memphis; cf. Assurbanipal's claim that 'I made [Egypt's] guards stronger than before' (e.g., B col. i.88–90; §9) in 667. Taharqa's successor Tanwetamani in 664 attacked 'the Assyrian troops present in Memphis' (e.g., B col. ii.8; §13). Note also the distinction between 'children of rebellion' at Memphis and 'chiefs' in the Delta in Tanwetamani's Dream Stela (Eide 1994, no. 29 lines 16–17).

13 LET 55'–69'; C col. iii.1'–15' (§21); A col. ii.8–19 (§15), with Onasch (1994) 151–54; Kahn (2006) 257–61. The conspiracy narrative has Necho restored to Sais, but the lists of kings in Prism A and C call him king of Memphis and Sais. Athribis as 'fief' of crown prince:

Assurbanipal's annals revert to calling Egyptian towns by their original names. By 666, Necho thus had Assyrian military backing for his claim to be legitimate ruler of Egypt.

Moreover, some potential rivals for power had been removed. Šarru-lū-dari of Pelousion was in prison in Nineveh, the only plotter to be punished. The third conspirator, Pekrur, was not even arrested, and his city of Pi-Soped went unpunished, unlike Sais and Pelousion: the implication may be that his subjects had thrown him out and that he had gone into exile. In the so-called Dream Stela which gives Taharqa's successor Tanwetamani's account of his own campaign into Egypt, Pekrur is the only Egyptian ruler mentioned by name, and he proclaims the right of the king of Kush to kill any of the Delta chiefs – a right which Tanwetamani graciously decides not to exercise.[14] I would suggest that Pekrur had fled south and was returning with the Kushite army, rather than speaking as an incumbent ruler. The defeat of Tanwetamani in 664 presumably put an end to Pekrur's career.

A third potential rival was removed from power by Psammetichos' appointment as ruler of Athribis. Conflict between Sais and Athribis is attested two generations earlier and may have been on-going.[15] Whoever ruled the city in 667 lost his position due to Assurbanipal's intervention, and we can perhaps infer what happened to him from a tiny change made in Assurbanipal's annals. In one version (Prism C), a list of six 'kings' of Egypt established by Assurbanipal gives the name of the ruler of Athribis as *Ni-her-a-u*, probably an Assyrian version of the Egyptian name rendered 'Inarōs' in Greek.[16] Prism A, composed a few years later, *ca.* 643, changed the name to *Bukkunanni'pi*, the Assyrian version of Egyptian Bakennefe.[17] Why change just this one name?

Kitchen (1986) 395; O'Connor (1983) 246–47. Onasch (1994) 51, 154, argues that the custom had been discontinued by 667.

14 Dream Stela: Eide (1994) no. 29, esp. lines 36–39 for Pekrur, called 'hereditary prince', suggesting a claim to the throne; in 728 the ruler of Pi-Soped was an ordinary 'Chief of the Ma' (no. 9, line 115).

15 Piye's Victory Stela speaks of 'hereditary prince Peteese' of Athribis supporting Piye against his opponent Tefnakht, king of Sais: Eide (1994) no. 9, lines 99, 107–126; cf. Onasch (1994) 44–48.

16 Prism C col. ii.80' (§16): the name is damaged in both extant copies but Jamie Novotny (personal communication) suggests this restoration of the name based on his own examination of the text: ^{m}*n[i!-he]r!?-a-u* in one copy (K 1974+) and [^{m}*ni*]*-her?-a-u* in the other (ND 5534). Quack (2006) esp. 501–502, endorsed by Holm (2007) 201–202, suggested *ni-har-a-u*, but Novotny points out that *har*, unlike *her*, is not an attested value of the character EZEN legible in ND 5534. Cf. Onasch (1994) 42.

17 A col. i.94 (§10); a hereditary prince by that name is attested in Athribis, *ca.* 700, probably the father of 'Inarōs', and the son of the Peteese noted above (n. 12): Onasch (1994) 48–51.

As we shall see, it is likely that Egypt had rebelled against Assyrian control *ca.* 644 and that the list of kings had in effect changed from a roster of loyal subjects to a roll-call of traitors. The replacement of the name 'Inarōs' must have served to exempt him from blame. This would have been worth doing only if he was still around in 643 but now on the side of Assyria. He may have lived in exile for twenty-odd years but seized the moment of Egypt's break with Assyria to make an alliance with Assurbanipal in a bid to regain power.[18]

Necho's position was thus much strengthened. Yet despite, or because of, the Assyrian occupation of their city, the priests at Memphis did not recognise his legitimacy. They continued to date events by the regnal years of Taharqa of Kush: deep into the reign of Psammetichos, the priests still called 665 Taharqa's year 26, not Necho's year 8.[19]

Necho's death and Psammetichos' accession are not mentioned in contemporary sources. A comparison of the competing narratives of the events of 664 in Assurbanipal's annals and in Tanwetamani's Dream Stela suggests that Tanwetamani won a pitched battle outside Memphis and occupied much of the city, but not the citadel where the Assyrian garrison was stationed. He was unable to attack any of the Delta towns and soon returned south, followed by an Assyrian relief force that went as far as Thebes and sacked this city.[20] If Necho was killed, it was presumably in the fighting at Memphis. In any case, his son succeeded him, and after the death of Taharqa and severe defeat of Tanwetamani even the priests at Memphis accepted Psammetichos as the new legitimate ruler.

---

18 Quack (2006) and Holm (2007) do not address the reasons for the change of name, but they point out that Inaros (and Pekrur) played leading roles in later Demotic historical fiction as heroes of Egyptian resistance to Assyria: the poorly preserved earliest version (mid-fifth century, in Aramaic) appears to feature Inaros as an ally of Taharqa conveying a threatening message to Necho, Holm (2007) 197–99.

19 Kahn (2006) 257, 261; Perdu (2004) 109 with n. 89. The evidence consists of Apis burial stelae: Serapeum Stela IM 2640 *et al.* for 667 (trsl. in Ritner 2009, no. 165), and IM3733 for 665. *Pace* Kahn, we need not infer re-conquest(s), merely a refusal to accept Assyrian-backed Necho.

20 For the Dream Stela (DS) see again Eide (1994) no. 29. Both sides agree that there was a pitched battle outside Memphis, each of course claiming victory for himself (DS, lines 16–17; LET recto 1–3). Partial control of Memphis is implied by contrast between DS 17–14, which lists temples at Memphis visited by the king, and Prisms B col. ii.8–19 (§13–14), C col. iii.22'–36' (§23–4), and A col. ii.22–33 (§17–18), which speak of him besieging the garrison (variations in F col. i.34–42; §5). DS 24–26 concede that no action against Delta chiefs was possible, but 27–41 claim that they offered submission voluntarily, after which the king went home; the Prisms claim that Tanwetamani fled as soon as the Assyrian army crossed the border, and that the Delta chiefs were very grateful.

In Lower Egypt, then, Psammetichos' position was well-established from the start: inherited from his father and from a dynasty which had been the main challengers to Kush for two generations, now backed by Assyrian military power and Memphite priestly authority, and without competition from the main local rivals faced by Necho. Moreover, he aimed at consensus, using both an advisory council of officials and a deliberative council in which the 'nobles' were to be persuaded by the king's proposals. In the main cities, local ruling families retained their position and, in the first generation, their inherited titles. Only gradually did they become mere 'governors' appointed by the king.[21]

In Upper Egypt, at Thebes, Tanwetamani continued to be recognised as king until his eighth year (657),[22] but here too Psammetichos gradually extended his power by diplomatic means. In 656 he had his daughter Nitocris adopted as successor-elect of the God's Wife of Amun, the highest-ranking priestess in Thebes. The incumbent was a Kushite princess who stayed in post, alongside the highest-ranking male priests, who were also members of the Kushite royal family. The highest secular authority, Montuemhat, 'Count' of Thebes, who was also Governor of the South, i.e., *de facto* ruler of Upper Egypt on behalf of Taharqa and his successor, also retained his position until his death, after 649; reliefs in his tomb acknowledged Psammetichos as king. Only after his death did the Stewards who managed Nitocris' vast estates take over the role of Governor of the South.[23] Every effort was clearly made to avoid provoking either local rulers or the kings of Kush.

Psammetichos also had large military forces under his direct control, and as king of Egypt was able to mobilise many more. Soldiers were of two types: local militiamen and 'professional' troops known as the Ma, descended from the Libyan troops who had established themselves in the Delta since the Late Bronze Age. Several rulers in the Delta had the title 'Chief of the Ma' and were thus primarily military commanders. The Tefnakht of Sais whose expansion had provoked the intervention of Piye two generations earlier was himself

---

21 For the two councils and gradual centralisation, see esp. Agut-Labordère (2013) 974–85; the broad picture of non-confrontational centralisation of power is shared by many accounts of Psammetichos' reign: e.g., Lloyd (2000, 1983); Spalinger (1976) 138–40, (1978b).

22 Kitchen (1986) 399–400; Kahn (2006) 266 with nn. 90–91.

23 See the Nitocris Adoption Stela (Cairo JdE 36327; trsl. Ritner 2009, no. 171), and, e.g., Leclant (1961) esp. 55, 179: image and name of Psammetichos in Montuemhat's tomb, 272–75: power relations at Thebes, 279 n.1: Montuemhat's assumption of near-royal privileges; also O'Connor (1983) 251; Kitchen (1986) 399–405; Morkot (2000) 237–38. Montuemhat is usually called 'mayor' of Thebes in English (and 'prince' by Leclant), but his title *ḥ3ty-ʿ* is in other contexts usually rendered as 'count'.

a 'Chief of the Ma', and although Psammetichos, as king, no longer had that title, he would still have had the soldiers. According to Herodotos, the professional soldiers of Egypt in his own day were divided into two groups called Hermotybies and Kalasiries, of which the former were stationed in the western Delta and the latter in the eastern Delta and Thebes: it looks as if the Hermotybies derived from the 'Ma' directly under the control of the kings of Sais, and the Kalasiries from the 'Ma' stationed elsewhere. If so, Herodotos' numbers imply that Psammetichos' own professional forces amounted to forty percent of the total across Egypt.[24]

Centralisation of the remaining Ma/Kalasiries under his command appears to have proceeded in typically gradual and diplomatic fashion. The ruler of Pharbaitos in the eastern Delta recognised Psammetichos as king but himself retained the full titulature of 'prince, count, great chief and army leader' until at least 656, as did others, but in the next generation the title disappears.[25] The only known military campaign from the first half of his reign illustrates how carefully Psammetichos trod. The fragmentary record of this episode has the king convene a meeting of chiefs and local mayors from the western Delta and neighbouring Libyan tribes in 655, in order to establish whether there were grounds for war against the 'Tehenu' Libyans. This was followed by a proclamation that no one should resist the king because Ptah of Memphis had granted him authority, and finally, in the next year, a muster of all the Ma and militia troops, who defeated the Tehenu.[26] At a minimum, this shows Psammetichos ensuring consensus in favour of war and actively soliciting and advertising the support of the priests at Memphis. One may even suspect that the offences of the Tehenu were a convenient excuse to make war and thereby establish a precedent and a legitimation – the blessing of Ptah, rather than merely the

---

24 Tefnakht, 'chief of the Ma': Eide (1994) no. 9, lines 28, 80, 126. Hdt 2.164–68: 160,000 Hermotybies in 6 areas of which Sais, Chemmis, Papremis and Prosopitis cover Psammetichos' own territory, and Busiris and 'half of Natho' are immediately adjacent; 250,000 Kalasiries from 11 areas covering the rest of the Delta and Thebes. The absolute numbers are unreliable, but the proportions may be significant. Documentary evidence for Hermotybies (*rmt-ḏm*, 'ḏm-man') and Kalasiries (*gl-šr*, 'brave man'): Fischer-Bovet (2013).

25 Kitchen (1986) 401–402, 492 (Pedikhonsu); cf. 400, 491 (Pimay/Pmui of Busiris); also Chevereau (1985) 64, 66, 227; Agut-Labordère (2013) 974–77; Perdu (2004) 104–106.

26 Stela: Goedicke (1962); English translation: Der Manuelian (1994) 323–32; the latter rejects on linguistic grounds Goedicke's suggestion (43, 47) that the text may distinguish 'mustered' from 'recruited' troops and that the latter may be mercenaries. This notion is repeated by Haider ((1988) 196–97, 210, 229, (1996) 96, (2004) 448) and Schlotzhauer and Weber (2012, 14–15) despite the lack of archaeological evidence for a Greek presence pre-625 which their meticulous studies establish.

backing of Assyria – for uniting all local forces, professional and militia, under the king's command.

Despite a common assumption that Psammetichos made himself independent from Assyria at some point before making himself king of Upper Egypt, and that he recruited outside help in order to do so,[27] I would suggest that his ability to achieve remarkable diplomatic successes relied on continued Assyrian backing and the military threat that this posed to any opponents. Such an arrangement was mutually beneficial because Assyria's main interest in Egypt was as a buffer zone between its provinces in the Levant and expansionist Kush in Sudan.[28] Psammetichos' consolidation of control over the Delta and extension of power in Upper Egypt served that purpose well, and Assurbanipal had every reason to support it. It would have made sense to maintain the Assyrian garrison in Memphis as well. The Assyrian royal annals in fact offer some evidence that Psammetichos did not break off relations with Assyria until some twenty years into his reign.

Assurbanipal's Prism A, composed *ca.* 643, reported the death of Gyges (Gūgu), king of Lydia, at the hands of Cimmerian invaders, and interpreted this event as divine punishment: Gyges had not kept up diplomatic relations with Assurbanipal and 'he sent his forces to aid Psammetichos, the king of Egypt, who had cast off the yoke of my lordly majesty'.[29] All previous versions of the annals from the mid-660s to 645 included only the story of Gyges' first diplomatic overtures to Assurbanipal, and we must conclude that Gyges died *ca.* 644. It does not necessarily follow that Gyges sent aid to Psammetichos shortly before his death, since a royal inscription would not mention an act of defiance unless it had already been punished, potentially many years after the offence had been committed.[30] However, the same text indirectly confirms

27 So, e.g., Spalinger (1976) 137, 143 (dated to 663/2); Sullivan (2011) 44; Bettalli (1995) 58; Lloyd (1988) 133–35 (662–658); Onasch (1994) 158 (before 656); Kitchen (1986) 406 (654); Goedicke (1962) 49 (652); Porten (1968) 12 (*ca.* 650). Later dates seem possible to Vittmann (2003) 34–35: Gyges' troops arrived *ca.* 660, but they were not necessarily used against Assyria; Braun (1982) 37: 'could have been as late as the 640s'; Haider (1988) 172, 174–75 (645 at latest). Some Greek historians assume that the troops were recruited already in 664: e.g., Hale (2013) 184; Trundle (2013) 330; Bowden (1996) 36.

28 As argued by Spalinger (1974), and widely accepted, e.g., O'Connor (1983) 245; Perdu (2004) 100.

29 Prism A col. ii.113–15 (RINAP 5.1 no. 11; cf. Borger §25).

30 Fuchs (2010) 420–21; Cogan and Tadmor (1977); Spalinger (1978a). Compare the case, discussed in Novotny and Watanabe (2018), of the Arab kings Iauta' and Abi-yate', who are mentioned from Prism B (649–648) onwards, but whose involvement in the Babylonian rebellion of 652–648 is not revealed until *ca.* 643, when Prism A (vii.82–ix.102) is able to report their capture.

that Psammetichos had still been loyal to Assyria recently, during the revolt of Assurbanipal's brother Šamaš-šuma-ukīn, king of Babylon, from 652–648. Prism A claims that Šamaš-šuma-ukīn had incited rebellion not only in Babylon, Elam, and other nations in the east, but also among 'the kings of … the land Amurru and Meluḫḫa … whom I had installed' (col. iii.103–4; §33). Amurru is the Levantine coast, while Neo-Assyrian texts regularly refer to Meluḫḫa in combination with Makan to mean Kush and Egypt, respectively.[31] The fact that the Prism does not mention *Makan*/Egypt here must be a deliberate omission, and indicates that alone among Assyria's western subjects and allies, Egypt had *not* joined the revolt of 652–648.

The almost equally surprising inclusion of Kush in the list confirms Egypt's continuing loyalty: a 'rebellion' by Kush – which was never actually subject to Assyria – could only have taken the form of an attack on Egypt, and if an attack on Egypt counted as a 'rebellion', then Assurbanipal must still have regarded Egypt as an Assyrian subject at the time of the revolt. Psammetichos thus did not declare Egypt independent until after 648.

Another indication that Egypt's revolt was recent is Prism A's wholesale revision of the story of Assurbanipal's interventions in Egypt and above all the narrative of the 'conspiracy' in which Necho and others had allegedly been involved. The first few iterations of the story followed the normal pattern in Assyrian annals of abbreviation and omission of older material as more recent events came to the fore,[32] but later versions show more radical changes and some elements were not abridged but expanded as the situation in Egypt changed and brought old news back into the headlines.[33] The Prism B-narratives, composed during the revolt of Šamaš-šuma-ukīn, presented the suppression of Necho's plot, not as a separate episode, but as a matter of reprisals for disloyalty shown during the campaign of 667, and mentioned only one plotter, Šarru-lū-dari, without making any reference to Necho's involvement or his subsequent restoration.[34] These striking omissions are likely to

31 As in Prism B col. i.48–9 (§7); C col. ii.4'–5' (§13); F col. i.35 (§5); A col. i.52–3 (§7). One of Esarhaddon's inscriptions makes the equation explicit: RINAP 4, no. 34 obv. 7'–8'.

32 The oldest fragment (fr. 82-5-22, 10; Onasch 1994, 91–92) includes the name of an Egyptian king who betrayed the conspiracy, omitted from the next version, Prism E_1 (E_2 has a lacuna) and all later accounts. The next version, LET, adds events of 664.

33 Spalinger (1974) offers an analysis in a similar vein, but it has in some respects been overtaken by more recent editions of the texts and does not note a number of differences that seem to me significant. Cf. Onasch (1994) 83–90, 126–27, for stylistic and historical differences. Cogan (2014) seeks explanations in distinctive literary and personal preferences of the author, not historical factors.

34 Prism B col. i.48–ii.37 (§7–14), with the conspiracy covered in a few lines at i.91–ii.3 (§10–11). The prism also omits what in earlier versions was the second part of the campaign of 667.

reflect the continuing loyalty of Psammetichos and Egypt at this time: previous disloyalty was not a topic on which to dwell with regard to a currently rare and precious ally.

The next versions, Prisms C, Kh, and G, were composed in 647–646 after the great revolt and duly added a grisly description of the deprivations suffered by the rebels.[35] The account of the first Egyptian campaign now included long lists of kings subject to Assyria in the Levant, Cyprus, and the Delta, and it restored the entire twenty-year old conspiracy narrative. These changes served to reassert Assyrian imperial power after the victory over Šamaš-šuma-ukīn, and to reinforce the message that disloyalty would be punished: only Prism Kh records the detail that Šarru-lū-dari ended his life in Assyrian prison.[36]

Three years later, Prism A rewrote the history of Necho's conspiracy.[37] The list of six named Delta kings found in Prism C now grew to twenty names, including at least four kings in Upper Egypt, culminating with Montuemhat, 'king' of Thebes (A col. i.90–109; §10). This was not because the scribe liked long lists or wanted to display his learning: he replaced the names of the rulers of the Levant and Cyprus by a brisk '22 kings' (A col. i.69; §8). The Egyptians' names and cities were added because they were now part of a denunciation of traitors to Assyria. In Prism A, Assurbanipal said that *all* the kings, 'as many as I had appointed' had broken their oath (col. i.118), and 'those kings' were thrown in chains (i.130). Not only the people of three named towns but also those of 'the rest of the cities' were massacred, flayed, and impaled (col. ii.1). Finally, 'those 20 kings who had constantly sought out evil (deeds) against the troops of Assyria' were taken to Nineveh (ii.5–7), where 'among them' only Necho was pardoned and returned to power (ii.8; §12). Prism A differed from versions down to 646, I suggest, because it projected a new and current state of

35 Prism C col. viii.2'''–ix.10''; Kh col. viii.1'–79'; G col. viii.1'''–36'''' (C §62–4).

36 Lists: C col. ii.25'–52' (§14; Levant and Cyprus), C col. ii.76'–85' (§16; Delta). Conspiracy: C col. ii.3''–iii.15'; Šarru-lū-dari's punishment: Prism Kh (RINAP 5.1 no. 7) col. ii.50''–53'' (§20); the episode falls in lacunae in C and G (RINAP 5.1 nos. 6, 8). The list of Levantine and Cypriot kings is virtually identical to a list of kings forced to contribute wood and stone for Esarhaddon's Arsenal (RINAP 4, nos. 1 col. v.54–73a; 5 col. vi.6b'–15'; in an earlier version just '22 kings': no. 2 col. iv.54–v.12), which also advertised the reach of the king's power. Prism C, unlike most others, boasts that Assurbanipal's empire was larger than his predecessors' by 'a distance of one month (and) 20 days (journey) into the midst of the sea and on dry land' (Kh/RINAP 5.1 no. 7 col. i.95'–99' [§12]; lacuna in C and G). The same phrase is used at this time in Prism I (Novotny 2014, no. 1, col. iii.30'–34'; *ca.* 648) and Prism T (Novotny 2014, no. 2, col. iv.22–26; *ca.* 645).

37 In between, Prism F of 645 severely truncated all earlier history in favour of a detailed account of the latest campaign against Elam, to the extent of altogether omitting Assurbanipal's first expedition to Egypt in 667, the conspiracy, and Necho (col. i.34–54; §5). In most respects, Prism A followed Prism F, which makes its elaborate coverage of Egypt even more significant.

'revolt' in Egypt – now including Upper Egypt as well as the Delta – back into the events of 667.

A final sign of Egypt's loyalty to Assyria until after 646 may be seen in an inscription on a statue of Psammetichos' general Djedptahiufankh, dated 'no later than 640'. The general calls himself 'His Majesty's mouth among the contingents of Asiatics, leading their troops when he was sent by the king to repel the foreign enemy'.[38] Djedptahiufankh also calls himself more broadly 'commander of foreigners', but the word for 'foreigners' used here, *chasut* (*ẖ3styw*), is not the same as the term later applied to troops that included Karian and Greek soldiers, and used in Ptolemaic Egypt to mean 'Greeks': *haunebut* (*ḥ3w.nbwt*).[39] 'Commander of foreigners' is a title attested already before 715, when it referred to soldiers from neighbouring countries such as Libya or Palestine.[40] Since Djedptahiufankh's statue was found at Thebes in Upper Egypt, the most likely foreign enemy would be Kush, and the most likely 'Asiatic' allies would be forces sent by Assyria, in 652–648.[41]

All of this means that the mention of Gyges' aid for Psammetichos occurs in Prism A, and only here, not because scribes with long memories decided to treat his recent death as punishment for an old offence, but because Gyges' death did indeed, coincidentally, come within a year or two of his backing Egypt's bid for independence. Assurbanipal's prayer for revenge was answered almost at once; the hand of god was an obvious explanation.[42]

Psammetichos and Gyges were not the only 'rebels' at this time. Prism A reported in detail a revolt by former Arab allies (col. vii.82–ix.102), and Assurbanipal's campaign of reprisals extended as far as Phoenicia, where he punished two cities for failing to pay their taxes (col. ix.115–28; §81–2). A later text reveals that Gyges' neighbour Mussi, king of Tabal, had made an alliance

38 The statue is Cairo JE 36949; see De Meulenaere (1965); cf. Chevereau (1985) no. 109. Date: Josephson and Eldamaty (1999) 88.

39 Cheverau (1985) nos. 114 (Potasimto, who also commands other 'foreigners', *ẖ3styw*, and appears in a Greek graffito at Abu Simbel as 'commander of foreign-language speakers' during an expedition of 593), 117 (Hor), 186 (Bakenrenef), 187 (Iufa); also, under Amasis, no. 178 (Udjahorresnet).

40 'Commanders of foreigners' under the XXIInd Dynasty: Chevereau (1985) nos. 40, 44, 61.

41 So De Meulenaere (1965) 22–25 (III, lines 2–3), 29–31 (*pace* Vittmann 2003, 199). A fragmentary inscription alludes to an expedition against Kush around this time: Sauneron and Yoyotte (1952) 201 n. 3; Habachi (1974). Sauneron and Yoyotte (1952) argued that the Judean 'allies' sent to help 'Psammetichos' against Kush in the *Letter of Aristeas* 13 relate to this campaign.

42 Prism A col. ii.111–25 (§25); the prism also reported Gyges' submission to Assyria in more detail (ii.95–110; §24) than other versions except the two earliest (E_1 frs. 11.11–12.31'; E_2 frs. 6.48'-9.3').

with the Cimmerians against Assyria (IIT 141b–6a), presumably at the time of the invasion that killed Gyges. These episodes show us glimpses of widespread defection in 645–644. We may see a crisis reflected also in the rest of Prism A's revisions: expanded versions of the subjection of Tabal and Arvad, and above all a completely rewritten, much longer account of Šamaš-šuma-ukīn's revolt. The latter went so far as to claim that the gods had communicated, in a dream, their decree that 'whoever makes war on Assurbanipal, king of Assyria, will suffer a terrible death'.[43]

Despite these threats, no attempt was made to recover Egypt, and five years later Assurbanipal had tacitly given up his claim to rule it. Three late inscriptions, from 639 and *ca.* 638, reduced the Egyptian campaigns to a single expedition against Thebes in 664, which they presented as a mere plundering raid, without a hint of imperial claims or ambitions in the area. They did not repeat the moralising story of Gyges' death.[44] Whatever Assyrian threat to Egypt existed in 643, it apparently evaporated within a few years, mainly no doubt because Cimmerian invasions threatened Assyria itself (IIT 146b–161). Psammetichos thus probably needed no further military aid to resist Assyria, and aid from Lydia would in any case not be forthcoming, because by 643 the Cimmerian danger had forced Gyges' successor to switch allegiance again, back to Assurbanipal (Prism A col. ii.120–5; §25).

In sum, the contemporary evidence makes no mention of Greek or Karian soldiers in Egypt before 645, and strongly suggests that Psammetichos' internal and international position and policies at the time were such that he had no need of such hired troops. Only when he made himself independent of Assyria *ca.* 644 did he require new external sources of support. If the troops sent to him from Lydia were soldiers recruited among Gyges' Ionian and Karian neighbours, as is likely, they will have been the first troops from the Aegean to serve the king of Egypt. A bilingual Karian and hieroglyphic inscription on a statue of Neith at Sais may hint at Karians sent by Gyges *ca.* 645: it has the cartouche of Psammetichos I, and is dedicated by a man and woman with Egyptian

43 Arvad and Tabal: A col. ii.63–94 (§20–23); among the additions is a list of the names of all 10 sons of the king of Arvad. New revolt narrative: A col. iii.70–iv.109 (§32–42; divine decree at iii.118–27; §34). Note also that Prism A prefaced its narrative of the Arab wars with an account of earlier campaigns against Arabs which Prism F had omitted altogether, and made a point of inserting here, too, some explicit moralising about the dire consequences of disloyalty to Assurbanipal (A col. ix.68–74; §78).

44 Prism H fr. 2.7'–14'; IIT 82; RINAP 5.1 no. 19, line 17'b; contrast Prism F col. i.40–42 (§5), which also focused on the sack of Thebes, but boasted of the submission of local 'kings' and thus asserted claims to imperial control. Prisms H and IIT do make claims to imperial rule in other regions. Gyges' submission, but not his death, is noted at IIT 86b–9; relevant parts of H do not survive.

names, but the man's parent's names are apparently Karian, which suggests a second-generation immigrant in Sais before 610.[45] Intriguing are the 100 or so Karian graffiti found in the open areas of Montuemhat's tomb at Thebes, along with a few Greek graffiti. If these texts were left by visitors who came to pay their respects to a figure of personal importance to them, they might date soon after Montuemhat's death, in the 640s. But the graffiti may very well have been left by much later visitors to this spectacular tomb.[46] In any case, both the Assyrian threat to Egypt and Egypt's alliance with Lydia evaporated almost immediately, and the troops sent by Gyges need not have been followed by any other Aegean mission to Egypt for twenty years.

2 Egypt versus Babylon: A New Market for Aegean Soldiers

With the collapse of the Assyrian Empire in the 620s, both Egypt and Babylon exploited the power vacuum to mount imperial campaigns of their own, and from *ca.* 625 onward Greek troops began to take their place in this international arena. Herodotos listed a few episodes of Egyptian expansion, but without mentioning Greek involvement, or indeed revealing that the Neo-Babylonian Empire was Egypt's opponent. He noted an alleged twenty-nine-year siege of 'Azotos in Syria' (Ashdod in Palestine) because 'this city held out longest under siege of all the cities we know' (2.157), but from contemporary sources we know that Psammetichos' power extended much further afield by the end of his reign. Biblical texts say that the entire Levant from the border of Egypt to Carchemish on the Euphrates 'pertained to the king of Egypt' (2 *Kings* 24.7), and imply that at some point before Psammetichos' death in 610 Judah had become tribute-paying vassal state of Egypt.[47] Contemporary documents

45 Adiego (2007) 33 (E.Sa 2 in his catalogue); discussion in Vittmann (2003) 162, on 175 he rightly dismisses the attempt to date a stela from Memphis with the name Pigres [*Pikre*, E.Me 3] to the start of the reign, on the basis of a Greek story about a Karian of that name, discussed below.

46 Adiego (2007) 95–106 (E.Th 1–53) includes many of these graffiti, which have not been fully published; cf. Kaplan (2003) 6 n. 29; Vittmann (2003) 164; Masson (1994), with the 'intervention' of J.D. Ray. Greek graffiti: noted in passing by Leclant (1961) 179. Conceivably, Karian troops came to Thebes when Ibi was installed as Chief Steward of Nitocris in 639 and needed 'to restore order' (Cairo JdE 36158, translated in Ritner 2009, 588–92). Sauneron and Yoyotte (1952) 201 n. 4, saw an allusion to a military or diplomatic mission to Kush in another inscription of Ibi.

47 Judah as vassal state before 604: 2 *Kings* 23.33–35; 2 *Chron.* 36.3–4. King Josiah is said to have opposed Necho II during his campaign of 609 (2 *Kings* 23.29–30; 2 *Chron.* 35.20–24), but this may be part of his portrayal as a great reformer and independent ruler, and

report that Egypt stationed officials and imposed taxes on trade in Phoenician cities by 612.[48] Accordingly, when Nabopolassar of Babylon attacked Assyria in 616, an Egyptian army went to the Assyrians' rescue to fend off this new threat to territories that were by now already under Egyptian control.[49]

Herodotos mentions that Psammetichos' successor Necho II (r. 610–595) built trireme fleets for the Mediterranean and the Red Sea, 'which he used when needed' (2.159). It is only from the *Babylonian Chronicle* that we find out a fleet was 'needed' for far-flung campaigns against further expansion by Neo-Babylonian armies, now led by Nebuchadnezzar as crown prince and then as king. In 610 and 609, Egyptian troops tried unsuccessfully to help the Assyrians defend their last stronghold at Harran. With Assyrian power destroyed, Egypt alone tried but failed to stop Babylonian expansion at Carchemish on the Euphrates in 606 and 605.[50] By the end of 604, Nebuchadnezzar had penetrated as far as Ashkelon in Palestine, and in 601 he attempted to invade Egypt itself but was defeated. Herodotos refers to the event as a great battle against 'the Syrians' at 'Magdolos' (Migdol, on the eastern border of Egypt), after which Necho II's forces retook Gaza.[51] Further Babylonian conquests, including the capture of Jerusalem in 597, seem to have been unopposed by Egypt, which presumably concentrated on self-defence at this time.

Psammetichos II (595–589) launched a great expedition against Kush in order to secure Egypt's southern border, while his successor Apries (589–567) briefly tried to renew the fight with Babylon over the Levant, sending a fleet to Phoenicia (Hdt. 2.161) and an army to Judah (Jer. 37.5–11), which provoked a second Babylonian capture of Jerusalem in 587. After this, there is little sign of military activity until 570, when an expedition into Libya sparked off a civil war between Apries and his general Amasis, ending in 567 with a failed attempt by Babylon to invade Egypt with a view to restoring Apries to power.[52] For the rest

in any case Necho's ability to impose a new king and tribute on Judah, despite having been defeated in Assyria, suggest pre-existing subordination rather than a new conquest (Na'aman 1991, 40–41, 51–55).

48 Phoenicia: Vittmann (2003) 37: Apis stela of 613; Na'aman (1991) 51–52, also cites a stela of Psammetichos I at Arwad, and of Necho II at Sidon. Palestine: Vittmann (2003) 39, 88: Sakkara letter; note also the story that Psammetichos paid Scythian invaders to leave Palestine (Hdt. 1.103).

49 *ABC* no. 3.10. A preliminary version of a new translation is on-line at https://www.livius.org/sources/about/mesopotamian-chronicles/.

50 *ABC* nos. 3.61–62 (610), 66–69 (609); no. 4.16–18, 24–26 (606); no. 5.1–8 (605).

51 Hdt. 2.159; the *Babylonian Chronicle* presents it as a draw, but tellingly its next entry reports that next year Nebuchadnezzar 'stayed home' for the first time and reorganised his army (*ABC* no. 5, rev. 5–8; cf. obv. 12–23; rev. 2–4).

52 Victory Stela of Amasis, from Elephantine: Jansen-Winkeln (2014); see further below.

of Amasis' long reign, military activity appears to have been minimal, although Herodotos does claim that this king 'was the first man to capture Cyprus and force it to pay tribute' (2.182.2).

Egypt thus experienced a period of expansion, *ca.* 625–600, followed by a largely defensive period, *ca.* 600–525, each of which in its own way created new demands for manpower. The wars to gain and retain control of the Levantine coast required soldiers who were available for long campaigns, and troops for permanent garrisons in occupied territory. The forces sent out in 606, for instance, conducted a four-month siege of Kimuhu on the Euphrates, then campaigned across the river, and wintered in Carchemish.[53] Such lengthy absences from home would have been problematic for Egyptian troops, since the Hermotybians and Kalasirians, despite forming a hereditary 'warrior class', were not full-time soldiers but had land to cultivate while they remained on stand-by for military service.[54] In the second phase, the acute Babylonian threat encouraged the employment of standing forces permanently stationed in border forts to guard against invasion. Herodotos' anecdote about half of the Egyptian army deserting to Kush when they had not been relieved after three years' continuous service at the southern border in Elephantine (2.30), while of doubtful historicity, illustrates the difficulty of maintaining a strong border defence on a traditional basis.[55] In order to meet these new needs, Egypt had to hire additional troops.

Not coincidentally, around 625 we find the first evidence for Greek troops in Egypt since the one-off intervention *ca.* 644. A famous Greek inscription at Abu Simbel, left by soldiers who took part in Psammetichos II's campaign against Kush, was signed by a number of newcomers to Egypt who still identified themselves by their Ionian home towns, but they were led by 'Psammetichos son of Theokles', presumably a second-generation immigrant, judging by his father's Greek name. Theokles' arrival might take us back to *ca.* 625. The equally famous Egyptian basalt block-statue dedicated by Pedon at Priene in the early sixth century may also imply military service towards the end of Psammetichos I's reign.[56]

---

53 According to Babylonian chronicles: *ABC* no. 4.16–18, 24–26; 5.1–8.

54 For their economic position, see Fischer-Bovet (2013). Herodotos mentions their ownership of land; only those who served as the king's bodyguard were provided with food and drink (2.168).

55 A similar story in Diod. Sic. 1.67.3–7 is set instead during Psammetichos' campaigns in 'Syria'.

56 Among numerous discussions, see Agut-Labordère (2012); Vittmann (2003) 200–206. Both also discuss a Wahibre-em-achet who may have been a soldier and had Greek parents: see now the full discussion (with additional material for his dossier) in Villing (2018).

At many sites in Egypt and the Levant, East Greek pottery begins to be found from *ca.* 625 onward. Not all such pottery need imply the presence of Greeks, but where Greek painted tableware and household wares such as cooking pots and lamps occur together in forts, they strongly suggest the presence of Greek soldiers. Herodotos mentions Stratopeda, 'The Camps', on the eastern border – probably the same place as biblical Migdol, meaning 'the Fortress' – as the first station of Greek soldiers in Egypt (1.154.1). A 25-acre site including a 10-acre fort at Tell Qedua, about 10 km southwest of Pelousion, may have been (one of) these 'Camps'. The earliest Greek finds were Chian, Samian, and Lesbian transport amphorae dating to *ca.* 625–600 and nearby cremation burials with such amphorae as grave goods. Very little Greek table- and cook-ware was found, alongside much larger quantities of local Egyptian and Levantine pottery, indicating that the fort housed mainly non-Greek troops.[57]

Herodotos also says that Psammetichos installed a garrison at Daphnai near Pelousion (2.30.2). Recent analysis of the nineteenth-century finds at this site (Tell Dafana/Defenneh) has shown that it was indeed founded by Psammetichos, but not as a mere fort: it was 'a classical temple town functioning as a frontier post'. The earliest Greek painted pottery, in the Middle Wild Goat II (MWGII) style, dates to *ca.* 625–600 and was manufactured in Ionia, especially Miletos. A substantial amount was found, again alongside larger amounts of Egyptian and Levantine material, but mainly in storage rooms attached to the temple, and in some cases specially made for dedication. This suggests the presence of Greeks but does not show that these Greeks were resident here. Daphnai may, therefore, have been a temporary base for Greek and other troops at the start and end of expeditions into the Levant, and probably did not feature 'a settled community' of Greeks.[58]

Elsewhere in Egypt, at Memphis, Mendes, Thebes, and Edfu, the earliest Greek pottery again dates after 625, and we cannot be sure that it implies the

57 Oren (1984) 24–30: Greek trade amphorae in fort and burials (date: 'late seventh century', and parallel with finds at Naukratis and Daphnai, i.e., 625–600, see below); the other Greek wares are limited to local copies of an East Greek cup and a Greek cooking pot at the fort (27; figs. 23:2; 42) and two East Greek drinking cups at nearby site T.94 (figs. 23:1, 4; 43; 51). The cremation burials could be Phoenician rather than Greek: Oren (1984) 40 n. 25. Identity of site: Oren (1984) 38 (tacitly rejected by Haider 1988, 203–204; Schlotzhauer and Weber 2012, 14–15; Weber 2012, 244–45). On the uses of Greek pottery in Egypt generally, see Villing (2013).

58 Leclère and Spencer (2014) esp. 1–4, 9–10, 135–36: interpretation of site; 118–24: pottery; see also Villing (2013) 92–93; Weber (2001) 131–33, (2012) 216–44: the date for the earliest pottery is 'about 600 BC', but it includes MWGII material, so 625–600, on the most widely accepted dating). Herodotos' references to the garrison (2.30, 107.1) do not suggest that it was manned by Greeks.

presence of Greeks, let alone Greek soldiers.[59] At Naukratis – which recent research has confirmed as a new foundation of Psammetichos I, and as a town that was both Egyptian and Greek, rather than a Greek settlement – the earliest closely dateable imported pottery starts *ca.* 620.[60]

The same Ionian, late seventh-century painted ware is the first Greek pottery for a long time to appear in the southern Levant, at quite a few sites, of which at least two can be identified as garrisoned forts. Finds of MWGII pottery at Ashkelon and two sites on the route inland from here are particularly important, because they were sacked in Nov/Dec 604 by a Neo-Babylonian invasion, confirming the date of the pottery.[61] The forts are at Mesad Hashavyahu just north of Ashdod in Palestine, where Greek finds constitute about half of the material and Greek soldiers were evidently numerous, and at Tel Kabri just south of Tyre in Phoenicia, where Greek soldiers were a small minority.[62] The fact that Greek soldiers are attested in the Levant at the time when Egypt dominated the area, but not afterwards, and that Mesad Hashavyahu was near Ashdod, which took up much of Psammetichos' military effort, suggests that they were present as garrisons installed by Egypt. Likewise, an Ionian bronze shield and Greek greaves left in the ruins of Carchemish in 605 surely belonged to a Greek in the service of defeated Egypt, not victorious Babylon.[63]

One channel through which Ionian and Karian troops were recruited may have been a renewed alliance with Lydia, made in or before 605 in the face

59 Weber (2012) 196–289, (2001) 130, 135–36, 139–41; cf. 129: at Sais from *ca.* 600.

60 Thomas (2015); Villing (2015, 2017). Schlotzhauer ((2001) 127–29, (2012) 21–72), dates earliest finds to *ca.* 620, but notes a 'Bird Bowl' of 640–600, a Milesian amphora of 630–610 (34) and the Ionian Cups (*Knickrandschalen*: 45–54); also (2001) 118–19 n. 53: three supposed MWGI fragments are of later date); Möller (2000) 89–181.

61 Waldbaum and Magness (1997) (pottery from Ashkelon, Tel-Miqne/Ekron and Tel Batash/Timnah, with reference also to the forts at Tel Kabri and Mesad Hashavyahu, below); see 37–40 for the date; for archaeological evidence see also: Niemeier (2001); Fantalkin (2001); Boardman (1999).

62 Meshad Hashavyahu: Fantalkin (2001); Tel Kabri: Kempinski (2002) 74–87, 223–52; Niemeier (2002) notes the different proportions of Greek and non-Greek material at the two sites.

63 Carchemish: illustrated and discussed in, e.g., Niemeier (2001) 19–20; Alexandra Villing (pers. comm.) reports that a new reconstruction of the shield from Carchemish at the British Museum confirms its Ionian workmanship. Mesad Hashavyahu: ostraca show that the commander, scribe, and some other leading men in the fort were Judeans, presumably serving either as vassals of Egypt or also as mercenaries alongside the Greeks (see Na'aman 1991, 45–47; Fantalkin 2001, 128–47; Niemeier 2002). The same situation apparently existed at an inland Judean fort at Arad, where ostraca (but not pottery) show the presence of *Kittim*, 'Greeks' or 'Cypriots' receiving rations, presumably as soldiers: Na'aman (1991) 47–48; Aharoni (1981).

of Babylonian expansion (Jer. 46:9). After his victory over Nebuchadnezzar in 601, Necho II made direct contact with Miletos by sending the clothes he had worn on the day of battle to be dedicated in the temple of Apollo at Branchidai (Hdt. 2.159). The purpose was surely to let Milesians, their Ionian neighbours and the Karians in their hinterland know that Egypt was finally winning again, and so encourage them to join his service. Two of the soldiers who put their names to the Abu Simbel graffito eight years later may be tentatively identified as brothers from a prominent Milesian family who had taken up Necho's invitation.[64]

Greeks and Karians were not the only outsiders recruited into Egyptian armies. Phoenician soldiers in Egypt are attested by graffiti at Abu Simbel alongside the Greek and Karian inscriptions, and by the existence of a separate 'Tyrian Camp' at Memphis.[65] After the sack of Jerusalem in 587, large numbers of Judean soldiers sought refuge in Egypt, and found employment as garrison troops at Memphis, in the border forts at Migdol and Daphnai, and 'in the land of Pathros', which probably refers to the southern border and the fort at Elephantine, where a Judean garrison is attested by detailed documentary evidence in the fifth century when Egypt was under Persian control.[66] An official who prided himself on preventing the desertion of foreign troops from Elephantine during Apries' reign specified 'Asiatics' as well as 'northerners' (*Haunebut*), i.e., Greeks and Karians, and an unidentified third group.[67] When Apries mobilised in defence of the border in 582, he involved 'the army of the state' as well as 'foreigners', and for the latter a generic term was used, rather than *Haunebut* to denote 'Aegean' troops in particular.[68]

If Apries employed 30,000 hired soldiers, as Herodotos claimed (2.163), they will have included Judeans, Phoenicians, and others, rather than exclusively Karians and Ionians. Apries' association with huge numbers of 'mercenaries' is, however, doubtful. Herodotos' narrative has the Battle of Momemphis in 570 fought by Apries 'leading his mercenaries' (*epikouroi*) against the rebel Amasis 'leading all Egyptians', i.e., the entire Egyptian army of Hermotybies

64 Haider (2004) 460–61, on Python and Archon, sons of Amoibichos.

65 Schmitz (2010); Vittmann (2003) 61–66. For various mercenary communities, see Kaplan (2003) esp. 7–9.

66 Jeremiah 43–46, esp. 44.1, 46.14. Elephantine: Vittmann (2003) 84–104, esp. 88 for a letter implying its existence before the Persian conquest; Porten (1968) esp. 8–16. See n. 63 above for Judean troops at Mesad Hashavyahu, possibly as hired soldiers rather than 'vassals' of Egypt.

67 Nesuhor inscription: Louvre A90; Schäfer (1904); Bassir (2016).

68 Stela of Apries from Daphnai: Abd el-Maksoud and Valbelle (2013) 4, cols. 4–6. The emphasis on compiling 'lists of men', forming units and appointing officers suggests a general militia levy.

and Kalasiries (2.163–69). Amasis' Victory Stela shows that this is precisely how he liked to paint himself. The Stela depicts him as the legitimate king, residing in Sais, 'minding the interests of the whole state' (§2), when he came under attack from Apries' 'ships filled with countless *Haunebut*' who intended to 'conquer the whole of Egypt' (§3–4). A lengthy portrayal follows of Amasis and his army as patriotic defenders, while Apries is a 'rabid' aggressor, a mere scavenging dog (§5–10). In this context, the claim that Apries was supported only by 'mercenaries' while Amasis had the whole region behind him is clearly propagandistic.[69] Other evidence suggests that Apries had support in large parts of the kingdom, and indeed that foreign troops were inclined to desert him.[70] Herodotos' version accepts that Apries was the legitimate ruler and Amasis a usurper, but presents his mutiny as justified and his reign as idyllic. The notion that Apries' only supporters were hired foreigners reflects the same pro-Amasis outlook.

Whatever the number and status of Greek and other foreign troops under Apries, they were essential to Amasis. He mobilised troops from 'distant regions which are in the midst of the sea' (perhaps Cyprus, or Samos) against Apries' attempt to return to power in 567,[71] and transferred the Karian and Ionian troops from the border forts to Memphis, 'making them his personal guard against the Egyptians', as Herodotos said in another context (2.154.3). We seem to have a glimpse here of an alternative tradition, accusing Amasis of excessive reliance on a mercenary power base, just as Amasis had accused Apries. With this transfer to Memphis, if not before, Karian and Ionian troops apparently served to uphold the position of the ruler against internal challenges, rather than as border garrisons or expeditionary forces. Whether they were kept on after the Battle of Pelousion and the Persian conquest of Egypt in 525 is not clear, but around 350 Memphis still had a distinct Greek Quarter (*Hellenikon*) and a Karian Quarter (*Karikon*), occupied by 'Hellenomemphites' and 'Karomemphites', perhaps descendants of Amasis'

---

69 A new edition of the stela (Jansen-Winkeln 2014) makes some older discussions irrelevant (e.g., the text no longer says that the mercenaries came from an 'island', but that they were on the 'Canobic branch'), but the overall interpretation of events by Leahy (1988) still seems valid.

70 Leahy (1988) 187–88, 197–99: Apries was still recognised as king in Hermopolis and Thebes towards the end of Amasis' first year (570); his burial as legitimate king (Hdt. 2.169; noted also in Amasis' stela, §18) is a concession to support for him in 567. Deserting mercenaries: see n. 67 above.

71 Leahy (1988) 191; Reyes (1994) 75–76. Note also his alliance with Greeks in Libya: Hdt. 2.181.

troops or more probably wider communities that originated from the military settlements of the sixth century.[72]

The period during which soldiers from the Aegean were hired by Egypt thus lasted only a century, *ca.* 625–525; the claim that they were employed in their tens of thousands is unreliable because it derives from a propagandistic source; and the impression that Ionians and Karians were the only foreign troops hired is misleading, since Judean and Phoenician troops are also attested, and the archaeological record suggests that Greeks formed only a small minority of troops in many a town or fort. Nevertheless, Ionian and Karian soldiers did become a significant presence in sixth-century Memphis, and perhaps elsewhere in Egypt, as the kings of Egypt increasingly needed permanent standing forces alongside the traditional 'warrior class' and wider militia armies. Whether the hiring of Greeks and Karians was motivated also by a need or preference for a specific set of hoplite skills that other foreign troops could not provide is the next question to consider.

3 Why Karians and Ionians?

We are poorly informed about the armament and tactics of the troops deployed by Egypt, Babylon, and their neighbours, which makes it easy to imagine that the Greek stereotype of the 'barbarian archer' as typical of 'Asia' might be true, and that hoplites capable of hand-to-hand, close-order combat were genuinely a rarity. However, such evidence as we have tends to contradict this assumption and suggests that heavy infantry soldiers armed with shield and spear were common in all areas where Egypt hired troops and indeed formed the bulk of the 'native' Egyptian army itself.

Assyrian reliefs of the siege of Lachish in 701 depict the armament of Judean soldiers. Among the spoils carried out of the city are large round shields and tall spears. The same shields are used by defenders on the city walls, some of

72 Aristagoras of Miletos *FGrH* 608 F 9ab. Austin (1970, 22) notes the special grain tax levied by the Persian empire to feed 'the Persians who lived in the White Fort at Memphis and their *epikouroi*' (Hdt. 3.91.3), who may have included Greeks. The Greek graffiti in the temple of Seti I at Abydos (the 'Memnoneion') cited by Kaplan (2002, 238) and Fischer-Bovet (2014, 30–31) are few and do not identify the writers as mercenaries, except in one instance (no. 405), which involves 'Cretan *epikouroi* under Amyrtaios', king *ca.* 400, not under Persian rule (Pedrizet and Lefèbvre 1919, viii–ix). The Persians moved at least some Karians from Memphis to Mesopotamia: Waerzeggers (2006).

whom wear tall conical helmets almost certainly made of bronze.[73] These images support the Biblical portrayal of eighth-century (and earlier) Judean armies as armed with shields and spears, as well as helmets and body armour, while the tribe of Benjamin alone provided archers and slingers.[74] An Assyrian relief that probably portrays Syrians and Israelites represents the few that are not fleeing or being trampled by Assyrian mounted troops as standing their ground in pairs, side-by-side with shields and spears raised.[75]

Phoenicians in Assyrian reliefs are equipped with round bossed shields and spears but wear no helmets or armour. The same is true of most armed figures marching in lines on Phoenician silver bowls, but some of these carry pairs of spears, presumably javelins, and others are archers.[76] Herodotos said that Phoenician marines in 480 were armed with shields 'without a rim', as indeed they are in the images of *ca.* 700, and wore 'linen corselets' and helmets 'in a style very close to the Greek', i.e., bronze and crested (7.89.1).

As for the Egyptian warrior class, Herodotos reports that the Hermotybians and Kalasirians who served as marines on Egyptian ships in the Persian fleet of 480 wore helmets (obscurely described as 'plaited'), carried 'hollow shields with large rims', that 'the majority of them wore cuirasses, and they carried long swords' (7.89.2). The Persians later adopted the Egyptian type of cuirass (1.135), which was made of iron scales (7.61). These Egyptian troops were deployed among the select troops retained by the Persian army for the campaign of 479, when they are again described as 'sword-bearers' (9.32). 'Hollow shields (*aspides*) with large rims' suggest round, convex shields like those of hoplites, but even bigger; an iron scale-corselet would be not much lighter than a bronze cuirass; and a long sword is even more of a short-range hand-weapon

73 Barnett *et al.* (1998) pls. 328, 330, 332, 334, 349–51: shields and conical helmets of defenders; pls. 332–33, 335, 337, 341: spoils, incl. shields *ca.* 45% of the bearer's height, *ca.* 75 cm in diameter. The helmets are identical to the Assyrian type, which was made of bronze or iron: Dezsö (2001) 18–55.

74 See 2 *Chron.* 26:14: army of Uzziah, *ca.* 750, had 'shields (*meginnah*), spears, helmets, corselets (*shiryon*), bows and slings'; cf. 2 *Chron.* 32:5 (Hezekiah, shields); 1 *Chron.* 12:9, 25, 35; 2 *Chron.* 11:12, 14:8, 25:5 (earlier armies); 1 *Chron.* 8:40, 12:2; 2 *Chron.* 14:8 (Benjamin's archers, slingers).

75 Albenda (1986) pls. 94–103 (Room 5, slabs 25–10, slabs 1–2 in door O, slabs 9–8); standing pairs appear in slab 1 of Door O, and slab 8. Admittedly, their shields are small, and they wear no armour. On Assyrian reliefs and the usefulness of their images for military history, Nadali (2019); Echeverría (2019) 327–28.

76 Reliefs: Barnett *et al.* (1998) pls. 134–37, 139 (in combat, incl. an archer); pls. 38–41, 46–47, also 151 (ships, walls lined with round shields); cf. Russell (1991) 160–64; Nadali (2019). Most soldiers on Cypro-Phoenician bowls wear the same kind of tunic, longer at the back than at the front, as Phoenicians in Assyrian reliefs: Markoe (1985) 242–43 (Cy1), 284–85 (E3), 290–301 (E5–9), 304–306 (E11), 308–309 (E13), 320–23 (G4).

than a thrusting spear. Even if one is sceptical of Herodotos' evidence for the Persian army, this must be how Egyptian soldiers were equipped by the 450s, when thousands of Athenians saw them during their Egyptian expedition. The idea that Persians copied Egyptian equipment, not the other way round, is significant.[77] Herodotos went so far as declare it his personal opinion that Greeks had learnt from Egypt (presumably in the heroic age) to use 'shield (*aspis*) and helmet' (4.180.4), while Plato believed in parallel invention: in Asia the Egyptians were the first to fight with spear and shield, just as the Greeks were in Europe (*Ti.* 24b).

Xenophon described an Egyptian contingent in the Persian army at Kounaxa as consisting of 'hoplites (*hoplitai*) with wooden shields reaching down to their feet' (*Anab.* 1.8.9; cf. 2.1.6). They fought with 'long spears' (*Cyrop.* 6.2.10, 7.1.33). In his historical novel *The Education of Cyrus*, he pictured the Egyptians, not the Greeks, in the army of Croesus as offering the only serious resistance to Cyrus by fighting in a deep, dense phalanx and 'pushing' the Persians back with their large shields and spears.[78] They were defeated in the end only because they were outnumbered and surrounded. Presumably, Xenophon's detailed and vivid narrative was based to a degree on what he had seen at Kounaxa. Hekataios of Abdera seems to have had an equally great regard for Egyptian soldiers, judging them 'unbeatable in daring and skill'.[79] It is, therefore, no coincidence that Iphikrates' so-called reform of hoplite equipment by lengthening spears and swords occurred during a campaign in Egypt in the 370s: it must have been an *ad hoc* measure to match the size of Egyptian weapons.[80]

Strictly speaking, none of this tells us how the Saite army operated in the seventh and sixth centuries, and contemporary evidence is extremely limited.

77 Note also Hdt. 7.91: Cilician swords (*xiphoi*) are described as 'very similar to Egyptian swords (*machairai*)'; 7.63: Assyrians and Babylonians ('Chaldaeans') are said to have 'shields and spears and hand-weapons (*encheiridia*) very similar to the Egyptian ones'. This description seems to apply to the whole clause rather than just to 'hand-weapons', and the latter are here presumably to refer to their swords, although the term more usually refers to 'daggers'.

78 Xen. *Cyrop.* 6.3.20, 23: '*hoplitai* drawn up 100 deep', 7.1.30: '*phalanx*', 7.1.33–34: shields 'rest against the shoulders and assist in pushing'; all this could have been attributed to the Greeks in the same army (6.2.10), but it is not. See Sekunda (2014) 127; Jędraszek (2011).

79 Diod. Sic. 1.73.9; for Hekataios of Abdera (*FGrH* 264) as his source, see, e.g., Vasunia (2001) 230–32. At 1.67.6, Diodoros/Hekataios refers to the spears and shields of Egyptian soldiers under Psammetichos I, and at 1.67.4 contrasts them with *psiloi*, i.e., he envisaged them as hoplites.

80 Diod. Sic. 15.44.2–4; Nepos 11.1.3–4; Anderson (1970) 129–32. Cf. Sekunda (2014) and the critical survey of the historiography in Konijnendijk (2014) 88–92. Fischer-Bovet (2014, 37–42) stresses the importance of Egyptian troops (not a 'decaying class', *pace* Austin 1970, 18; Möller 2000, 34).

Egyptian representations of soldiers stop in the twelfth century; Assyrian representations of campaigns in Egypt are rare and only show Kushite troops as the enemy. In one of Assurbanipal's reliefs, the Kushites are armed with round shields, spears, and scale-armour corselets, but Egyptians need not have been armed in the same way.[81] Late Bronze Age reliefs show Libyan invaders, the ancestors of Psammetichos' Ma-troops, armed with bows or very long swords and little else, but we cannot assume that this did not change for half a millennium.[82]

A Biblical prophet pictured Necho II's army at Carchemish in 605 in the course of a sarcastic exhortation, meant as a warning that Egypt was doomed to defeat by Babylon, and this may be a reasonably accurate description of the contemporary Egyptian army:

> Array shields (*magen*) and bucklers (*tsinnah*) and draw near for battle. Harness the horses and mount, you horsemen. Take your stations with your helmet (*koba*); polish your spear (*romach*); put on your scale-armour (*has-siryon*) ... Come up, horses; shine, chariots; advance, warriors, Kushites, Libyans who handle the shield (*magen*), Lydians (*Ludim*) who bend the bow.
>
> Jer. 46.3–4, 9

The Egyptian infantry was thus equipped with shields, while both infantry and cavalry had spears, helmets, and body-armour. Only the Lydians, among Egypt's allies, were (mounted) archers. Less detailed is the evocation of Amasis' army in his Victory Stela, where his council agrees that he should fight Apries: 'Your horses are numerous, your archer corps is infinite, no one knows the number of your bodyguard, and there is no shortage of picked recruits!'[83] The separation of the archer corps from the bodyguard and the 'recruits' implies that the latter were not archers. Similarly, the title 'Commander of the Archers' appears only once in the Saite period, while all other infantry commanders simply lead 'the army' (*mšˤ*) or 'the troops' (*mnfyt*), which suggests that most infantry were not archers.[84]

81 See Barnett (1976) pl. 36, for Assurbanipal's relief showing the sack of Thebes (?); the scale armour of the Kushite defenders is hard to see in photographs but clear in the original (British Museum 124928); other Kushites in Albenda (1986) pl. 94–95 (Sargon II); Nadali (2006) fig. 2ab (Esarhaddon). Cf. Nadali (2019).

82 See O'Connor (1990) esp. 47–60, 81–89; Spalinger (2005) 235–63, for these representations.

83 My translation, from the German translation by Jansen-Winkeln (2014) §8.

84 Title: Chevereau (1985) no. 109: Djedptahiufankh, discussed above as commander of 'foreign' troops, 'Asiatics'; see Agut-Labordère (2013) 988. The fourth-century 2 *Chronicles*

A demotic papyrus written in the late sixth century but referring to events of 630 mentions in passing a group of Kalasirians who come armed with spears and shields.[85] Intriguingly, several Phoenician bowls of *ca.* 710–675 show an image of Pharaoh smiting his enemies in the conventional manner, but with the addition of a bearded attendant with a feather in his hair who carries a dead body on his right arm while holding a spear. I would suggest that this is an image of what the Victory Stela of Piye called a 'plume-wearing Chief of the Ma', representing Pharaoh's bodyguard of 1,000 (Hdt. 2.168) or the army at large. If so, spears were indeed their main weapon even before Psammetichos came to power; whether they also carried shields we cannot tell because the attendant's left hand waves a fan instead.[86] Actual finds of military equipment are rare, but fragments of iron scale armour were found at Tell Dafana, along with bronze and iron arrowheads, and an iron dagger.[87] On balance, this limited Archaic evidence tends to confirm the Classical Greek picture of Egyptian soldiers armed with spears, shields and metal body armour, much like Greek hoplites, and thus in principle capable of hoplite-style tactics.

Karian and Greek hoplites did have a distinctive double-handle system that enabled them to carry heavier shields offering more protection, but this was surely a marginal advantage at best. Similarly, even if one were to accept that this type of shield entailed an especially dense battle order – a notoriously disputed matter – this would constitute a difference of degree rather than a difference in kind between hoplites and other troops armed for close combat. Depending on one's views of the development of hoplite phalanx tactics, it is even possible that Greeks learnt these tactics in Egypt before they emerged in the Greek cities. Whereas at home Greek armies of the seventh and sixth centuries may still have consisted of war bands that combined hoplites, horsemen and light infantry, and fought in relatively open and fluid order,[88] in

(35:23; cited by Agut-Labordère 2013, 988) mentions Egyptian archers shooting King Josiah in 609, but this detail does not feature in the older 2 *Kings* (23:29).

85 *P. Rylands* 9 (*Petition of Petiese*), col. XI.11–12: see Vittmann (1998); cf. Winnicki (1977) 260.

86 Phoenician bowls: Markoe (1985) 46; 170–71 and 244–45 (Cy2, Louvre AO 20134, Idalion); 177–78 and 256–59 (Cy8, Met. Mus. 74.51.4554, Kourion); 188–91 and 274–77 (E1, Villa Giulia 61574, Praeneste); 198–99 and 303 (E10, Musee du Petit Palais, Paris inv. Dutuit 170, Pontecagnano); also, fragmentary, 177 and 254–55 (Cy7, Met Mus 74.51.4556, Kourion). In the only later example, from 675–625 (Period IV), the attendant has bow, quiver, and helmet: Markoe (1985), 174–75, 251 (Cy5, BM 1892/5–19/1, Salamis). The attendant figure is discussed but not explained by Schneider-Hermann (1945–8).

87 Leclère and Spencer (2014) 46, 69–74; pl. 26; cf. Yadin (1963) 354, for tenth-century scale armour.

88 As suggested by Van Wees (*W&V*, *GWMR* 166–83). On cavalry see Konijnedijk (this volume); on light-armed see Kucewicz, Lloyd and Konijnedijk (this volume).

foreign service hoplites would probably have found themselves in exclusive heavy infantry units that could and would operate as closed phalanxes.

We cannot even be sure that infantry service rather than naval service was the primary use of Greek and Karian troops. Given the geography of Egypt, ships were an integral part of warfare, and not just as transports. Amasis described Apries' forces as 'ships filled with *Haunebut*'. He listed his own forces as if they were cavalry and infantry, and himself mounted a chariot before battle, yet the fight turns out to take place in the river at Momemphis: 'many ships capsized and fighting men fell in the water; they could be seen floating like fish'.[89] Herodotos remarked on the 'slipways of the ships' constructed for the soldiers settled at Stratopeda/Migdol (2.154.5). Necho II's building of two fleets of triremes, with crews of 200 men per ship, created a surge in demand for manpower, not necessarily for specialist oarsmen, but for men who could fight as well as row. It is no coincidence that some commanders of *Haunebut* were also 'admiral of the royal fleet'.[90]

All in all, it seems unlikely that hoplite armour and tactics counted for much in Egypt's decision to hire Karians and Greeks. The supposed uniqueness of the 'men of bronze' in Herodotos' story, which has persuaded so many modern scholars of the superiority of the hoplite,[91] is belied by our other evidence. Insofar as special skill rather than mere manpower was a factor in the recruitment of Greek troops, their appeal probably lay at least as much in their maritime expertise as in their heavy armour.

Otherwise, the main reason why Greeks and Karians met a significant part of Egypt's manpower needs, alongside troops recruited from regions much closer to Egypt, was probably simple availability. The alliance between Egypt and Lydia made Ionians and Karians easily accessible, since the Lydian

---

89 Jansen-Winkeln (2014) §10–13; the Babylonian invasion of 567 also includes a fleet and ends with Apries drowned (§14–16) – not strangled by an angry populace (Hdt. 2.169).

90 See Agut-Labordère (2013) 990–94; Kaplan (2003) 3; Lloyd (2000). Iancu (2016) 13 n. 8, rightly stresses that not all Greek mercenaries were necessarily hoplites. Radner (2012) esp. 475–77, notes a document referring to two men of Egyptian origin living in Nineveh in 692 who were 'ship's captains', i.e., in Sennacherib's navy; she suggests that they had been taken captive in Sennacherib's campaign of 701 and imply the existence and presence of an Egyptian Mediterranean fleet at this time. If so, Necho II may not have created a Mediterranean fleet but upgraded it to a trireme fleet.

91 Fischer-Bovet (2014) 19–21; Hale (2013) 190; Perdu (2010) 143; Schwartz (2009) 135; Haider (2004) 452, (2001) 199; Murray (1993) 231; Chevereau (1985) 311, 321–22; Spalinger (1978b) 19: Egyptians 'not up to the standard of the Karian or Ionian mercenaries'; Garlan (1975) 94: 'technical superiority of the hoplite armour and formation'. Lloyd (1975) 16, sees superior equipment, but not yet phalanx tactics (cf. Lloyd 1983, 289). Kaplan (2003) 12, rightly rejects the idea of hoplite superiority.

kingdom already made use of both as 'allied' troops (see below). Moreover, the lower date of 625–600 for the beginning of regular hiring of troops from the Aegean means that this development coincided with a period of high mobility and apparent manpower surpluses in the Greek world. It was at this time that Miletos and other Ionian cities also began to settle on the farther coasts of the Black Sea, while Phokaia and Samos explored the far Western Mediterranean. Greeks first settled in Libya *ca.* 630, followed by a mass influx of Cycladic islanders and Peloponnesians in the 570s. As shown by their presence at Naukratis, and the spread of East Greek pottery in Egypt and the Levant, traders joined settlers and soldiers for hire, together constituting a huge outward movement, mainly from Ionia which until then had played only a small role overseas. The driving force may have been that the Greek world experienced a crisis of exploitation and inequality around 600,[92] to which migration was a common solution. Whatever the cause of this widespread mobility, the result was that, just when Egypt needed more naval and military manpower, thousands of Greeks happened to be looking for opportunities.

4 Fact and Fiction in the Greek Tradition

The stark difference between the origin of Greek military service in Egypt according to Herodotos and the history reconstructed above from largely contemporary evidence makes it tempting simply to dismiss Greek stories as fiction. However, it is worth asking what purpose such stories served: we may gain some insight into the self-image of the Greek and Karian soldiers among whom these tales presumably originated, and we may even be able to discern elements of historical fact amid the myth-making.[93]

4.1 *A Story from the Karian Quarter at Memphis*

Ironically, one of the tales that sounds most legendary is likely to be based on a historical event that brought Karian troops to Egypt twenty years before any Ionian or other Greek soldiers arrived. The story sounds a bit like Herodotos, and is only attested in Polyainos' *Stratagems*, compiled in the time of Marcus

92 See Van Wees (1999, 2008, 2009). Kaplan (2002) 240–41, dating the phenomenon earlier, sees civil war among elite factions as the main cause.

93 Lloyd ((1975) 15–16, 89–116, (1988) 114–35) offers detailed discussions of the relevant stories in Herodotos, but he assumes a far greater element of historical fact than I would accept.

Aurelius, so that it could easily be dismissed as a 'late' invention.[94] Closer inspection shows that it is an independent and early tradition.

> Psammetichos overthrew Tementhes, king of Egypt. When Tementhes sought an oracle about his kingship, the god Ammon instructed him to be on his guard against cocks. Since Psammetichos had a companion, Pigres the Karian, and learned from him that the Karians were the first to put crests on their helmets, he understood the utterance, and after recruiting many Karians as mercenaries led them against Memphis and won a battle near the temple of Isis outside the palace, about five stades distant. After these Karians a part of Memphis is called *Karomemphitai*.
>
> Polyaenus, *Strat.* 7.3

The final sentence shows that this is in essence a foundation story for the Karian Quarter at Memphis, and suggests that Polyainos' source was Aristagoras of Miletos, whose *Aigyptiaka* explained the meaning of the term 'Karomemphitai' (n. 72 above). The similarity with Herodotos lies in the oracle predicting that Aegean soldiers will play a decisive role in the conflict. In every other way, however, the story is fundamentally different.

Most remarkable is the appearance of 'Tementhes, king of Egypt', who is generally and rightly identified as Tanwetamani/Tanutamon of Kush, a figure unknown to Herodotos and indeed wholly unknown to later Greek tradition.[95] Significantly, the oracle is given to him, not to Psammetichos: Tanwetamani's own Dream Stela claimed that his attempt to take back Memphis and the Delta was inspired by a dream sent to him during incubation in the temple of Amun ('Ammon') at Thebes. A pitched battle outside Memphis in 664, also unknown to Herodotos, is mentioned in contemporary texts, as we have seen. The story presupposes that Karians were not new arrivals in Egypt – Pigres was already Psammetichos' advisor – but were now recruited in larger numbers and became the founders of the Karian Quarter at Memphis, under Psammetichos, rather than a century later under Amasis. Aristagoras' story was clearly independent of Herodotos and had some basis in fact. While his Karian tale was certainly not the whole truth – it ignored the central role of the Assyrians in these events – it is hard to see how anyone could have known about the obscure 'Tementhes', who was no longer recognised as king anywhere in Egypt

94 Lloyd (1988) 133. Haider (1988) 179–84, rightly treats it as an independent tradition.

95 'Tementhes' did not make it into Manetho's 25th Dynasty, which has only 'Sabakos', 'Sebichos' (Shebiqu/Shabataqa) and 'Tarkos' (Taharqa).

by the time of the next expedition in 644, unless Karians were indeed involved in the fight against him in 664.[96]

Given that this campaign was led by Assyria, rather than by Psammetichos, the most likely reason for a Karian presence would be that they had been sent by Gyges, who at that stage had just become Assyria's ally.[97] Assyrian annals do not mention his sending of troops, but they had no reason to record an ally doing simply what was expected of him. The implication that Gyges on this occasion sent only Karians, not Greeks, fits well with the Greek belief that it was Karians who 'invented' mercenary service. This is attested not only by later authors but by the contemporary poet Archilochos, *ca.* 680–640, who said: 'I shall be called an ally (*epikouros*), like a Karian' (fr. 216). Anyone could be an 'ally' at times, but the suggestion is that Karians served as 'allies' so often that they became what we would call 'mercenaries', which Greeks of Archilochos' generation seem to have regarded as a disreputable way of life. A Greek proverb, 'the Karian runs your risks', and an Archaic tradition that Karians invented hoplite shields and crested helmets, both point in the same direction. An early Karian expedition to Egypt may, therefore, be historical.[98]

The lack of any other evidence for Karians in Egypt at this early stage suggests that the expedition of 664 was a one-off, but it became famous because it ended with the notorious sack of Thebes. The memory of this exploit later became the basis for the foundation legend of Memphis' Karian Quarter, which took both Gyges and the Assyrians out of the picture and turned the Karians into independent agents who had served Psammetichos I, allowing them to boast that they had been instrumental in founding the Saite dynasty and had received their area of Memphis as a reward at the very beginning of its reign.

---

96 Karians adopted Egyptian language and culture more than Greeks did (Kaplan 2003, 17–18), which may help explain their detailed knowledge.

97 Gyges' diplomatic overtures must date before 664, since the story is included in Prisms E_1 and E_2, which do not yet include the second Egyptian campaign of that year. This episode always comes after the first Egyptian campaign in the Prisms, but this need not mean that it dated after 667, since Assurbanipal's annals did not follow a strict chronological order. The Gyges-story is in E, LET, B and C sandwiched between the conquest of Qirbit and the submission of Arwad, which are known to have occurred before 667: Grayson (1980) 230, 232, 236; so possibly it, too, dates to the beginning of Assurbanipal's reign. Haider (1988) 179–84, endorsed by Niemeier (2001) 17, (2002) 228; Rollinger (2001, 252) suggests that the story implies Karian and Greek troops serving in the Assyrian army.

98 See Kaplan (2002) 233–34; Bettalli (1995) 105–107, 109–11 (with n. 31 for references to the proverb); Snodgrass (1964) 182–86. First mercenaries: schol. ad Plato *Laches* 187b; Ael. *NA* 12.30; invention of shield emblems and handles, helmet crests: Hdt. 1.171.4, cf. 7.93; Strabo 14.2.27–28; crests: Alkaios fr. 22 B, 58 D; handles: Anakreon fr. 91B, 81.

If this was the story told by Karians at Memphis, one would expect Herodotos to have been aware of it, especially given his own part-Karian background. The *Histories* do indeed feature a tacit polemic against it. When Herodotos said that at the Ionian and Karian settlement at Stratopeda 'the slipways of their ships and the foundations of their houses' were still visible in his own day (2.154.5), he was not offering gratuitous tourist information but citing evidence that the soldiers really had lived here at some point, and had not, as the Karians claimed, always resided at Memphis. His comment 'Amasis is the one who built the temple for Isis at Memphis' (τῇ Ἴσι τε τὸ ἐν Μέμφι ἱρὸν Ἄμασις ἐστὶ ὁ ἐξοικοδομήσας, 2.176.2) – rather than simply 'Amasis built the temple' – subtly refutes the Karian story that a battle had been fought at this temple about 100 years before Amasis. One may even wonder why Herodotos consistently referred to 'Ionians and Karians' in Egypt,[99] never simplifying to 'Ionians' or 'Greeks', nor expanding to include the 'Asiatic' soldiers of whose presence he was aware (2.112). Perhaps he firmly linked the two in an implicit denial of the Karian boast that they had arrived earlier than the Ionians. A polemic of this kind would surely not originate with Herodotos himself, but with two communities of soldiers offering and challenging rival claims to superiority.[100]

4.2 *A Milesian Tale*

In his description of places along the Bolbitine branch of the Nile, Strabo comes to 'the watchtower of Perseus and the fort of the Milesians', and he explains:

> Milesians, sailing in 30 ships in the time of Psammetichos (he lived at the time of Kyaxares the Mede),[101] landed at the Bolbitine mouth, then disembarked and fortified the aforesaid settlement. Later, they sailed upstream into the province of Sais, defeated Inaros in a naval battle, and founded the city Naukratis.
>
> Strabo 17.1.18

99 Six times: Hdt. 2.152.4–5, 2.154.1 and 3, 2.163, 3.11 ('Hellenes and Karians').

100 Kaplan (2003, 13) suggests that Ionians and Karians formed 'a well-integrated and cohesive fighting unit' and that their separation in Greek and Karian Quarters was 'imposed' later. Herodotos' constant linking of the two groups is our only evidence for such integration, unless one takes '*Pelekos Oudamou*', 'Axe son of Nobody' in the Greek graffito at Abu Simbel as a genuine Karian name, rather than as an obviously joking reference to the tool used to cut the inscription.

101 This dating to the time of Kyaxares (*ca.* 625–585) suggests an awareness of the real date at which Naukratis was established, late in Psammetichos I's reign. Psammetichos II also lived at the time of Kyaxares, but Strabo can hardly mean this king, since he knows only one Psammetichos (cf. 17.1.18, 25), and our Psammetichos II and III were in any case known as Psammis and Psammenitos (Herodotos) or Psammouthis and Psammecherites (Manetho).

Naukratis did not stand on the Bolbitine branch of the Nile but further west on the Canobic branch, as Strabo was well aware (17.1.22),[102] so presumably he had in mind two separate Milesian expeditions. The initial force founded 'the fort of the Milesians' at the mouth of the Bolbitine branch, which led to Psammetichos' capital Sais; later, either the garrison of this fort or a second force sent from Miletos itself sailed down the Canobic branch of the river to defeat Inaros and settle at Naukratis. Although these expeditions have sometimes been taken as aggressive actions against Psammetichos,[103] the tradition of Greek support for this king and his successors makes it much more likely that they were understood as actions in his defence. The coastal fort, from this perspective, would have been established to guard the Bolbitine mouth and protect Sais from attack by sea, while the second expedition repelled a rebel assault on Psammetichos' western border.

A sequel to this second expedition featured in Aristagoras of Miletos' *Aigyptiaka*, which mentioned several cities along the Canobic branch, including Gynaikospolis, 'The Woman's/Wife's city' (*FGrH* 608 F8). Aristagoras gave three etymologies for the name, of which the most interesting is that the city was so called because its men were 'women': 'when the Naucratites sailed up the river and were prevented from disembarking by the rest of the Egyptians, these people were stricken with cowardice and did not stop them'. This is obviously fanciful, but it is premised on the same idea as Strabo's Milesian expedition, namely that Naukratis was founded in the face of hostility. Aristagoras of Miletos may, therefore, have been the source of Strabo's story, too.[104]

We have one further detail of this version of events. Diodoros says that the decisive battle between Psammetichos and his 11 rivals took place at Momemphis (1.66.12). This city lay south of Gynaikospolis on the Canobic branch (Strabo 17.1.22) and was mentioned, in an unknown context, by Aristagoras (*FGrH* 608 F 12), who may once again have been the ultimate source. If so, we can piece together a story according to which the Milesians sailed out from their fort and into the next branch of the river down to Momemphis

102 The 'watchtower of Perseus' actually also stood on the Canobic branch and was already in the sixth century known as a landmark delimiting the extreme western border of Egypt (Hdt. 2.15.1).

103 As suggested by Drijvers (1999) 21. The customary dismissal (e.g., Bowden 1996, 25; Gorman 2001, 56) of this story as somehow derived from the fifth-century Athenian (not Milesian) expedition to Egypt in support of (not against) Inaros son of (not enemy of) Psammetichos is weak.

104 Similarly, e.g., Braun (1982) 37; Drijvers (1999) 21. Haider (1988, 154–55, 191–99) takes Strabo's source to be Artemidoros of Ephesos, cited as 'saying the same things' as Aristagoras (608 F 6, 8, 10). of the latter. The other etymologies were that the city's women in general or the wife of the city's governor in particular had defeated unspecified 'enemies' or 'the king' in the absence of the men.

where they helped Psammetichos defeat Inaros, a leading opponent, and then settled at Naukratis. Miletos' claim to have founded Naukratis is well-attested in later sources,[105] but this is clearly false: Naukratis was not a Greek colony but an Egyptian town with a Greek quarter, even if Milesians were prominent among its earliest Greek inhabitants.[106]

Nevertheless, we have reason to think that this legend, too, was derived from a historical event, namely the expedition of troops sent by Gyges to support Psammetichos in his revolt from Assyria. As in the Karian story, the naming of a historical but thoroughly obscure figure as Psammetichos' enemy, in this case Inaros, is an important clue. We argued above that Inaros was a ruler of Athribis ousted by Psammetichos, and that his name was removed from the list of Egyptian traitors in Assurbanipal's Prism A because he took sides with Assyria at the time of Psammetichos' revolt. If this is correct, Milesian ships might genuinely have fought and defeated this Inaros at Momemphis around 644. The establishment of a fort at the river mouth to protect Sais from attack, too, is plausible at a time when Assyria and its allies posed a direct threat, but it would be surprising after 625, when Psammetichos' military activity was all in the Levant.

What is more, Milesians apparently had a special connection with Gyges, because we hear that Abydos on the Hellespont was founded by Miletos 'by permission of Gyges, king of the Lydians, for he was in control of this territory and the entire Troad' (Strabo 13.1.22). Other Milesian settlements on the coasts of the Hellespont, Propontis, and Paphlagonia were also within the borders of the Lydian Empire, and presumably also required the permission of Gyges or his successors.[107] After initially waging war against Miletos (Hdt. 1.15), the Lydian king may have concluded an alliance with this city and subsequently

105 Eusebius' *Chronicon* 88b Helm (placing the foundation at a period when Miletos supposedly 'ruled the sea', at the impossibly early date of 749–732); Suda s.v. *Naukratis*: 'this city was founded in Egypt by the Milesians when they held the thalassocracy' (cf. Diod. Sic. 7.11); schol. ad Theocritum 17.98; *CIG* 2878: 'mother-city of many great cities in the Black Sea and Egypt' (195 CE); Peek (1966) (*ca.* 200 BCE). An impossibly early date for Naukratis was also assumed by Polycharmos of Naukratis (*FGrH* 640 F1) who dated an event at Naukratis in 688. Cf. Austin (1970) 22–23, 58–59.

106 Milesian pottery is common at Naukratis and many of the earliest pieces are dedications to Apollo, whose temple was built by Miletos, so Milesians played a prominent (but not exclusive) role among the earliest settlers: e.g., Schlotzhauer (2012) esp. 71–72 (also 61–62 on pottery from the Troad found at Naukratis, which may have been made at Abydos or another Milesian 'colony' on the Hellespont).

107 Hellespont and Propontis: Anaximenes of Lampsacos *FGrH* 72 F 26; Strabo 14.1.6. A 'Milesian expedition' founded several cities in Paphlagonia (Pseudo-Scymnos 1001–8 Diller; with Strabo 12.3.10) and *Milet* 1.3 no. 185 (Apollonia-on-the-Rhyndakos); cf. Ehrhardt (1983) 44–47, 52–55.

used Milesians in leading roles among settlers and garrisons sent out to secure the borders of his kingdom, and finally in 644 to sail out and serve under his new ally Psammetichos.[108]

On balance, it seems likely that Milesian ships did go to Egypt at the time of Psammetichos' rebellion and fought on his side, but they did so as allies or mercenaries in the service of Gyges rather than as independent agents. They did not found Naukratis, or even merely settle there, until twenty years or so later, but the memory of a dramatic earlier historical event became the basis of a legend which allowed the Milesians to claim superiority over their fellow-Greeks at Naukratis as founders of the town and as owners by right of conquest.

Again, one would expect Herodotos to have been aware of this Milesian tale if it was current at Naukratis, and again we can detect signs of an implicit polemic against it, in the notorious errors in his account of Naukratis. Although Greeks lived in this town from *ca.* 620 onward, Herodotos gives the impression that they first arrived some fifty years later: he says that Amasis 'gave those who came to Egypt the city of Naukratis to live in' (2.178.1). He states unambiguously that the Greek temples at Naukratis were built on land granted by Amasis (2.178.1–3), although archaeological evidence shows that several existed much earlier. He also dates the lifetime of the courtesan Rhodopis of Naukratis to the reign of Amasis (2.134–35), despite her appearance in the poems of Sappho, *ca.* 600. These were not mistakes but political statements. Herodotos argued that only nine Greek cities, including his native Halicarnassos but not including Miletos or Samos, had jointly founded 'the biggest and most famous and most used' of the temples, the Hellenion, and had the sole right to appoint the 'directors of the trading post' (2.178.3). This right was probably bestowed by Amasis when he gave Naukratis a monopoly on import trade into Egypt (2.179). It was in the interest of the beneficiaries to deny stories about an earlier foundation that supported Milesian claims of priority, and to insist that rights granted by Amasis were no innovation but the original dispensation by the city's 'true' founder.[109] Herodotos was sufficiently invested in the matter to endorse this claim, at the cost of creating several anachronisms.

---

108 In this capacity, too, he may have used Karians at an earlier stage, since Kios (Aristotle fr. 519 Rose) and Sesamos (scholion on Apollonius of Rhodes, *Argon* 943) are said to have been inhabited by Karians before the Milesians arrived. Cf. Herda (2013) 448–49 (further citing Pliny *NH* 6.20 for Karians at Tanais; Pomp. Mela 2.22; Arrian, *Periplus* 35 for a 'Karian harbour' on the west coast of the Black Sea), who regards these settlements as independent Karian enterprises.

109 Similarly Murray (1993) 230–31; cf. Möller (2000) 192–96, on Amasis reorganising Naukratis.

4.3 *A Story from the Greek Quarter at Memphis*

The story about 'men of bronze' emerging from the sea can be understood as the Ionian Greeks' answer to the claims made by the foundation legend of the Karian Quarter and the Milesian tale of Naukratis. Ionian soldiers in Memphis wanted to claim their own link with Psammetichos I and their own decisive role in founding the Saite dynasty, but unlike the Karians and Milesians did not have a major historical action to build on. Instead, they developed a story at the fictional end of the foundation-legend spectrum.

Psammetichos' original Ionian troops – as well as the Karians, who were brought into this story to acknowledge their prominent presence at Memphis while denying that they arrived any earlier than the Greeks – were imagined as sea-borne raiders 'sailing in search of plunder', whom 'necessity caused to be driven off course to Egypt' (2.152.4). One can see why sixth-century Ionian soldiers might have imagined their predecessors in this way, since raiding was still a common Ionian activity at the time, and Herodotos hints that Egypt was a particular target for raiders from Samos in the years before 525.[110] One may be tempted to rationalise this story, and to posit that Greeks were genuinely a key source of manpower for an Egyptian ruler seeking to recruit soldiers because they often raided Egypt.[111] Yet this is the opposite of what Herodotos said: the point of his story was precisely that no Greek or Karian soldiers had ever been seen in Egypt before, that the raiders did not intend to go to Egypt but were blown 'off course', and that it was only the intervention of 'necessity' that took them to Psammetichos. The story thus took typical Ionian adventurers but elevated them to founding fathers driven by fate and an oracle to become not just the first Greek soldiers in Egyptian service, but 'the first foreigners (*alloglōssoi*) to settle in Egypt' (Hdt. 2.154.4).[112]

---

110 Hdt 3.39.4: Polykrates of Samos allows his subjects to raid his 'friends', which suggests above all his 'guest-friend' Amasis of Egypt (2.182; 3.40–43); cf. Van Wees (2013) 31–32; Carty (2015) 169–73.

111 Hale (2013) esp. 184; Sullivan (2011) 43–45; Rawlings (2007) 67–68; Luraghi (2006) 35–36: 'a telling example of how easily raiders could turn into mercenaries', 41; Raaflaub (2004) 209; *GWMR* 42: 'raiders found themselves hired as mercenaries'; Kaplan (2002) 237–38; Bettalli (1995) 52; Braun (1982) 37; Austin (1970) 17–18. Iancu (2016, 12–13) rightly warns against treating this story as a model.

112 The story is sometimes treated as a generic fiction based on a well-known tale in the *Odyssey* about Greeks raiding Egypt (*Od.* 14.243–86; cf. 17.424–44), but this tale, too, is in some ways the opposite of Herodotos' story. In the *Odyssey*, the raiders do not arrive in Egypt by chance, but set out to raid there, and Egypt is a well-known destination for travellers. Nor are the raiders in epic welcomed as potential allies, but the local king and his townsmen massacre them all, bar one who lives to tell the tale. Sullivan (2011, 41–42) suggests that the 'many goods' gained by the son of Castor in this story when he stays in

The detail that their first place of settlement was Stratopeda presumably means that this was the major place where Ionians were initially posted – not in 664 but from 625 onward – and it is mentioned to counter the Karians' claim that they had arrived in Egypt earlier and had always lived at Memphis. Moreover, Stratopeda is presented as a prize: Psammetichos gave his new friends, on top of the other rewards he had promised, 'places to live' (2.154.1). This has been taken to mean that he granted them farmland, and thus tied them to Egypt permanently.[113] But these 'places to live' were 'Camps' consisting of 'houses' and dockyards, not agricultural villages. In reality, living in these forts was the job, not the reward, but in legend it became the reason why the raiders stayed and assumed a role as founding heroes.

One odd detail of the story is likely to have been a mid-fifth century addition rather than part of a sixth-century version: Psammetichos' eleven rivals 'stripped him of most of his power and forbade him to leave the marshes to visit the rest of Egypt' (1.151.3). In other words, he remained king, but his territory was now strictly limited to the coastal marshland. Quite apart from the fact that the real Psammetichos was never in this position, even as a fictional story this is peculiar: the logic of restricting his territory without deposing or exiling him is far from obvious and never explained. However, another king of Egypt did find himself in exactly this situation: Amyrtaios, 'the king in the marshes', who for some time around 450 maintained his independence in the Delta when Persia controlled the rest of Egypt (Thuc. 1.110.2, 112.3–4; Hdt. 2.140.2; 3.15.3). Psammetichos' strange position in the tale was surely modelled on Amyrtaios', and his location in the marshes was a topical twist to the story at the time Herodotos heard it from someone in Memphis or Athens. Earlier versions may simply have had Psammetichos governing in Sais and recruiting help to fend off his eleven 'envious' rivals, as Diodoros had it (1.66.8–10).

On one other point, Diodoros' version probably also reflects an earlier version of the story than the one told by Herodotos, namely the account of how Egypt came to be ruled by twelve kings, a 'Dodecarchy', including Psammetichos, in the first place. Diodoros relates how the 'Ethiopian' ruler Sabakos suddenly departed in deference to a warning from the gods and how this led to 'anarchy in Egypt for two years', 'mobs engaging in riots and internal bloodshed',

---

Egypt after his raid are rewards for military service; Carty (2015, 155–58) sees the son of Castor as a slave in military service, which she argues was the norm in Egypt. However, the text says that he 'gathered many goods among the Egyptians, because they all gave' (14.285–86). These are not spoils or pay, but gifts from hosts, as so often in the *Odyssey*.

113 So Sullivan (2011) esp. 46, 49–50, also taking Pedon's 'city' as 'land', in an otherwise judicious and comprehensive discussion of how Archaic mercenaries were rewarded for their service.

until twelve leading men made a pact to rule jointly as kings (1.66.1–2). This is fiction, but it makes good narrative sense and is close to historical chronology insofar as it has Psammetichos become king almost immediately after the expulsion of the last Kushite ruler. The two years of anarchy may allude to 666–664, when the Assyrians imposed Necho I on Memphis, but the priests of Ptah refused to recognise his legitimacy.

Herodotos' version of events, by contrast, is meandering, internally inconsistent and chronologically impossible because it features two more kings between the abdication of Sabakos and the accession of Psammetichos and the Dodecarchs. First Anysis from Anysis (sic), a blind man who had spent fifty years in hiding 'in the marshes', resumed his earlier reign (2.137.1, 140). Then Sethos took power, a priest of Hephaistos (i.e., Ptah at Memphis), who was so pious that without the aid of professional soldiers, relying only on militia troops and divine intervention, he repelled an Assyrian army sent by Sennacherib (2.141). Next,

> when the Egyptians were liberated (*eleutherōthentes*) after the priest of Hephaistos had been king, they established 12 kings, dividing the whole of Egypt into 12 parts, for they were unable to live without a king for any length of time.
>
> Hdt. 2.147.2

The remark that the Egyptians were 'liberated' is bizarre. In what sense were they 'freed' from a ruler who combined deep piety and great military success, especially if they had no wish to be free of kingship as such? 'Liberation' only makes sense if, in the story as told to Herodotos, it referred to the departure of Sabakos, which left the Egyptians 'free' from their Kushite conquerors. I would suggest that originally the story was as Diodoros told it,[114] but that Herodotos'

114 In other respects, Diodoros' version of events is unlikely to reflect Archaic legend. His story is that Psammetichos' championing of openness, free trade with Greeks and Phoenicians, and foreign alliances – in a kingdom that had previously killed or enslaved all foreign arrivals – enabled him to 'send for mercenaries from Karia and Ionia' when he needed them (1.66.8–9, 12; 1.67.8–11). This is no alternative tradition but a new spin reflecting Classical ideas about sound foreign policy (see, e.g., Thucydides' *Archaeology*; Van Wees 2017) and the new openness of Egypt after the Macedonian conquest. The idea probably derived from Hekataios of Abdera (see n. 79 above), as did two other details: (a) Hekataios seems to have believed that Athens was a 'colony' of Sais (Diod. Sic. 1.28), so Athens was made one of Psammetichos' allies (Diod. Sic. 1.67.8); (b) Psammetichos gave his Greek troops preferential treatment during Syrian campaigns, causing desertion by Egyptians (Diod. Sic. 1.67.3–7), which is reminiscent of the status of Macedonian and Greek troops in Egypt after Alexander.

efforts to incorporate everything he knew about Egyptian kings into a single historical narrative led him to insert two free-floating tales about legendary rulers here.[115]

Part of the original story will have been the detail that Psammetichos had lived in exile 'in Syria' after Sabakos 'killed his father Necho' but was brought back by 'the Egyptians of the province of Sais' when Sabakos left (Hdt. 2.152.1). This again makes no chronological sense in Herodotos' account, where after this return from exile two other kings reigned before the Dodecarchy was established, but in the original story the people of Sais brought Psammetichos back precisely in order to make him one of the twelve new kings at once.

The sixth-century version of the tale told in the Greek Quarter at Memphis thus managed not only to roll three foundation legends into one – Ionian raiders were fated to become the first foreigners to live in Egypt, founders of Stratopeda, ancestors of the Hellenomemphites – but also to justify the Ionians' alleged role as kingmakers of the Saite dynasty. This role posed ethical problems for Greeks who generally did not approve of kingship, let alone of monarchical rule seized by military force. Hence the story included multiple justifications for supporting Psammetichos. First, the Egyptians proved congenitally incapable of handling freedom 'for any length of time', so that imposing kingship on them – as opposed to Greeks – was simply necessary. Secondly, Psammetichos had suffered extreme provocation: his father had been murdered by Sabakos and he himself had been forced into exile, yet he still had no intention of becoming sole king until his colleagues deprived him of his position (2.151.3), 'a thorough, violent insult' (1.152.3) that warranted 'revenge'.[116] Thirdly, he had been invited to assume power by the people of Sais. Fourthly, his rule was divinely ordained by oracle, and his supporters were fulfilling their own oracular destiny. Thus Ionian soldiers at Memphis created a distinctive and carefully thought-out story to compete with the Karians and Milesians. Some of its details may seem plausible, but Egyptian history and archaeology show that it was fundamentally false, and we must treat it as the legend it is.

4.4 *Tales from the Temple of Ptah*

A final problem is the vast gap between Psammetichos' real career as an agent of the Assyrian Empire who diplomatically unified the kingdom and eventually

115 See Appendix below for further discussion.

116 See Fisher (1992) 344, for this meaning of the verb *perihybrizein* here and elsewhere in Herodotos. Diodoros, by contrast, makes no attempt to justify Psammetichos' actions but suggests that a small group of oligarchs, like the Dodecarchs, will inevitably end up fighting one another for power.

made it independent and the notion that he emerged as sole leader from an oligarchy of Twelve by his own and his foreign soldiers' efforts. The Ionians premised their legend on this notion, but they surely did not themselves invent such a complex back-story with such a tenuous relation to history. More probably, they adapted a pre-existing legend about Psammetichos to their own ends, and this older legend probably originated with the priests of Ptah, the most powerful authority in Memphis, and Herodotos' chief Egyptian source.[117]

The priests of Ptah, as we have seen, refused to recognise Assyrian-backed Necho I and instead accepted the kings of Kush as legitimate rulers of all Egypt. Hence their version of history presented the Kushite dynasty as a single fifty-year reign by 'Sabakos', who is clearly based on Shabaqa, the first Kushite king to make Memphis his capital. Sabakos, they said, had conquered Egypt but ruled justly; he had voluntarily abdicated and left the kingdom in pious obedience to divine signs. This fiction enabled the priests to account for the end of Kushite rule without so much as mentioning Assyrian invasions or Necho I.[118]

The downside was that, without Necho or Assyrians, it was hard to account for the accession of Psammetichos I, whose legitimacy they did accept, while the story of Sabakos' departure made it hard to acknowledge Tanwetamani's continuing reign in the south. The solution was to posit that Psammetichos was one of twelve men picked by the people of Egypt to step into a power vacuum and to divide the kingdom between them. An oracle foretold that one of the twelve was destined to rule alone, and the high priest of Ptah at Memphis – who else? – would eventually fulfil the oracle, in effect anointing Psammetichos as king (Hdt. 2.147.2–151.1; Diod. Sic. 1.65.2–66.7).

The story went on to portray the Dodecarchs as governing in complete harmony, and collaborating to build the 'Labyrinth', a temple more impressive than the Pyramids and all the temples of Greece put together, in Herodotos' opinion (2.147.3–148.3; Diod. Sic. 1.66.1–6). The priests of Ptah thus projected a time of mutual toleration and peaceful co-operation such as the contemporary evidence shows did indeed exist between Psammetichos, the leading men of Upper Egypt, and the kings of Kush. This idyll lasted 'some time' (Hdt. 2.151.1); Diodoros makes it fifteen years (1.66.2, 7). This date puts the

117 Hdt. 2.147.1 with 2.3; see Lloyd (1975) 15, 89–116, (1988) 4, 114–16. Haider (1988, 176–77) assumes that the story was a Greek invention.

118 Hdt. 2.137, 139; Diod. Sic. 1.65.2–8. The priestly origin of the story is confirmed by the detail that Sabakos' dream told him that he could extend his reign beyond fifty years if he 'cut in half' all the priests in Egypt, an atrocity which he piously refused to commit.

end of the Dodecarchy in 649, in the middle of the Babylonian revolt against Assyria, when, as we have seen, Egypt came under attack from Kush – a time at which peaceful relations within Egypt may well have been interrupted, as the ruling elite of Upper Egypt with its Kushite connections parted ways with its Lower and Middle Egyptian peers.[119] Having the harmonious Dodecarchy end after fifteen years may therefore have been the priests' way of acknowledging this break and placing here the moment when they intervened to anoint Psammetichos sole ruler.

As Herodotos tells it, the priest made a silly mistake and brought out only eleven golden libation bowls for twelve kings, marking the one left without as the future monarch (1.147.4, 151). One may imagine that in the original version the high priest deliberately brought out too few libation vessels, as a means of divining which of the twelve candidates Ptah wished to be king. How the story ended before the Greeks of Memphis inserted their Ionian raiders into it is hard to tell. Obvious possibilities are that Psammetichos after receiving Ptah's blessing duly 'smote' all his enemies in the traditional manner of Pharaohs, or that the god's blessing itself was enough to make the other eleven kings accept Psammetichos as the chosen one and for the monarchy to be re-established without any fighting at all.

Either way, the story of the Dodecarchy was no reflection of genuine long-term and deep political division within the Delta[120] that could only be resolved by military action. It was a stylised version of the harmonious relations across the kingdom and with Kush, despite a lack of formal unification, during the early years of Psammetichos' reign, devised by the priests at Memphis to account for the end of Kushite rule and the rise of the new dynasty without having to mention the long period of Assyrian domination.[121] This idealising construct rather than historical events provided the framework for the Ionians' foundation legend.

---

119 Haider (1988) 177–78, suggests that the '15 years' cover 671–656, from the first Assyrian invasion to the unification of Egypt under Psammetichos, but overlooks the fact that the priests at Memphis regarded Taharqa as legitimate sole king down to 664 (and does not explain the two-year 'anarchy').

120 So, e.g., Lloyd (1988) 119 (cf. 2007, 353), stating explicitly what is generally assumed tacitly.

121 This construct was not accepted in the history of Egypt compiled by the Egyptian priest Manetho in the Hellenistic period, which recognised three 'Ethiopian' kings (not Tanwetamani) as the XXVth Dynasty, sandwiched between XXIVth and XXVIth Dynasties from Sais, and had Psammetichos I simply inherited the kingship from Necho I and earlier forebears (*FGrH* 609 F 2, 3a–c, 28).

5 Greek and Karian Soldiers in Egypt: A Reassessment

The fictional or fictionalised nature of all these stories means that some of the things they seem to tell us about Greek and Karian military service abroad are questionable. The Ionian raiders-legend has fed into a general modern perception that raiding and mercenary service went hand-in-hand in Mediterranean history, yet Greeks who raided Egypt were surely not often, if ever, converted into mercenaries. Quite apart from the practical difficulty of arranging to hire hit-and-run raiders, the long-term garrison service typically available in Egypt was a very different prospect from the quick profit sought by plunderers.[122] Military service abroad was not so much a complement as an alternative to raiding or overseas migration: more stable, if typically less profitable, than the former, and a less permanent commitment than the latter.[123]

The suggestion of the various stories that the reward for loyal service was typically land reflects their nature as foundation legends. In reality, the places where soldiers 'settled' were forts and garrison posts where they lived and worked, not places to which they retired to lead a farming life. If foreign troops were hired because they were available full-time and not hampered by agricultural responsibilities as the Egyptian warrior class was, it would defeat the object to give them land as payment. Instead, they will have been paid in rations, uncoined silver, and an occasional share of spoils.[124] Some may have received or acquired land eventually, but others surely went back to Greece. Even Pedon, so successful that he was given a town to administer, seems to have left Egypt and returned home in old age.[125] Instead of a single large migration of Greeks that established a self-sustaining community of

122 Interdependence between mercenary service and piracy: Horden and Purcell (2000) 386–87; they cite Gabbert (1986), but this study argues that the use of 'pirates' in warfare was a phase, *ca.* 320–250, not a permanent phenomenon, and involved recruiting locally active pirates or bandits to help fight a specific war – not appointing raiders of one's own coast into standing border guards.

123 Carty (2015) 149–74, strikingly argues that Greek and other foreign troops in Egypt were slaves bought, rather than mercenaries hired, on the grounds that Near Eastern states often employed 'captives' in their armies (esp. 160–65), but I would object that forcible drafting of defeated enemies is a quite different custom from the hypothetical buying of chattel slaves from foreign raiders.

124 See Sullivan (2011) for all three forms of payment other than land.

125 The inscription on Pedon's statue says that 'he brought it out of Egypt', though this could refer to a temporary visit. His receipt of 'a golden bracelet and a city' suggests an appointment as governor of a town, or commander of the town's garrison, along with a token of office: e.g., Agut-Labordère (2012); Vittmann (2003) 203–206.

mercenaries, we should thus envisage a continuous flow of Greeks into and out of Egypt.[126]

From 625–525, then, military service in Egypt provided vital opportunities for many Greeks, perhaps on a par with trade, raiding, and overseas settlement in economic terms. In this respect, the recent reappraisal of Archaic Greek mercenary service as a mass phenomenon, not confined to small bands of elite adventurers, is justified.[127] On the other hand, our conclusion that Greek military service in Egypt was the product of particular circumstances that arose at a relatively late date, requires reconsideration of the view that Greeks, as hoplites, were in demand as soldiers everywhere in the Middle East. The evidence for the employment of Greek troops by the Assyrian Empire has often been set out and cannot be discussed in full here. However, many if not all instances may concern Greeks living within the Assyrian Empire – in Cilicia, Syria, or Cyprus – rather than hired as 'foreign' mercenaries, and the only time that 'Ionians' undeniably did serve in an Assyrian campaign was as naval crews, alongside Phoenicians.[128]

Assyria's successor and Egypt's main opponent, the Neo-Babylonian Empire, probably made little if any use of Greek mercenaries. The only evidence for Greeks in Babylonian service is the famous fragment of Alkaios welcoming back his brother Antimenidas from 'the ends of the earth', where he killed a giant warrior as 'an ally of the Babylonians' (fr. 350 Page). The absence of any other evidence contrasts sharply with the situation in Egypt. It may well be that Antimenidas volunteered his services during one of Nebuchadnezzar's campaigns in the Levant – perhaps even after deserting from Egyptian service as Phanes was to do some eighty years later (Hdt. 3.4) – but it does not follow that the Babylonian Empire employed the services of mercenaries on a regular basis.[129] Herodotos saw seventh- and sixth-century wars in the Levant purely from an Egyptian point of view, and never even mentioned the Neo-Babylonian Empire as Egypt's opponent: if Greeks had regularly fought in Babylonian armies, it would surely have been a different story, or set of stories.

126 See Murray (1993) 232–35, *contra* Austin (1970) 17–18.

127 Esp. Luraghi (2006); see also Hale (2013); Sullivan (2011); Raaflaub (2004) 209 (who despite a general focus on the role of the elite stresses wide participation, *contra* Kaplan 2002); *GWMR* 40–42, 71–75.

128 Among numerous discussions I would single out Luraghi (2006) and Rollinger (2001) as especially important. Greeks and Phoenicians as naval crews under Sennacherib in 694: RINAP 3.2 no. 46.48–62a (reading 'Ionians' rather than 'Cypriots').

129 Fantalkin and Lytle (2016) analyse what Alkaios does and does not say, and they point out the lack of archaeological evidence for Greeks in the Neo-Babylonian empire; cf. Carty (2015) 150–51.

The Achaemenid Persian Empire also did not employ Greek mercenaries, until *ca.* 440. Herodotos says that 'a few Arcadians who had no livelihood' volunteered their services as mercenaries to Xerxes after the Greeks were driven back from Thermopylai and Artemision (8.26). Successful expansionist armies will often have received such offers, but even if they were accepted (Herodotos does not tell us in this instance), it does not amount to the use of mercenaries as a regular part of the armed forces. When we eventually do find Greek mercenaries in Persian service, the Persians also employed hired soldiers from other parts of the world, including Assyria and eastern Anatolia.[130] Xenophon and others may give the impression that only Greek hoplite mercenaries mattered, but this is surely no more than patriotic Greek bias.[131] The Persians did of course mobilise many Greeks who lived within the territory of their empire, but these Greek subjects always served in the navy rather than the army. From the reign of Cambyses (530–522) to the start of the Ionian Revolt in 499, Greeks from mainland Anatolia, and especially Greek states merely 'allied' to Persia – Miletos, Samos, Chios, Lesbos – provided an important part of the Persian fleet. Even in 480, Xerxes levied only naval personnel, not hoplites, among his Greek subjects.[132]

Finally, Egypt's ally and the Persians' enemy, the Lydian kingdom, did hire Greeks, and indeed seems to have pioneered mercenary service. Apart from Gyges' use of Karians and Milesians, we hear from Alkaios, *ca.* 600, that 'the Lydians gave us 2,000 staters to try to enter the sacred city, although they had never received any benefit from us and did not even know us' (fr. 69). This is the very definition of 'mercenary' service: a purely contractual relation with foreign soldiers, offering pay in precious metal for specified services.[133] The 'staters' here may still be weights of uncoined metal (electrum), but as has

130 In 430, Pissouthnes, satrap of Sardis, has 'Arcadian *and barbarian*' mercenaries (*epikouroi*, Thuc. 3.34.2–4); 700 *epikouroi* supporting Samian oligarchs in 440 (Thuc. 1.115.4) are also his (Diod. Sic. 12.27.4); see further Parke (1933) 14–15 n. 1; Bettalli (1995) 115–21. Xenophon's *Anabasis* mentions many non-Greek mercenaries in Persian service (4.3.4: Armenians, Mardians, Chaldaeans; 4.4.18: Chalybes, Taochoi; 7.8.15: Assyrian hoplites and Hyrcanian cavalry, stationed near Pergamon).

131 See Rop (2019) for full critical discussion of this attitude in the Classical sources.

132 Hdt. 7.61–99, 185. Detailed discussion in Van Wees (2013) 30–35, 147–48, (2017) 55–57. Parke (1933) 6 n. 2 cites the Greeks with Cambyses in Egypt mentioned at Hdt. 3.25.2, 7 as 'the only instance of Greeks in Persian service during the sixth century' (i.e., as hired troops; ditto Bettalli 1995, 84), but as Austin (1970, 15, 58 n. 2) already pointed out, they are clearly the Ionian and Aiolian (naval) troops mobilised as subjects ('patrimonial slaves', 2.1; 3.1) or as allies (Mytilene: 3.13; Samos: 3.44–45).

133 Although it is possible that the soldiers here are exiles and the 'sacred city' is their home town of Mytilene, so that the Lydians were sponsoring them to effect regime change.

often been noted, the Lydian habit of paying mercenaries was likely a key factor in their development of coined money around 600. Herodotos explicitly says that the army with which Croesus fought the Persians in 546 'consisted of his mercenary troops' (*xenikos*), dismissed at the end of the campaign (1.77.4), and that the next year, in a rebellion against the Persians, Pactyes the Lydian took 'all the gold of the Sardians' and 'hired *epikouroi*' (1.154). Later sources add anecdotes about Croesus sending out recruiters and borrowing money from rich individuals to raise funds for hiring troops in Greece.[134] If the Karians were credited with the invention of 'mercenary' service, it was presumably because the Lydians hired them.

One anecdote claims that Greek hoplites were so important to the Lydians that, when his Greek troops did not arrive in time, Croesus 'armed the Lydians in the Greek manner' and by this stratagem defeated a Persian cavalry force (Polyaenus, *Strat.* 7.8.1). However, Herodotos, despite characterising Lydia as a cavalry power (1.79.3; cf. 1.27, 80), reveals that Lydian infantry 'had equipment very close to the Greek type' (7.74). The 'stratagem' thus evidently consisted of having regular Lydian hoplite infantry, instead of Lydian horsemen, fight the Persian cavalry. Conversely, another anecdote has Croesus' father Alyattes keen to have Greek cavalry (from Kolophon) among his allies, and even to seize their horses.[135] On the whole, the picture is that foreign military manpower was much used in Lydia, but that Ionians and Karians were hired primarily because they were immediate neighbours, rather than because they offered military skills that the Lydians themselves did not have.[136]

In the Archaic period, then, Greek hoplites could find employment as mercenaries in Lydia from *ca.* 650–545 and in Egypt from *ca.* 625–525, because these two kingdoms were under threat from even larger and more expansionist empires and needed additional manpower for long campaigns, long-term garrison service, and naval crews. However, these greater expansionist powers

134 Recruiter sent to Peloponnese: Diod. Sic. 9.32 (= Ephoros 70 F 58a–d; Duris 76 F 20); Carty (2015, 161–62) sees a reference to buying slaves rather than hiring free men here. Borrowing hiring funds from wealthy individuals: Nikolaus of Damascus FGrH 90 F 65; Aelian 4.27. Note also the story that Gyges seized power in Lydia, *ca.* 685–680, with the aid of a Karian friend (perhaps a fellow-member of Kandaules' bodyguard: Hdt. 1.8.1) who brought further forces: Plut. *Quaest. Graec.* 45 (*Mor.* 302a).

135 Polyaenus, *Strat.* 7.2. Lydian equipment: Mimnermos, *ca.* 600, described the troops of king Gyges as having charged into battle 'while fencing themselves in with hollow shields' (*koileis aspisi phraxamenoi*, fr. 13a West), the very words used by Tyrtaios to describe the advance of the Dorian tribes in Sparta (fr. 19.7 West). See also Brouwers (2013) 98 and this volume.

136 Rop (2019) esp. 231–32, draws a similar conclusion about Greek military service in Persia and Egypt in the fourth century.

themselves – Babylonians, Medes, and Persians – made little or no use of hired foreign soldiers. Opportunities for Archaic Greek mercenaries were thus limited, both chronologically and geographically, and we have no reason to believe that they were sought out for their particular arms, armour or combat tactics. The rest of the world was not in awe of the 'men of bronze' as much as the Greeks themselves were.[137]

6 Appendix: Herodotos' Insertions into the Priests' Story – Why and How?

The story of Anysis is minimal. No sooner does he become king than Sabakos invades and he is forced to 'flee into the marshes' (2.137.1). He spends fifty years here, sustained by people who secretly bring him supplies, and during that time creates an island, Elbo, from earth and ashes. 'This island no one was able to discover before Amyrtaios, but for more than 700 years the kings who lived before Amyrtaios were not capable of finding it' (2.140.2). After Sabakos leaves, Anysis returns to power and that is the end of his story. The island of earth and ashes probably has some mythological significance,[138] but whatever its meaning, it was clearly a story told by the supporters of Amyrtaios, himself a 'king in the marshes' around 450 (Thuc. 1.109–10). Amyrtaios claimed to have rediscovered a special place created by a predecessor many centuries ago. Whether as a result of the Athenian expedition in support of Amyrtaios, or from a visit to Memphis and the Delta around this time, Herodotos will have heard this story and felt obliged to find a place for it in the history he was told by the priests of Ptah. Since Sabakos was the only invader who featured in this priestly history, Herodotos had little choice but to place Anysis' exile during his reign, despite the implication that Psammetichos then also dated nearly 700 years before Herodotos' own day, a serious chronological inconsistency.

The story of Sethos is more substantial. He was a 'priest of Hephaistos' (Ptah), who took away the land assigned to the Egyptian soldier-class and

137 Comments and corrections by Jamie Novotny, Alexandra Villing, Karen Radner, Aideen Carty, Sotirios Peithis and the editors have helped improve this chapter a great deal, but I alone remain responsible for its flaws and errors. The paper was written in 2017, and although it has been revised, it has not been consistently updated. The recent PhD-thesis by Lola Legrand-Sibeoni, 'Problèmes de la guerre en Grèce ancienne' (Paris Nanterre 2021) includes a fine discussion of much of the same material and questions as are addressed here. Special thanks to Jamie Novotny for enabling me to read RINAP 5.1 in advance of publication.

138 Lloyd 1988, 91–92.

inflicted other dishonours upon them as well. When Sennacherib invaded with his 'Arabians and Assyrians', the soldiers refused to fight, and Sethos mobilised volunteers instead: shopkeepers, manual labourers and market traders. In a dream, Ptah promised Sethos his help, and during the night field mice ate the enemy's quivers, bows and shield handles so that Sethos' unwarlike militia was able to kill many unarmed in flight. A stone image of this king stood in the temple of Ptah, holding a 'mouse' (probably an ichneumon) in the palm of his hand, with an inscription allegedly exhorting the viewer to piety (2.141). This too was clearly a story that stood outside the priests' account of Sabakos and Psammetichos, which as we have seen avoided all mention of Assyria. It was a separate tale, told to explain an unusual statue in the temple,[139] and to glorify the priesthood of Ptah at the expense of the military class. The idea that Memphis might deploy a non-professional militia against an invading enemy is paralleled in the Victory Stela of Piye, where the Kushite army outside Memphis faces 'an army of artisans, overseers of builders, and sailors'.[140] Herodotos conscientiously tried to fit this tale into the history of Egypt as well, and this time he placed it as near as possible to its actual chronological position. Its correct place, in 701, the only year in which Sennacherib came near Egypt,[141] would have fallen in the middle of the supposed fifty-year reign of Sabakos. Since the Egyptian story left no room for it there, Herodotos put it in the nearest available slot, after the return of Anysis, despite the cost to the coherence of the priests' narrative.

Bibliography

Abd el-Maksoud, M. and Valbelle, D. (2013) 'Une stele de l'an 7 d'Apriès découverte sur le site de Tell Défenneh'. *REgypt* 64: 1–13.

Adiego, I.J. (2007) *The Carian Language*. Leiden.

Agut-Labordère, D. (2012) 'Plus que des mercenaires! L'intégration des hommes de guerre grecs au service de la monarchie saïte', in Martinez-Seve (2012) 293–306.

Agut-Labordère, D. (2013) 'The Saite Period: The Emergence of a Mediterranean Power', in Moreno García (2013) 965–1028.

139 Lloyd 1988, 104–105.

140 Eide 1994, no. 9, line 86.

141 By his own account he defeated Egyptian forces at Eltekeh in Palestine (RINAP 3.1 no. 4.32–58, and parallel passages in nos. 15–19, 21–3, 26, 34). A similar story about his abrupt retreat in 701, supposedly caused by divine intervention, features in biblical accounts (2 *Kings* 19:35–6; 2 *Chronicles*. 3:21–2; *Isaiah* 37:36); the two stories are combined by Josephus *AJ* 10.1.1–5.

Aharoni, Y. (1981) *Arad Inscriptions*. Jerusalem.

Albenda, P. (1986) *The Palace of Sargon, King of Assyria. Monumental Wall Reliefs at Dur-Sharrukin, from Original Drawings Made at the Time of their Discovery in 1843–1844 by Botta and Flandin*. Paris.

Anderson, J.K. (1970) *Military Theory and Practice in the Age of Xenophon*. Berkeley, CA.

Austin, M.M. (1970) *Greece and Egypt in the Archaic Age*. Cambridge.

Baker, H., Kaniuth, K., and Otto, A. (eds) (2012) *Stories of Long Ago. Festschrift für Michael D. Roaf*. Münster.

Balot, R.K., Forsdyke, S., and Foster, E. (eds) (2017) *The Oxford Handbook of Thucydides*. New York.

Barnett, R.D. (1976) *Sculptures from the North Palace of Ashurbanipal at Nineveh*. London.

Barnett, R.D., Bleibtreu, E., Turner, G.E., and Collon, D. (1998) *Sculptures from the Southwest Palace of Sennacherib at Nineveh, Volume II: Plates*. London.

Bassir, H. (2016) 'Neshor at Elephantine in Late Saite Egypt'. *Journal of Egyptian History* 9: 66–95.

Bettalli, M. (1995) *I mercenari nel mondo Greco*. Pisa.

Boardman, J. (1999) *The Greeks Overseas: Their Early Colonies and Trade*. 4th edn. London.

Borger, R. (1996) *Beiträge zum Inschriftenwerk Assurbanipals*. Wiesbaden.

Bowden, H. (1996) 'The Greek Settlement and Sanctuaries at Naucratis: Herodotus and Archaeology', in Hansen and Raaflaub (1996) 17–37.

Braun, T.F.R.G. (1982) 'The Greeks in Egypt', in *CAH* 3.3: 32–56.

Brélaz, C. and Ducrey, P. (eds) (2008) *Sécurité collective et ordre public dans les sociétés anciennes*. Geneva.

Brock, R. and Hodkinson, S. (eds) (2000) *Alternatives to Athens. Varieties of Political Organization and Community in Ancient Greece*. Oxford.

Brouwers, J.J. (2013) *Henchmen of Ares. Warriors and Warfare in Early Greece*. Rotterdam.

Burliga, B. (ed) (2011) *Xenophon: Greece, Persia and Beyond*. Gdańsk.

Carty, A. (2015) *Polycrates, Tyrant of Samos. New Light on Archaic Greece*. Stuttgart.

Chevereau, P.-M. (1985) *Prosopographie des cadres militaires égyptiens de la basse époque: carriers militaires et carrières sacerdotales en Egypte du XIe au IIe siècle avant J.-C.* Paris.

Cogan, M. (2014) 'The Author of Ashurbanipal Prism A (Rassam): An Inquiry into his Plan and Purpose, with a Note on his Persona'. *Orient* 49: 69–83.

Cogan, M. and Tadmor, H. (1977) 'Gyges and Ashurbanipal: A Study in Literary Transmission'. *Orientalia* 46: 65–85.

De Meulenaere, H. (1963) 'La statue du général Djed-ptah-iouf-ankh, Caire JE 36949'. *BIAO* 63: 19–32.

Der Manuelian, P. (1994) *Living in the Past*. London.

Dezsö, T. (2001) *Near Eastern Helmets of the Iron Age*. Oxford.

Drijvers, J.W. (1999) 'Strabo 17.1.18 (801C): Inaros, the Milesians and Naucratis'. *Mnemos.* 52.1: 16–22.
Echeverría, F. (2019) 'Epilogue: Ancient Sieges and Modern Perspectives', in *Sieges* 321–41.
Ehrhardt, N. (1983) *Milet und seine Kolonien: Vergleichende Untersuchung der kultischen und politischen Einrichtungen*. Frankfurt.
Eide, T. (1994) Fontes Historiae Nubiorum. *Textual Sources for the History of the Middle Nile Region between the Eighth Century BC and the Sixth Century AD*. Vol. 1: *From the Eighth to the Mid-Fifth Century BC*. Bergen.
Fantalkin, A. (2001) 'Mezad Hashavyahu: Its Material Culture and Historical Background'. *Tel Aviv* 28.1: 3–16.
Fantalkin, A. and Lytle, E. (2016) 'Alcaeus and Antimenidas: Reassessing the Evidence for Greek Mercenaries in the Neo-Babylonian Army'. *Klio* 98.1: 90–117.
Fischer-Bovet, C. (2013) 'Egyptian Warriors: The *Machimoi* of Herodotus and the Ptolemaic Army'. *CQ* 63.1: 209–36.
Fischer-Bovet, C. (2014) *Army and Society in Ptolemaic Egypt*. Cambridge.
Fisher, N.R.E. (1992) *Hybris*. Warminster.
Fuchs, A. (2010) 'Gyges, Assurbanipal und Dugdamme/Lygdamis: absurde Kontakte zwischen Anatolian und Ninive', in Rollinger (2010) 409–27.
Gabbert, J.J. (1986) 'Piracy in the Early Hellenistic Period: A Career Open to Talents'. *G&R* 3: 156–63.
Garlan, Y. (1975) *War in the Ancient World*. London.
Gianotta, M.E., Innocente, L. and Gusmani R. (eds) (1994) *La decifrazione del cario*. Rome.
Goedicke, H. (1962) 'Psammetik I. und die Libyer'. *MDAI(K)* 18: 26–49.
Gorman, v. (2001) *Miletos, The Ornament of Ionia. A History of the City to 400 B.C.E.* Ann Arbor, MI.
Gorman, V. and Robinson, E. (eds) (2002) *Oikistes: Studies in Constitutions, Colonies, and Military Power in the Ancient World Offered in Honor of A.J. Graham*. Leiden.
Grayson, A.K. (1980) 'The Chronology of the Reign of Ashurbanipal'. *Zeitschrift für Assyriologie* 70: 227–45.
Grimal, N.-C. (1981) *Études sur la propagande royale égyptienne 1: La stèle triomphale de Pi ('Ankh)Y au Musée du Caire*. Cairo.
Habachi, L. (1974) 'Psammétique II dans la région de la première cataracte'. *OA* 13.4: 317–26.
Haider, P.W. (1988) *Griechenland, Nordafrika: Ihre Beziehungen zwischen 1500 und 600 v. Chr.* Darmstadt.
Haider, P.W. (1996) 'Griechen im Vorderen Orient und in Ägypten bis ca. 590 v. Chr.', in Ulf (1996) 59–115.
Haider, P.W. (2001) 'Epigraphische Quellen zur Integration von Griechen in die ägyptische Gesellschaft der Saïtenzeit', in Höckmann and Kreikenbom (2001) 197–215.

Haider, P.W. (2004) 'Kontakte zwischen Griechen und Agyptern und ihre Auswirkungen auf die archaisch-griechische Welt', in Rollinger and Ulf (2004) 447–91.
Hale, J.R. (2013) 'Not Patriots, Not Farmers, Not Amateurs: Greek Soldiers of Fortune and the Origins of Hoplite Warfare', in *MoB* 176–93.
Hansen, M.H. and Raaflaub, K. (eds) (1996) *More Studies in the Ancient Greek* Polis. Stuttgart.
Herda, A. (2013) 'Greek (and our) Views on the Karians', in Mouton, Rutherford and Yakubovich (2013) 421–506.
Höckmann, U. (ed) (2012) *Griechische Keramik des 7. und 6. Jhs. v. Chr. aus Naukratis und anderen Orten in Agypten*. Worms.
Höckmann, U. and Kreikenbom, D. (eds) (2001) *Naukratis: die Beziehungen zu Ostgriechenland, Ägypten und Zypern in archaischer Zeit*. Möhnesee.
Holm, T.L. (2007) 'The Sheikh Fadl Inscription in its Literary and Historical Context'. *Aramaic Studies* 5.2: 193–224.
Hopwood, K. (ed) (1999) *Organized Crime in Antiquity*. Swansea.
Horden, P. and Purcell, N. (2000) *The Corrupting Sea. A Study of Mediterranean History*. Oxford.
Iancu, L.M. (2016) 'Greek and Other Aegean Mercenaries in the Archaic Age: Aristocrats, Common People, or Both?'. *StudHerc* 20.2: 9–29.
Jansen-Winkeln, K. (2014) 'Die Siegesstele des Amasis'. *ZAS* 141.2: 132–53.
Jędraszek, S. (2011) 'The Egyptian Phalanx in the *Cyropedia*', in Burliga (2011) 63–73.
Josephson, J.A. and Eldamaty, M.M. (1999) *Statues of the XXVth and XXVIth Dynasties*. Cairo.
Kahn, D. (2006) 'The Assyrian Invasions of Egypt (673–663 B.C.) and the Final Expulsion of the Kushites'. *Studien zur Altägyptischen Kultur* 34: 251–67.
Kaplan, P. (2002) 'The Social Status of the Mercenary in Archaic Greece', in Gorman and Robinson (2002) 229–43.
Kaplan, P. (2003) 'Cross-Cultural Contacts among Mercenary Communities in Saite and Persian Egypt'. *MHR* 18.1: 1–31.
Kempinski, A. (2002) *Tel Kabri: The 1986–1993 Excavation Seasons*. Tel Aviv.
Kitchen, K.A. (1986) *The Third Intermediate Period in Egypt* (1100–650). 2nd edn. Warminster.
Konijnendijk, R. (2014) 'Iphikrates the Innovator and the Historiography of Lechaion', in Sekunda and Burliga (2014) 84–94.
Leahy, A. (1988) 'The Earliest Dated Monument of Amasis and the End of the Reign of Apries'. *JEA* 74: 183–99.
Leclant, D. (1961) *Montouemhat: quatrième prophète d'Amon, prince de la ville*. Cairo.
Leclère, F. and Spencer, J. (2014) *Tell Dafana Reconsidered: The Archaeology of an Egyptian Frontier Town*. London.
Lichtheim, M. (2006) *Ancient Egyptian Literature. Volume III: The Late Period*. 2nd edn. Berkeley, CA.

Lloyd, A.B. (1975) *Herodotus Book II: Introduction*. Leiden.
Lloyd, A.B. (1983) 'The Late Period, 664–323 BC', in Trigger (1983) 279–348.
Lloyd, A.B. (1988) *Herodotus Book II: Commentary 99–182*. Leiden.
Lloyd, A.B. (2000) 'The Saite Navy', in Oliver (2000) 81–91.
Lloyd, A.B. (ed) (2010) *A Companion to Ancient Egypt*. Blackwell.
Luraghi, N. (2006) 'Traders, Pirates, Warriors: The Proto-History of Greek Mercenary Soldiers in the Eastern Mediterranean'. *Phoenix* 60: 21–47.
Markoe, G. (1985) *Phoenician Bronze and Silver Bowls From Cyprus and the Mediterranean*. Berkeley, CA.
Martinez-Seve, L. (ed) (2012) *Les diasporas grecques du VIIIe à la fin du IIIe siècle av. J.-C.* (Special issue of *Pallas*). Toulouse.
Masson, O. (1994) 'Les inscriptions carienne du tombeau de Montouemhat (Thèbes)', in Gianotta, Innocente and Gusmani (1994) 191–94.
Mazarakis-Ainian, A., Alexandridou, A. and Charalambidou, X. (eds) (2017) *Regional Stories – Towards a New Perception of the Early Greek World*. Volos.
Möller, A. (2000) *Naukratis: Trade in Archaic Greece*. Oxford.
Moreno Garcia, J.C. (ed) (2013) *Ancient Egyptian Administration*. Leiden.
Morkot, R. (2000) *The Black Pharaohs. Egypt's Nubian Rulers*. Leiden.
Morkot, R. (2013) 'From Conquered to Conqueror: The Organisation of Nubia in the New Kingdom and the Kushite Administration of Egypt', in Moreno Garcia (2013) 911–63.
Mouton, A., Rutherford, I., and Yakubovich, I. (eds) (2013) *Luwian Identities: Culture, Language and Religion Between Anatolia and the Aegean*. Leiden.
Murray, O. (1993) *Early Greece*. 2nd edn. London.
Na'aman, N. (1991) 'The Kingdom of Judah Under Josiah'. *Tel Aviv* 18: 1–69.
Nadali, D. (2006) 'Esarhaddon's Glazed Bricks from Nimrud: The Egyptian Campaign Depicted'. *Iraq* 68: 109–19.
Nadali, D. (2019) 'Images of Assyrian Sieges: What They Show, What We Know, What Can We Say', in *Sieges* 53–68.
Niemeier, W.D. (2001) 'Archaic Greeks in the Orient: Textual and Archaeological Evidence'. *BASO* 322: 11–32.
Niemeier, W.D. (2002) 'Greek Mercenaries at Tel Kabri and Other Sites in the Levant'. *Tel Aviv* 29.2: 328–31.
Novotny, J. (2014) *Selected Royal Inscriptions of Assurbanipal: L3, L4, LET, Prism I, Prism T, and Related Texts*. Helsinki.
Novotny, J. and Watanabe, C.E. (2018) 'Revisiting the Identities of the Four Foreigners Represented on Ashurbanipal Relief BM ME 124945–6: Unravelling the Mystery of an Unrecorded Event', in Yamada (2018) 93–115.
O'Connor, D. (1983) 'New Kingdom and Third Intermediate Period', in Trigger (1983) 183–278.

Oliver, G.J., Brock, R., Cornell, T.J., and Hodkinson, S. (eds) (2000) *The Sea in Antiquity*. Oxford.

Onasch, H.-U. (1994) *Die assyrischen Eroberungen Agyptens*. Wiesbaden.

Oren, E. (1984) 'Migdol: A New Fortress on the Edge of the Eastern Nile Delta'. *BASO* 256: 7–44.

Parke, H.W. (1933) *Greek Mercenary Soldiers from the Earliest Times to the Battle of Ipsus*. Oxford.

Pedrizet, P. and Lefèbvre, G. (1919) *Inscriptiones Graecae Aegypti, Vol. III Les graffites grecs du Memnonion d'Abydos: nos. 1–658*. Strasbourg.

Peek, W., 'Ein milesisches Polyandrion', *WS* 79 (1966) 218–30.

Perdu, O. (2002) 'De Stéphinatès à Néchao ou les débuts de la XXVIe dynastie'. *Comptes rendus des séances de l'Académie des Inscriptions et Belles-Lettres* 146.4: 1215–244.

Perdu, O. (2004) 'La chefferie de Sébennytos de Piankhy à Psammétique Ier'. *REgypt* 55: 95–111.

Perdu, O. (2010) 'Saites and Persians', in Lloyd (2010) 140–58.

Pope, J. (2014) *The Double Kingdom under Taharqo*. Leiden.

Porten, B. (1968) *Archives from Elephantine. The Life of an Ancient Jewish Military Colony*. Berkeley, CA.

Quack, J.F. (2006) 'Inaros, Held von Athribis', in Rollinger (2006) 499–505.

Quack, J.F. (2011) 'Zum Datum der persischen Eroberung Ägyptens unter Kambyses'. *Journal of Egyptian History* 4: 228–46.

Raaflaub, K.A. (2004) 'Archaic Greek Aristocrats as Carriers of Cultural Interaction', in Rollinger and Ulf (2004) 197–217.

Raaflaub, K.A. and Van Wees, H. (eds) (2009) *A Companion to Archaic Greece*. Oxford.

Radner, K. (2012) 'After Eltekeh: Royal Hostages from Egypt at the Assyrian Court', in Baker, Kaniuth and Otto (2012) 471–79.

Rawlings, L. (2007) *The Ancient Greeks at War*. Manchester.

Reyes, A.T. (1994) *Archaic Cyprus. A Study of the Textual and Archaeological Evidence*. Oxford.

Ritner, R.K. (2009) *The Libyan Anarchy: Inscriptions from Egypt's Third Intermediate Period*. Atlanta.

Robinson, D. and Goddio, F. (eds) (2015) *Thonis Heracleion in Context: The Maritime Economy of the Egyptian Late Period*. Oxford.

Rollinger, R. (2001) 'The Ancient Greeks and the Impact of the Ancient Near East: Textual Evidence and Historical Perspective (ca. 750–650 B.C.)', in Whiting (2001) 233–64.

Rollinger, R. (ed) (2006) *Altertum und Mittelmeerraum: die antike Welt diesseits und jenseits der Levante. Festschrift für Peter W. Haider zum 60. Geburtstag*. Stuttgart.

Rollinger, R. (ed) (2010) *Interkulturalität in der Alten Welt: Vorderasien, Hellas, Ägypten und die vielfältigen Ebenen des Kontakts*. Wiesbaden.

Rollinger, R. and Ulf, C. (eds) (2004) *Commerce and Monetary Systems in the Ancient World: Means of Transmission and Cultural Interaction*. Melammu Symposia 5. Stuttgart.

Rop, J. (2019) *Greek Military Service in the Ancient Near East 401–330 BCE*. Cambridge.

Russell, J. M. (1991) *Sennacherib's Palace Without A Rival at Nineveh*. Chicago.

Sauneron, S. and Yoyotte, J. (1952) 'La campagne nubienne de Psammétique II et sa signification historique'. *BIAO* 50: 157–207.

Schäfer, H. (1904) 'Die Auswanderung der Krieger unter Psammetich I und der Söldneraufstand in Elephantine unter Apries'. *Klio* 4: 152–63.

Schlotzhauer, U. (2012) 'Untersuchungen zur archaischen griechischen Keramik aus Naukratis', in Höckmann (2012) 21–72.

Schlotzhauer, U. and Weber, S. (2012) 'Einleitung', in Höckmann (2012) 13–21.

Schmitz, P.C. (2010) 'The Phoenician Contingent in the Campaign of Psammetichus II against Kush'. *The Journal of Egyptian History* 3.2: 321–37.

Schneider-Hermann, G. (1945–8) 'Over de figuur achter den zegevierenden Pharao'. *JVEG* 10: 355–68.

Schwartz, A. (2009), *Reinstating the Hoplite: Arms, Armour and Phalanx Fighting in Archaic and Classical Greece*. Stuttgart.

Sekunda, N. (2014) 'The Chronology of the Iphicratean Peltast Reform', in Sekunda and Burliga (2014) 126–44.

Sekunda, N. and Burliga, B. (eds) (2014) *Iphicrates, Peltasts and Lechaeum*. Gdańsk.

Snodgrass, A.M. (1964) 'Carian Armourers: The Growth of a Tradition'. *JHS* 84: 107–18.

Spalinger, A.J. (1974) 'Esarhaddon and Egypt: An Analysis of the First Invasion of Egypt'. *Orientalia* 43: 295–326.

Spalinger, A.J. (1976) 'Psammetichus, King of Egypt: I'. *JARCE* 13: 133–47.

Spalinger, A.J. (1978a) 'The Date of the Death of Gyges and its Historical Implications'. *JAOS* 98: 400–409.

Spalinger, A.J. (1978b) 'The Concept of the Monarchy during the Saite Epoch – An Essay of Synthesis'. *Orientalia* 47.1: 12–36.

Spalinger, A.J. (2005) *War in Ancient Egypt: The New Kingdom*. Malden, MA.

Sullivan, B.M. (2011) 'Paying Archaic Greek Mercenaries: The View from Egypt and the Near East'. *CJ* 107.1: 31–59.

Thomas, R. (2015) 'Naukratis, "Mistress of Ships", in Context', in Robinson and Goddio (2015) 247–66.

Török, L. (2009) *Between Two Worlds. The Frontier Region between Ancient Nubia and Egypt, 3700 BC–AD 500*. Leiden.

Trigger, B. (ed) (1983) *Ancient Egypt: A Social History*. Cambridge.

Trundle, M. (2013) 'The Business of War: Mercenaries', in *WCW* 330–50.

Tsingarida, A. and Viviers, D. (eds) (2013) *Pottery Markets in the Ancient Greek Word (8th–1st Centuries B.C.)*. Brussels.

Ulf, C. (ed) (1996) *Wege zur Genese griechischer Identität: Die Bedeutung der früharchaischen Zeit*. Berlin.

Van Wees, H. (1999) 'The Mafia of Early Greece. Violent Exploitation in the Seventh and Sixth Centuries BC', in Hopwood (1999) 1–51.

Van Wees, H. (2008) '"*Stasis*, Destroyer of Men": Mass, Elite, Political Violence and Security in Archaic Greece', in Brélaz and Ducrey (2008) 1–48.

Van Wees, H. (2009) 'The Economy', in Raaflaub and Van Wees (2009) 444–67.

Van Wees, H. (2013), *Ships and Silver, Taxes and Tribute: A Fiscal History of Archaic Athens*. London.

Van Wees, H. (2017) 'Thucydides on Early Greek history', in Balot, Forsdyke, and Foster (2017) 39–62.

Vasunia, P. (2001) *The Gift of the Nile. Hellenizing Egypt from Aeschylus to Alexander*. Berkeley, CA.

Villing, A. (2013) 'Egypt as a "Market" for Greek Pottery: Some Thoughts on Production, Consumption and Distribution in an Intercultural Environment', in Tsingarida and Viviers (2013) 73–102.

Villing, A. (2015) 'Egyptian-Greek Exchange in the Late Period: The View from Nokradj-Naukratis', in Robinson and Goddio (2015) 229–46.

Villing, A. (2017) 'Greece and Egypt: Reconsidering Early Contact and Exchange', in Mazarakis-Ainian, Alexandridou and Charalambidou (2017) 563–96.

Villing, A. (2018) 'Wahibreemakhet at Saqqara: The Tomb of a Greek in Egypt'. *ZAS* 145: 174–86.

Vittmann, G. (2003) *Ägypten und die Fremden im ersten vorchristlichen Jahrtausend*. Mainz.

Waerzeggers, C. (2006) 'The Carians of Borsippa'. *Iraq* 68: 1–22.

Waldbaum J.C. and Magness, J. (1997) 'The Chronology of Early Greek Pottery: New Evidence from Seventh-Century B.C. Destruction Levels in Israel'. *AJArch.* 101: 23–40.

Weber, S. (2001) 'Archaisch ostgriechische Keramik aus Ägypten außerhalb von Naukratis', in Höckmann and Kreikenbom (2001) 127–50.

Weber, S. (2012) 'Untersuchungen zur archaischen griechischen Keramik aus anderen agyptischen Fundorten', in Höckmann (2012) 195–432.

Whiting, R.M. (ed) (2001) *Mythology and Mythologies: Methodological Approaches to Intercultural Influences*. Melammu Symposia 2. Helsinki.

Winnicki, J.K. (1977) 'Die Kalasier der spätdynastischen und der ptolemäischen Zeit. Zu einem Problem der Ägyptischen Heeresgeschichte'. *Historia* 26: 257–68.

Yadin, Y. (1963) *The Art of Warfare in Biblical Lands in Light of Archaeological Discovery*. London.

Yamada, S. (ed) (2018) *Neo-Assyrian Sources in Context: Thematic Studies on Texts, History, and Culture*. State Archives of Assyria Studies, Vol. 28. University Park, PA.

Epilogue

Matthew A. Sears

It was a long time coming; perhaps too long. We are finally moving beyond the seemingly endless debates about Greek hoplites, namely how they fought, when they emerged, and what implications they had for the developing polis. We no longer need to be in the shadow of Hilda Lorimer, Victor Davis Hanson, and other proponents of the 'orthodox view' of the hoplite phalanx.[1] More than that, we no longer need to spend so much time on the 'revisionist' accounts of the phalanx, either, but we can learn from and expand upon their example in questioning the nature and supposed dominance of the hoplite.[2] Their own scholarly *othismos* lasted for decades; Donald Kagan and Gregory Viggiano's 2013 edited volume *Men of Bronze: Hoplite Warfare in Ancient Greece* defined the terms and presented the different sides as clearly as possible at the time. Now nearly a decade beyond even that collection of papers, students of ancient Greek warfare have an opportunity to turn to other things, to revise our understanding of Greek warfare in much broader terms.

Fortunately, as the contributions to the present volume have amply demonstrated, ancient Greek warfare itself involved plenty of other things. The preoccupations of modern scholarship, and ancient elite sources, might have obscured those other things, but they did not remove all trace. As the preceding chapters also reveal – in the process, pointing out directions for further research – these other things were not merely 'things in addition or ancillary to the phalanx,' but things of great importance to Greek warfare in their own right. The phalanx debate is not only stale; it perhaps pointed its lens in the wrong direction.

Consider Echeverría's chapter on assaults and sieges. Beyond making the case for assaults against cities being far more pervasive than clashes of hoplites, which should be obvious to anyone seriously considering the evidence from antiquity, he argues that even the category of 'siege warfare' fails to encompass the full range of experiences and implications involved in attacking, defending, and taking cities and territory. Instead, assaults and sieges are just a part of

1 The paradigmatic statements of the so-called 'orthodox' view: Lorimer (1947); *WWW*.

2 Representative revisionists include Krentz (1985, 1997); Van Wees (2000, *GWMR*). I think the revisionists have had the better of the argument. For a good picture of how the 'revisionist' position has evolved, and a convincing refutation of so-called 'orthodox' views, see Konijnendijk (2018).

 | DOI:10.1163/9789004501751_013

war, of one side campaigning against another, a nexus of interrelated activities of which pitched battles too play a part. This insight, like a successful assault, begins to break down walls once thought impenetrable, the disciplinary walls that force aspects of the ancient world into distinct categories: literary vs. material evidence, pitched battles vs. small raids, soldiers vs. civilians, war vs. peace, and even war vs. everything else. Disciplinary boundaries can serve a purpose, chiefly to foster specialised expertise. But they can also stifle new insights and new questions. Because the hoplite was self-evidently important, we had better hash it out until we get it right, or so we thought, and the best way to do that is to focus on everything there is to know about the hoplite and master the body of evidence, including the burgeoning field of 'experimental' evidence. All along, however, too many succumbed to the temptation to listen only to the hoplites, that is, those reasonably well-off (mainland) Greek men who wrote and listened to literary histories, and who dedicated armour at sanctuaries and bought vases depicting warriors. While we always knew this 'hoplite class' was a minority in the polis (even if they were a majority of *citizens* in many cases), too few specialists stopped to ask whether hoplites were not also of minority importance in the experience of war, let alone just how broad the experience of war was.

If I may be forgiven for appealing to a Greek philosopher, one who is more responsible than many others for our obsession with hoplites, we can find value in one of Aristotle's research methods – interdisciplinarity. He appreciated that to understand one subject – living well – we must also understand many others, even ones that appear unrelated. Seeing the way form and matter combine to make frogs different from cats can yield insights into the different ways human communities can be organised, which in turn can teach us about human nature and ultimately how best to live in accordance with that nature. No topic exists in a vacuum.[3] While we now reject many of Aristotle's illustrative examples and ethical conclusions, we should not dismiss his fundamental example of interdisciplinarity.

We cannot understand, say, Greek cavalry – about which Konijnendijk's contribution fills in many of the scholarly gaps and corrects the overemphasis on heavy infantry – without also understanding a whole host of other things, including many things apparently far removed from cavalry. In the same way, as students of classical antiquity are finally realising, and as several authors

3 Gotthelf (2012) is a good primer on Aristotle's broader philosophical methods as applied to biology.

demonstrate in this volume, we cannot understand the Greeks, including Greek warfare, without going much further afield.[4]

Professional classicists and ancient historians no longer subscribe to the idea of the 'Greek Miracle', that the Greeks emerged *ex nihilo* with wholly new ideas about politics, philosophy, society, culture, and, yes, warfare. The field of military history was particularly stubborn in clinging to the paradigm of Greek exceptionalism, largely because of the fixation on the hoplite. When scholars did go beyond the hoplite, it was to demonstrate how the hoplite influenced and interacted with other types of soldiers, or how the Greeks 'Hellenised' the warfare of non-Greeks, and, occasionally, how non-Greeks influenced Greek warfare, including even the hoplite.[5] A parochial notion of Greek warfare ever loomed as the backdrop to these studies. In this volume, Brouwers and Hall point out that 'Greek' warfare has been too narrowly defined, neglecting the Greeks who lived in Anatolia and southern Italy, respectively. Future studies must take their scholarly example even further, widening the net of Greekness but also problematising the notion of 'Greek' warfare in general. Perhaps the same Orientalist bias that affected the Greeks, defining 'Greek' and Greek warfare by what it was not, still hovers over some modern scholarship.[6] Now that we have firmly established just how intertwined, or 'culturally entangled', the Greeks were with their neighbours, how useful is it to talk about Greek warfare as a phenomenon in and of itself? I have certainly been guilty of boxing in the Greeks, if for no other reason than for scholarly and pedagogical convenience. Thanks in large part to the inspiration I have taken from this volume, I will be more careful about my categories in the future. Van Wees's contribution blazes a promising trail. Aside from offering a fascinating account of the Egyptian Psammetichus' revolt on its own terms, relying on an assortment of Near Eastern sources, he shows that Greek mercenaries were employed in the region not because they were prized as Greeks, or fought like Greeks, but because they just happened to be there or be the subjects of allied powers

4 Morley (2018) is a lively manifesto for classicists getting outside their disciplinary boxes and working together with specialists from a variety of fields and perspectives.

5 Rop (2019) for a recent consideration of this issue. The better part of a decade ago, I was reasonably pleased with my own efforts (Sears 2013) to illuminate the Thracians' influence on the Greeks (instead of the other way around), but looking back on that work, I am struck by how Atheno-centric, and elite-focused, my scholarship was.

6 More than three decades ago, Edith Hall (1989) demonstrated how powerful this Orientalist impulse was for Greek self-definition in the wake of the Persian Wars. This important line of inquiry has continued and expanded in the work of Bonfante (2011); Vlassopoulos (2013); and others.

like the Lydians. The same was true for many other peoples to whom classical scholars have paid scant attention.

Lloyd's chapter sets an important methodological example related to broadening our concepts and categories. Instead of relying on hard lines like the 'Late Bronze Age Collapse' and the 'Eighth-Century Revolution', Lloyd conceives of more permeable boundaries and gradual change. Evolutionary biology presents a useful analogy. While ideological opponents of evolution once remarked on the lack of 'transitional forms' in the fossil record, those who truly understand biology recognise that *all* forms are transitional. Most evolutionary biologists think Stephen Jay Gould went too far with his idea of punctuated equilibrium, which posited that most species remain relatively the same unless some major event or change forces rapid evolution.[7] Instead, evolution is held by most scientists to proceed slowly and incrementally, though rapid change does sometimes occur.[8] Likewise, many historians now look for continuity and gradual change between historical epochs, rejecting the rigid periodisation that tends to encourage a revolutionary view of history that looks so much like punctuated equilibrium. It is likely that both models have merits, and that in history, as in evolution, gradual change and sudden, transformative change both take place. The key is to be open to interrogating the nature and reality of historical periods – as we interrogate the interaction between Greeks and non-Greeks, including in the sphere of warfare – since for too long the idea of revolutionary change was predominant.

Lloyd also goes far beyond traditional military-historical topics to reveal that everything in a given society informs its ideas and practices of warfare. Broad cultural trends in the Early Iron Age and Archaic Age, such as the prestige associated with long-distance travel and exchange, are relevant to war and a culture that valued warrior imagery (even if those who used such imagery were not necessarily warriors). In a similar spirit, Martinez corrects the narrow focus on women and war by moving beyond women as merely war's victims (even if they often were subject to its horrors), but instead as individuals who had a range of experiences and attitudes with regard to war, just as war had implications for society beyond battles, sieges, and other traditional subjects of study. This type of analysis should be applied to non-citizens, enslaved persons, and others typically left out of the conversation beyond concessions that

7 Eldredge (2015), esp. 219–332, covers this debate in detail.

8 Gould (2002) is as good a place to start as any for those seeking an engaging and accessible introduction to his views on evolutionary biology.

they took part in war and combat, too.[9] Decades ago, scholars such as Peter Hunt showed that enslaved persons fought as soldiers far more frequently than the Greek historians tell us, and we of course know that enslavement happened in the wake of campaigns, not to mention the centrality of helotage for Spartan society and Sparta's ability to make war.[10] As enslaved persons are increasingly understood as having their own thoughts and agency, it is well past time to ask what they thought of war, and how they responded to it.[11] For decades, light-armed troops have afforded scholars a gateway to the experiences of the non-elite, marginalised, and even enslaved in ancient warfare, and it is gratifying to see the editors of this volume take this line of inquiry even further and in a more holistic direction than mere tactics and recruitment.

With all that said, the nuts and bolts are still important, especially since the nuts and bolts addressed in this volume are not the 'drums and trumpets' variety of too much military history.[12] Kucewicz sheds new light on Spartan commemorative practices at a time when military commemoration is a pressing issue in our own society.[13] If commemoration reflects and shapes society, and if Spartan commemoration was different than we once thought, this should inform how we think about commemoration today. If the burial of the Spartan war dead on the battlefield reflects a growing egalitarianism, like the undifferentiated list of names on Maya Lin's Vietnam Veterans' Memorial in Washington, DC, why did several Spartans continue to receive ever greater special honours, and why do we still think about modern wars in terms of the actions of a select few? Our own cynicism tends to obscure the fact that in the ancient world divine forces were accepted by nearly everyone as causes for everything, including the ups and downs of war, as Millington reminds us in his study of the cult of Ares. Moving beyond an analysis of Ares in Athenian literature, future scholarship should integrate the material evidence for the

9 Liston (2019) uses the evidence of bones to paint a vivid – and often horrifying – picture of the physical effects of war on soldiers, but also a variety of persons not typically given much attention by military historians.

10 Hunt (1998).

11 The terminology 'enslaved person' is starting to be preferred to 'slave', since it emphasises agency and mitigates against the idea that slavery is in a person's nature or a fundamental part of who they are. Raaflaub (2014) makes a start in this direction, though the focus is still on non-combatants merely as victims of war. Kamen (2016), while not dealing with warfare specifically, deals with enslaved persons as having their own agency in an Athenian context.

12 *NAGRW* offers an excellent selection of papers on new ways to approach the nuts and bolts of ancient warfare, both Greek and Roman.

13 The examples of Confederate monuments in the United States and Soviet monuments in Eastern Europe come to mind.

cult of Ares into the study of ancient warfare. In an Athenian context, for example, the recently published work of Andrew Stewart, Eric Driscoll, and Seth Estrin on the sculptures from the Temple of Ares, a building moved from the Attic countryside to the Agora during the Roman period, should prove enlightening.[14] Why was the temple moved? What does the temple's original context tell us, and how did its artistic program relate to the Greeks – or Athenians, at least – at war?

By way of concluding these brief reflections, I want to reiterate that this volume on going 'Beyond the Phalanx' is an important first step not just because, and not even primarily because, there are many non-phalanx topics that are worth studying. If that were the focus, this volume would not be a first step at all, but within in a well-established tradition stretching back at least thirty-five years to the first rumblings of the 'revisionists.' Rather, this volume challenges us to recognise that military historians were not only *too* focused on hoplites, but were even *wrongly* focused on hoplites. It is now the time to increase the pace of correcting that scholarly distortion (based, to be fair, on a distortion in much of the canonical ancient evidence itself) in order to bring military history in line with the scholarly fields of classics and ancient history, and to open up scholars and students to a fuller appreciation of the ancient world and those who lived in it – all of those who lived in it. Social scientific approaches might be one area in which to expand ancient military history. Psychological studies that go beyond the soldier's experience; economic analyses of war, death, and destruction that look outside of Athens; and theories of command and control that do not rely solely on elite literary portrayals of famous generals are all possible avenues.[15]

This challenge should prove an exciting one for military historians, who in recent years have taken great strides in their field, as this volume demonstrates. Many of us, myself included, went into this discipline because we were

---

14 Stewart *et al.* (2019). See also Stewart (2016) for more on the temple's history and the cult of Ares in Attika. It seems relatively certain that the original temple was dedicated to Athena, and the image of Ares was part of that temple's adornment. Augustus moved the temple to the Agora and transferred its dedication to Ares.

15 Those taking a cue from Crowley's (2012) analysis of the Athenian hoplite should now go beyond Athens and beyond soldiers; see Meineck and Konstan (2014); Heidenreich and Roth (2020); Tritle (2020). Those following Pritchard's (2018, 169–79) study of the role of public finances and war in Athens should now move past Athens and consider the broader economic implications of war. 'Leadership studies' are today all the rage. While scholars like Roisman (2017) consider the leadership styles of household names, we should also explore the factors that made a broader range of ancient leadership styles effective or ineffective, and why soldiers and non-combatants followed some orders and not others.

fascinated by ancient warfare as it has been traditionally conceived: the high drama of the battles of Thermopylai and Salamis, and the life-and-death struggle of the Ten Thousand making their way back to the sea from deep in the heart of the Persian Empire. Many of the readers of our books come to our work hoping to read about shield clashing against shield and wily generals outfoxing each other. Drawing on my own experience once again, I have found that the broader my conception of ancient warfare has become, the more fascinating I have found the ancient world. My job as a teacher and scholar is to convey that fascination to others, including those who come to my courses and scholarly writing because they liked the 2006 Zack Snyder film *300*. It may be instructive that the subject matter of *300* occupies but a few pages in Herodotos' sprawling nine-book exploration of the human condition, which spends far more time on the customs of the Egyptians than on Leonidas and the Three Hundred. While Herodotos made many 'digressions' from his supposed main subject of the Persian Wars, these digressions were vital for understanding the Persian Wars. Herodotos never uses the term 'Persian Wars', and certainly does not constrain the conflict within a narrow chronological or geographical scope, as today's textbooks do. We can be confident that our audiences are just as curious as those who listened to Herodotos.

Bibliography

Bonfante, L. (ed) (2011) *The Barbarians of Ancient Europe: Realities and Interactions*. Cambridge.

Crowley, J. (2012) *The Psychology of the Athenian Hoplite: The Culture of Combat in Classical Athens*. Cambridge.

Eldredge, N. (2015) *Eternal Ephemera: Adaptation and the Origin of Species from the Nineteenth Century through Punctuated Equilibrium and Beyond*. New York.

Gotthelf, A. (2012) *Teleology, First Principles, and Scientific Method in Aristotle's Biology*. Oxford.

Gould, S.J. (2002) *The Structure of Evolutionary Theory*. Cambridge, MA.

Hall, E. (1989) *Inventing the Barbarian: Greek Self-Definition through Tragedy*. Oxford.

Hamilton, C.D. and Krentz P. (eds) (1997) *Polis and Polemos: Essays on Politics, War, and History in Ancient Greece in Honor of Donald Kagan*. Claremont.

Heidenreich, S. and Roth, J. (2020) 'The Neurophysiology of Panic on the Ancient Battlefield', in *NAGRW* 127–38.

Hunt, P. (1998) *Slaves, Warfare, and Ideology in the Greek Historians*. Cambridge.

Kamen, D. (2016) 'Manumission and Slave-Allowances in Classical Athens'. *Historia* 65: 413–26.

Konijnendijk, R. (2018) *Classical Greek Tactics: A Cultural History*. Leiden.
Krentz, P. (1985) 'The Nature of Hoplite Battle'. *ClAnt* 4: 50–61.
Krentz, P. (1997) 'The Strategic Culture of Periclean Athens', in Hamilton and Krentz (1997) 55–72.
Liston, M.A. (2020) 'Skeletal Evidence for the Impact of Battle on Soldiers and Non-Combatants', in *NAGRW* 81–98.
Lorimer, H.L. (1947) 'The Hoplite Phalanx with Special Reference to the Poems of Archilochus and Tyrtaeus'. *ABSA* 42: 76–138.
Meineck, P. and Konstan, D. (eds) (2014) *Combat Trauma and the Ancient Greeks*. New York.
Morley, N. (2018) *Classics: Why It Matters*. New York.
Pritchard, D.M. (2018) *Athenian Democracy at War*. Cambridge.
Raaflaub, K.A. (2014) 'War and the City: The Brutality of War and the Impact on the Community', in Meineck and Konstan (2014) 15–46.
Roisman, J. (2017) *The Classical Art of Command: Eight Greek Generals Who Shaped the History of Warfare*. Oxford.
Rop, J. (2019) *Greek Military Service in the Ancient Near East 401–330 BCE*. Cambridge.
Sears, M.A. (2013) *Athens, Thrace, and the Shaping of Athenian Leadership*. Cambridge.
Stewart, A. (2016) 'The Borghese Ares Revisited: New Evidence from the Agora and a Reconstruction of Augustan Cult Group in the Temple of Ares'. *Hesperia* 85: 577–625.
Stewart, A., Driscoll, E, and Estrin, S. (2019) 'Classical Sculpture from the Agora. 2, The Friezes of the Temple of Ares (Temple of Athena Pallenis)'. *Hesperia* 88: 625–705.
Tritle, L.A. (2020) 'Soldiers' Home', in *NAGRW* 53–64.
Van Wees, H. (2000) 'The Development of the Hoplite Phalanx. Iconography and Reality in the Seventh Century', in *W&V* 125–66.
Vlassopoulos, K. (2013) *Greeks and Barbarians*. Cambridge.

Index

Abu Simbel 73, 308, 311
Achaemenids 351
Acharnania 236, 253
Achaia 19, 21
Aeschines 134
Agesilaos 171, 174–75, 176n28, 177, 178, 182–183, 187, 191–192, 247, 261
aggrandisers 22, 24–25, 48, 50, 51–52
agriculture 5, 45–46
Ahlberg, Gudrun 32
Aigina 33–34, 228
Aineias Taktikos 125–126, 128, 255
Aischylos 2, 151, 156–157
  Seven against Thebes 148, 151–156, 158, 161, 162–164
Aitolia 248, 261
Akragas (Agrigento) 40, 270, 273
Alkaios 30–31, 50, 52–53, 70–71, 333, 334
alphabet 28–29, 55–56
Amasis 294, 307–308, 311–312, 316, 318, 322, 325
Amathus Bowl 77
Amphipolis 255
Amyrtaios 327, 336
Anatolia 66–69
Antalkidas 261
Antonaccio, Carla M. 25
Anysis 328, 336
apophthegms, Laconian 84
Apries 293–294, 307, 311–312
Archer, Robin 75
archers 20, 26, 30, 40, 52, 206–214, 216–220, 223, 225–227, 252, 271, 313, 316
  Skythian 218–220
Archidamos 245, 255, 260
Archilochos 28, 30–31, 41, 42–43, 73, 205, 212–213
Ares 145, 148–150, 152–156, 157–164, 349–350
Argolid 21, 26–27, 38, 41
Argos 28, 41–42, 43–44, 47–48, 49
Aristagoras (author) 323
Aristocrats 49, 268–269, 277
Aristodemos 268–269
Aristophanes 135, 137
Aristotle 3, 48n153, 127, 147, 192, 194, 346
Arjan Bowl 77
arms and armour 1, 20–21, 26, 31, 38–40, 49–51, 206–207, 214, 221, 267, 270, 275n38
  dedication of 20, 27–28, 31, 39–40, 47, 49–51, 53, 148, 213, 216
army organisation 252
  Bronze Age 22–23
  composition 252, 267–274
  Early Iron Age 25
  Egypt 299–300
  numbers 253, 254
  Sparta 106, 110
  see also mobilisation
Artemisia 123
Ashdod 306, 310
Asine 41–42
Assurbanipal 297, 301–302, 304–305
Assyrians 2, 38, 50, 75, 301, 307, 315n77, 321, 328, 337
  Neo-Assyrian Empire 67, 75, 164, 295–296, 301–303, 306, 329–330, 333
Athens 2, 19–20, 24–28, 31–33, 43–44, 47, 48–49, 160, 163, 208–209, 217–220, 224–226, 266, 269–270, 282–283
  Athenians 236, 244, 248, 251, 253, 256
Attika 19–20, 236, 252

Babylonia 164
Babylonians 315n77
  Neo-Babylonian Empire 67, 307–308, 316, 333
Barry, William 125
battle 26–27, 185–189, 255–256
  frequency of 249–250, 253
Battle of the Champions 102, 104, 107
bell-cuirass: *see* cuirass
Bérard, Reine-Marie 107
bilingualism 68, 305
black-figure pottery 215–220
blazons: *see* shield
blockade 254–256, 257–259
Boiotia 38, 47, 248, 253
bow 316

Brasidas 236, 247, 253, 255, 257
Briant, Pierre 131
Bugh, Glenn 169
Burial 44n130
 cremation 24–25, 207–209
 with weapons 19–21, 26–28, 39, 47, 49–51, 206–207, 209
 see also war dead

Cambyses 334
campaigns 181–182, 236, 238, 245, 247, 249–250, 252
 objectives of 247–248, 250, 252–253, 259–262
 poliorcetic 249, 253, 254–259
Carthage 279, 281, 284
 Carthaginians 270, 272, 273, 276, 277–278, 282, 284–285
 see also Phoenicians
Cartledge, Paul 125
Carty, Aideen 332n123
Catling, Hector 207
causality 145–156, 160–161, 162–164
cavalry 26–27, 32–33, 38, 72–73, 75, 169–171, 175–176, 178–179, 180–181, 183–184, 188–200, 222, 271–272, 282–283, 316, 346–347
 Athenian 176–178, 180, 181, 191–193, 252
 Boiotian 171, 173, 175, 177–178, 179–180, 181, 186, 188, 193, 196, 199, 252
 Greek 335
 Lokrian 172, 199
 Lydian 335
 Macedonian 170, 172, 176n28, 177, 183n46, 188, 196
 Persian 170, 172–173, 174–175, 177, 178, 179–180, 183, 335
 Phokian 172, 199
 scholarship on 169–172, 189–190
 Spartan 171, 177, 192, 194, 196, 198–199
 Syracusan 173–176, 180, 183–184, 186–187, 191, 193–194, 196, 223, 272, 283
 Thessalian 171, 173, 176n28, 182, 194
 see also mounted warrior
Chalkidike 242, 247, 251, 253, 261
change, theory of 17–18, 27–28, 43, 45, 53–54
chariots 26–27, 33–34, 48, 75, 76, 91, 210
Chigi Vase 36, 73
Chios 243, 334
Christesen, Paul 94
Cimmerians 304–305
circumvallation: *see* walls
climate change 44–46
combined arms 188–189, 200, 252, 254
commemoration 349
 see also war dead
Cook, R.M. 72
Corinth 181, 183, 186, 191, 196, 199, 224, 251–252
 Corinthians 244, 253, 260, 261
Corinthian vase painting 211–213
 see also Protocorinthian
Crete 19–20, 26, 38, 47–48, 50, 226
Croesus 110, 315, 335
cuirass 76, 77
 bell cuirass 39, 47
 linothorax 70–71, 78, 314
Cumae 268–269, 281, 282–283
Cyprus 24, 47–48

Daphnai 309, 311
'Dark Age' 66
 see also Early Iron Age
Dayton, John 7
deceit 240–241
dedications 111n98, 148, 213, 216
Delion 256
Delphi 277
Demosthenes (general) 224–226, 244–245, 248, 256, 257, 261
Demosthenes (orator) 192n63
devastation 42, 176–182, 188, 199, 224, 236, 243, 249, 255, 261, 283–284
Diodoros 124–125, 183, 228, 273–274, 282, 327–328, 329n116
Dionysios of Syracuse 181, 187, 196, 252n50, 257
disciplinary boundaries 346
Djedptahiufankh 304
Dodecarchy 327–329, 330–331
D'Onofrio, Anna Maria 208
Ducat, Jean 135–136

Early Iron Age (EIA) 8, 17–21, 23–27, 39, 43–44, 47, 52–53, 206–211, 348

Egypt 73, 149, 164, 294–295, 299, 302, 306–308, 310, 312, 313, 316, 318–319, 327–328
Egyptians 2
Eleans 236
elites 22–25, 27–28, 46–47, 49, 51–53, 164, 206, 208–209
Epameinondas 3, 175
epistrateia: *see* invasion
epiteichismos 244–245
Eretria 41, 50–51
Esarhaddon 296
ethnicity 65
Etruscans 267, 268, 277–281, 282–283
Euboia 19–20, 30, 41, 47, 205–208, 212–213
Euboian gulf 19, 21–24, 26–27, 47
Euripides 134, 156–157
Phoenician Women 148, 156–164
Euxitheos 134
evolutionary biology 348
expansionism 243–244, 335–336
experimental research 346

feasting 22–25
Finley, Moses 97
Fisher, Nick 52
fleets 253
see also warfare, naval
fortifications 40–41, 49–50, 238–239
foundation legend 320–321, 326, 329, 331, 332
Foxhall, Lin 7

Gaca, Kathy 132
Gaebel, Robert 169–170
Garlan, Yvon 6, 239, 257, 259
garrisons 296, 298, 308–311, 322–323, 327, 332, 335
Gelon 225, 271–272, 282
Geometric period 26–28, 30–33, 43n128
Late Geometric vase painting (LG) 31–34, 36, 48–49, 206, 208, 209–211
Gordion 74
gradualist view 5–6, 5n17
Greek culture 64–66
Greek exceptionalism 293, 347
Greek warfare: *see* warfare, Greek
Greenhalgh, Peter 74
Grote, George 3–4
Grundy, George 5, 6
Gyges 68–69, 70, 72, 301, 304–306, 321, 324
gymnetes 36, 52–53, 95, 214–215
see also light-armed
Gytheion 248

Haider, Peter 330n117, 331n119
Hale, John R. 49
Hall, Jonathan 65, 213
Hanson, Victor Davis 6, 64, 191n60, 240, 249, 345
Harvey, David 125, 130
heavy infantry 163–164, 205, 211, 213–215, 218–221, 239, 252, 274, 313
Anatolian 69–72
Assyrian 2, 315n77
Babylonian 315n77
Egyptian 2, 314–317
Greek 317
Judean 313–314
Karian 317
Lydian 335
Persian 2
Phoenician 314
see also hoplite
hegemony 244
Hegetorides, daughter of 131
Hekataios 315, 328n114
Hellas 130
Hellenomemphites 312–313, 329, 331
helmet 313–314, 316
Corinthian 34, 36, 38–39, 76, 77, 214, 267
crest 69–70, 314
Ionian 71–72, 76
Kegelhelm 38–39
Spangenhelm 71
Hermotybies and Kalasiries 300, 311–312
Herodotos 2, 131, 138, 146–147, 215, 217, 221–222, 272, 282, 293–294, 311–312, 314–315, 319, 322, 325, 328, 333, 336–337, 351
Hesiod 46–48, 52
Works and Days 41, 47
hippobatai 38, 48, 74–75
hippodromoi psiloi (ἱπποδρόμοι ψιλοί) 194n69, 271–272
hippomachos 72

hippostrophos 38, 74
Hodkinson, Stephen 91, 94, 136
Homeric period 5–6
Homeric epics 2, 8, 28–31, 47, 52–53, 207, 210–211, 218, 243, 275n41
see also Iliad and *Odyssey*
hoplite 1, 2–3, 5–7, 49, 74, 163–164, 169–171, 178–179, 180, 183–184, 188–189, 221, 222–223, 227–229, 239, 266, 270–271, 273–274, 293–294, 313, 317, 318, 335, 345–346, 350
hoplite ideology 199, 222, 240, 252
hoplite mirage 239
hoplite revolution theory 2, 4–5, 40
see also heavy infantry; warfare, hoplite
Hornblower, Simon 128
horseman: *see* cavalry; mounted warrior
Hysiai, battle of 102

Iason of Pherai 244
Iconography 20–22, 29–30, 31–38, 50, 53, 170–171, 208, 209–210
Iliad 6, 20, 28, 30, 34, 41, 50, 145–146, 149–150, 164, 207, 210–211, 213, 214–215
Inaros 297–298, 322–324
institutionalisation 97, 111, 113–114
invasion 236, 248, 254–256
Ionians 305, 310–311, 312–313, 318–319, 322, 326, 329, 331, 335
Iphikrates 226, 227–228, 243, 315
Isokrates 228
Italic peoples 267, 275–277
Italy, Southern 266–267, 275–277
see also Magna Graecia

Jarva, Eero 70
javelin throwers 256
Jerusalem 307, 311
Judeans 310n63, 311, 313
justifications for attack 242–243

Kagan, Donald 5, 345
Karians 67, 69–70, 73, 78, 304, 305–306, 310–311, 312–313, 318, 319–322, 325n108, 326–327, 334–335
Karomemphites 312–313
Kephallenia 236, 243
Kern, Paul 239n12
Kızıbel burial mound 76
Kleomenes I 98, 247
Kleon 255
knight-and-squire: *see* mounted warrior
Knossos 20, 24–25, 26, 47
Konijnendijk, Roel 250n46
Korkyra 244, 261
Korkyrians 236, 245, 248
Kos 19
Kounaxa, battle of 315
Kramer-Hajos, Margaretha 21–23
krater 21–24, 26
Krentz, Peter 6–7, 35, 249
Kush 295, 301, 302, 308, 316, 328, 330–331
Kyaxares 221
Kynos 21–23

Lagia, Anna 209
Lakonia 251
Latacz, Joachim 5–6
Late Bronze Age 18–23, 43–44, 206, 208–209
Late Helladic IIIC (LH IIIC) 21–23, 26
Lechaion, battle of 227–228
Lefkandi 19–25, 27, 41, 47, 206–208
Toumba cemetery 23–24, 26, 47, 51, 206–208
leisure class 46–54, 282
Lelantine War 41, 212–213
Lendon, J.E. 242n25
Lesbos 71, 334
Leukas 245, 248
Leuktra, battle of 139–140, 181, 183, 188, 189, 229
Libyans 300, 316
light-armed 26, 30, 35–36, 42, 51, 182, 205–231, 252, 256–257, 271–272
light infantry 179, 256
linen corslets: *see* cuirass, *linothorax*
Lokrians 251
Lorimer, Hilda 5, 35, 345
Lydians 68, 70, 73, 75, 76, 316, 334–335
Lydian Kingdom 66–68, 72, 77–78, 110, 306, 310–311, 318–319, 324–325, 334
Lykia 76
Lysias 176

MacDowell, Douglas 137
Macedonia 248, 251, 260
machines: *see* sieges, machines and
Magna Graecia 266–267, 273–274, 281–283, 285–286
Manetho 331n121
manpower 112, 241, 308, 326
Marathon 2, 171, 183
Martin, Albert 169
mass burials: *see polyandria*
Mediterranean 7–8, 19, 25, 38–39, 44, 51, 214
 Mediterranean world 266
Megara 251–252, 253
 Megarians 197–198
 Megarid 236, 251
Memphis 295, 296, 298, 312–313, 320, 328, 330–331, 337
'men of bronze' 11, 293–294, 318, 326, 336
mercenaries 49–50, 73, 77, 180, 223, 225–226, 227–229, 252, 270–271, 272–273, 311–313, 321, 325, 332–336
Messenia
 First Messenian War 101, 112n91
 Second Messenian War 86, 97
Methone (Lakonia) 236, 251, 255, 256
Midas 68
Migdol (Stratopeda) 307, 309, 311, 318, 327
Miletos 244, 311, 319, 322–325, 334
Miltiades 248
Mimnermos 72–73, 78
missile troops: *see* light-armed
Mitrou 27
Meyer, Eduard 5
mobile defence 176–178
mobilisation 192–195, 241, 252, 253, 254, 299–300, 312
Momemphis 318, 323
monarchy 329
Morris, Ian 44, 71
Motya 270, 284
mounted warrior 74–75, 78
Müller, Karl Otfried 4
Mykonos 42
Mytilene 253, 255, 256

Naukratis 310, 319, 322–325
Naupaktos 245, 253
naval warfare: *see* warfare, naval
Naxos (island) 19
Neandria 68
Necho I 295–297, 298, 302–303, 328, 330
Necho II 306n47, 307, 311, 316, 318
Nereid Monument 128–129
Nikarete 134–135
Nikias 173, 269–270, 282
nudity 34–36, 42, 211–215
 heroic nudity 34–36

Oath of Plataia: *see* Plataia, Oath of
O'Brien, Stephen 22, 27
occupation 255
Odyssey 28, 30, 48, 207, 211, 275–276
Olympia 47–48, 50
Olynthos 133, 194, 244
Onasander 187–188
Orientalism 347
orthodoxy, hoplite 1, 40, 185, 266, 275, 286, 345n1

panoploi 36, 213–215
panoply, hoplite: *see* arms and armour
Papadopoulos, John 17
Paros 27, 28, 30, 32–33, 40, 49, 210, 212, 214, 248
Patroklos 48n154, 145–146
Pausanias (commander) 173
Pausanias (author) 41, 101–102, 103, 139, 163
pay 332, 334–335
Pavlides, Nicolette 108–109, 216
Pedon 308, 332
Peisistratos 217
 Peisistratids 172, 220
Peloponnese 172, 196–200
Peloponnesian League 110, 113
Peloponnesian War 6–7, 157, 170, 192–193, 210, 242–243, 245, 249
Pelousion, battle of 312
peltast 175, 178, 180, 224, 226–230, 257
periodisation 348
Persians 2, 170, 174–175, 242, 314–315
Persian Empire 333, 351
 see also Achaemenids
Persian Wars 3, 4, 157, 172–173, 197–198, 223, 271–272, 281

phalanx 1, 2, 7, 36–38, 40, 72–73, 179, 197, 210, 226, 227–228, 239, 315, 317–318, 345, 350
Pheia (Elis) 236
Phoenicians 55–56, 275–277, 277n49, 311, 313, 314, 333
  see also Carthaginians
Phokaians 277–278
Phokians 184, 248, 251
Phormion 253, 260
Phrygians 68, 75, 218
  Phrygian Kingdom 67, 78
picked troops 326
pictorial pottery: *see* iconography
piracy 277–280, 332n122
Plataia 130–131, 244, 247, 257, 258
  battle of 103, 170, 171, 173, 179–180, 183, 185n46, 198, 222, 225
  Oath of 107, 113
Plato 162, 183–184, 192, 229
Plutarch 84, 127, 135, 178–179, 212–213
poliorcetics 237–238, 241, 245, 248, 253, 254–259
  see also sieges
Polyainos 319–320
polyandria 28, 32–33, 40, 42, 48–49, 51, 83, 85–86n10, 96n44, 102, 104, 107, 111, 113, 210
  Kerameikos *polyandrion* 83, 104–105
Polybios 195, 196
Polykrates 217, 225, 244, 246
Potasimto 73
Potidaia 188, 244, 248, 251, 253, 256, 258–259
population 43–46
Post-palatial 19–25, 43
Priene 244
Prinias 26
Protoattic (PA) 31–32, 34
Protocorinthian (PC) 31–32, 34–38, 211–212
Protogeometric (PG) 24, 26, 43n128, 207–209
provisioning 243
Psammetichos (Psamtik) I 73, 293, 297, 298–301, 305, 306–307, 308, 310, 320, 322–324, 326–330, 336
Psammetichos (Psamtik) II 307
Psammetichos (son of Theokles) 73, 308
psychology 9, 350, 350n15
psiloi (ψιλοί) 222
Punic: *see* Carthaginians
Pylos 244, 253, 256
Pyrrhic: *see* war-dance

Raaflaub, Kurt 137–138
Radiocarbon 43n128
Radner, Karen 318n90
Rahe, Paul A. 112n91
raiding 32–33, 42, 46–47, 236, 252, 326, 332
rape: *see* sexual violence
ravaging: *see* devastation
Rawlings, Louis 7, 252
revisionism, hoplite 286, 345n2
Rhodes 19
Rhodians 253, 260
Romans 195, 267, 278–279
Roosevelt, Christopher 66

Sabakos/Shabaqa 327–329, 330, 336–337
sailor-warrior 22–23, 25, 27, 206, 225
Sais 295, 296, 305–306, 323, 329
Salamis 123, 281, 351
Šamaš-šuma-ukīn 302–303
Samos 103, 243, 244, 319, 334
sanctuary 162–163, 211–212
panhellenic 213, 274, 277
  see also Delphi; Olympia
Sappho 325
Sardis 67, 71–72
Scheidel, Walter 44
Seaman, Michael 249
Selinous 103
Sennacherib 328, 337
Sethos 336–337
sexual violence 42, 124, 132–133
shield 20–21, 3184, 313–316
  Argive/*porpax-antilabe/aspis* 2n4, 34–37, 39–40, 74, 75, 76, 78, 211, 214, 315, 317
  Boiotian 34–36, 211–212
  Dipylon 32, 34–36, 96n42, 211–212
  hoplon 2n4
  pelte 2n4
  Phalara 20, 39
  of Achilles 41, 77
  shield apron 77
  shield blazon 69–70

Sicily 40, 253, 260, 266–267, 269–271, 275–276, 284–285
siege engines: *see* sieges, machines and
sieges 30, 41, 236–238, 283–285, 256–259
  assault 251–252, 256–257
  impact on women 132–133
  machines and 256, 257
  mining 258
  mound 68, 72, 78
  ramp 258
  resources and 241, 245–246, 258
  sallies 255–256
  scholarship on 4, 6, 237, 238–241
  vocabulary of 246, 256–257
  with circumvallation walls 258
  see also poliorcetics
Sikels 275–276
Simonides 83
skytalides 96n42
Skythians 76
slingers 205, 210, 212, 223, 227, 256–257, 271
  see also light-armed
Smyrna (Old Smyrna) 61, 68, 78
Snodgrass, Anthony 27, 29, 43–44, 69–70, 163, 275n41, 276–277
Snyder, Zack 1, 351
society, hierarchical 267–268
Socrates 183–184
Solon 89
Solygeia, battle of 186, 196, 199, 262
Sophocles
  Antigone 161n58
Sparta 1, 2, 30, 171, 212, 216, 223, 226, 228, 243, 251, 253, 283, 349
  army organisation 106, 110, 194
  citizen militia 112–113
  egalitarianism 85, 96, 99–100, 112–114, 199
  foreign policy 198–199
  funerary practices 90–94, 98–101, 104, 108–109
  Lycurgan reforms 97–98
  royal funerary practices 104n67
  'Spartan mirage' 85
  syssitia 106–107
  see also cavalry, Spartan
Spartolos, battle of 184, 256
spears 20, 21, 26, 31–32, 36, 40, 50–51, 150, 206, 270–271, 313, 314, 316
Spence, Iain 169–170, 172, 176
Sphakteria 252
spoils 243
Strabo 41, 163, 191n71, 213n37, 280n59, 322–323
Stela of the Vultures 73
strategy 243–245, 248, 254, 262
  see also campaign; invasion
Stratopeda: *see* Migdol
sword 20, 21, 26, 30, 32, 34, 40, 51, 209, 314, 315, 316
sworn bands (*enomotiai*): *see* Sparta, army organisation
Syracuse 173–174, 180, 193–194, 196, 223–224, 251–252, 269–272, 275, 281, 283, 284–285

Taharqa 298
Tanwetamani/Tementhes 297, 298, 299, 320, 330
Taras (Tarentum) 194n71
Tatarlı 75
Tefnakht 295
Tegea 108
terrain 183–184, 186–187, 190, 223–224, 244–245
Thasos 258
Thebes (Egypt) 299, 304, 306, 320–321
Thebes (Greece) 22, 244
  Thebans 181, 251
Themistocles, Decree of 138–139
Thermopylai, battle of 83, 172–173, 198, 351
Thrasyboulos 135, 175, 191, 243, 257
Theognis 53
Thessaly 24–25
Thucydides 2, 41, 124–125, 130–131, 146–147, 162, 173–174, 177, 212–213, 222, 223–224, 228–229, 242, 243, 245–246, 246–247, 260, 272
Thyrea 251
  battle of: *see* Battle of the Champions
Tiryns 19–20, 27
tremblers (*tresantes*) 135–137, 139–140
triremes 307, 318
Troy 42, 49, 245, 246
  Trojans 75

Tyrtaios 2, 28, 30–31, 34, 36, 85, 86–87, 91, 93, 95n41, 150, 205, 213–215, 216, 220

Van Wees, Hans 7, 46, 52, 53, 103–104, 105–106, 207, 213, 218, 242n25, 246, 271
vengeance 245
Viggiano, Gregory 5, 345

walls 68, 250, 256–257, 258
war-dance 76
war dead 32, 50
  battlefield burial 97, 102–103, 104–105, 111
  en polemōi stelai 99, 101
  fights over 30, 210, 214–215
  numbers 95–96, 112
  repatriation 83–84, 88–89, 94
  status 86–87, 89–92, 94–95, 104
warfare 23–24, 42–43, 48–49, 51
  Greek 1, 7–8, 169, 238–239, 266
  Homeric 28–30, 36, 275n38
  hoplite 6, 28–29, 35, 169, 185, 189–190, 199–200, 239–240, 275–276
  ideology and 214–215, 239–240, 242–243, 246–247
  naval 21–22, 32, 48–49, 241, 277–281, 318
  predatory 240–241
  siege 236–238, 241, 249, 251, 262, 283–285, 345–46
warrior 7, 19, 21–23, 25, 27–28, 36, 42–53, 150, 158–159
'warrior class' (Egypt) 308, 313, 332, 336–337
Warrior-Trader 24, 206–207, 209
Weber, Max 4, 5
Whitley, James 64
Will, Edouard 237n4
women 42, 44, 153–154, 156n45, 348
  as 'fighters' 128–129
  see also women and war
women and war
  absence of husbands 134–135, 137
  economic impact of war 134–137
  impact of war 131–134
  in Sparta 126–128, 135–137, 139–140
  in tragedy 123n6, 134
  negative contribution 130–131
  'passive' roles 130
  social factors 138–139
  throwing tiles 124, 128–129
  war widows 137–138

Xenophon 2, 127, 130, 135, 146–147, 162, 174–175, 176, 179, 192, 222, 227–229,183, 192, 315, 334
Xerxes 3, 198, 334

Yamani: *see* Ionians

Zagora 41, 44

www.ingramcontent.com/pod-product-compliance
Lightning Source LLC
Chambersburg PA
CBHW071249180926
51613CB00002B/175
9789004501720